**Microstructural Science
Volume 12**

Corrosion, Microstructure, and Metallography

Microstructural Science
Volume 12

Corrosion, Microstructure, and Metallography

Proceedings of the Sixteenth Annual Technical Meeting of the International Metallographic Society

Edited by

Derek O. Northwood, Technical Meeting Chairman
The University of Windsor

William E. White, General Chairman
Petro-Canada, Inc.

George F. Vander Voort, Series Editor
Carpenter Technology Corporation

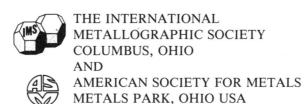

THE INTERNATIONAL
METALLOGRAPHIC SOCIETY
COLUMBUS, OHIO
AND
AMERICAN SOCIETY FOR METALS
METALS PARK, OHIO USA

Copyright © 1985
by the
AMERICAN SOCIETY FOR METALS
All rights reserved

No part of this book may be reproduced, stored in a retrieval system, or transmitted, in any form or by any means, electronic, mechanical, photocopying, recording, or otherwise, without the prior written permission of the publisher.

Nothing contained in this book is to be construed as a grant of any right of manufacture, sale, or use in connection with any method, process, apparatus, product, or composition, whether or not covered by letters patent or registered trademark, nor as a defense against liability for the infringement of letters patent or registered trademark.

Library of Congress Catalog Card No.: 84-72692

ISBN: 0-87170-193-6

SAN 204-7586

PRINTED IN THE UNITED STATES OF AMERICA

Contents

Contributors xi

Preface xv

Metallography of Corrosion and Corrosion Related Failures
Chairpersons: I. LeMay, J.E. Bennett, M.R. Louthan, P.J. Kenny, D.W. Stevens, A. Koul, and M.E. Blum

M.H. Rafiee
A General Metallographic Approach to Corrosion Failure Analysis 3

P.J. Kenny
Metallography and Failure Analysis of Corrosion Related Failures 11

I.D. Peggs and D.P. Skrastins
The Metallography of Electrochemical Corrosion Testing 25

H. Roper
Fatigue and Corrosion Fatigue Failure Surfaces of Concrete Reinforcement 37

A.K. Koul, R.V. Dainty, and C. Marchand
Metallurgical Investigation of a Turbine Blade and a Vane Failure From Two Marine Engines 45

G. Rondelli, B. Vicentini, and D. Sinigaglia
Investigation Into Precipitation Phenomena Following Heat Treatment of ELI Ferritic Stainless Steels and Their Influence on Mechanical and Corrosion Properties 73

E.W. Filer
Nickel-Beryllium Alloys Resistance to Sulfide Stress Cracking . . 89

H.J. Cleary
The Microstructure and Corrosion Resistance of 55% Al-Zn
Coatings on Sheet Steel 103

D.W. Hoeppner and I. Sherman
Fractographic Observation on Corrosion Fatigue and Fretting
Fracture Surfaces 115

M. Yanishevsky and D.W. Hoeppner
Corrosion Fatigue Behavior of Ti-6Al-4V in Simulated Body
Environments... 127

M.E. Blum
On-Site Metallographic Investigations Using the Replica
Technique.. 143

G.R. Smolik and D.V. Miley
Sulfidation Characteristics of Incoloy 800 151

P. Mahadevan and P. Breisacher
Nondestructive Room Temperature Tests for Metallic
Corrosion on Space Hardware 163

J.R. Keiser and A.R. Olsen
Corrosion Studies in Coal Liquefaction Plants 173

J.C. Thornley and J.K. Sutherland
The Identification of the Origins of Some Deposits in a Steam
Turbine 187

G.M. Raynaud and R.A. Rapp
In-Situ Observation of Whiskers, Pyramids and Pits During the
High Temperature Oxidation of Metals 197

P.F. Tortorelli and J.H. DeVan
Effects of a Flowing Lithium Environment on the Surface
Morphology and Composition of Austenitic Stainless Steel 213

K.J. Chittim and D.O. Northwood
Metallography of Corrosion and Hydriding of Zirconium Alloy
Nuclear Reactor Pressure Tubing........................... 227

W.J.D. Shaw
Surface Corrosion Comparisons of Some Aluminum Alloys in
3.5% NaCl Solution 243

H. Hindam and D.P. Whittle
Amelioration of Alloy Oxidation Behavior by Minor Additions
of Oxygen-Active Elements 263

Microstructure-Property Relationships
Chairpersons: R.J. Gray, P.M. French, M.R. Krishnadev, and
J.H. Richardson

T.A. Roth
Metallographic Applications in Interfacial Free Energy Studies .. 281

P.J. Blau
Relationships Between Knoop and Scratch Microindentation
Hardness and Implications for Abrasive Wear 293

A.K. Sinha, M.G. Hebsur, and J.J. Moore
The Effects of Aging Between 704 and 871C on
Microstructural Changes in Inconel X-750 315

M.T. Shehata, V. Moore, D.E. Parsons, and J.D. Boyd
Characterization of Nonmetallic Inclusions in Steel 329

R.J. Gray, R.K. Holbert, Jr., and T.H. Thrasher
Microstructural Analysis for Series 300 Stainless Steel Sheet
Welds and Tensile Samples 345

G. Petzow and H. Hofmann
The Precipitation Behavior of Martensitic Fe-Ni-W Alloys..... 371

R. Rungta, R.C. Rice, R.D. Buchheit, and D. Broek
An Investigation of Shell and Detail Cracking in Railroad
Rails 383

R.C. Wasielewski and M.R. Louthan, Jr.
Hydrogen Embrittlement of AISI 316 Stainless Steel 407

Microstructure-Fracture Relationships
Chairpersons: M. Ryvola and M.H. Rafiee

I.D. Peggs
Fractography and Metallography of Polymers 423

F. Smith and D.W. Hoeppner
Observations on Fatigue Crack Growth/Microstructure
Interactions Using Advanced Techniques 435

D. Eylon, C.M. Cooke, and F.H. Froes
Fatigue Crack Initiation in Shot-Peened Blended Elemental
Ti-6Al-4V Powder Compacts 445

D.C. Wu and D.W. Hoeppner
The Effects of Welding-Induced Residual Stresses and
Microstructural Alterations on the Fatigue-Crack Growth
Behavior of Commercially Pure Titanium 459

G. Hopple
Shear Band Failures in Threaded Titanium Alloy Fasteners 473

R. Chattopadhyay
Effect of Sulfide Inclusions on the Fracture of a Low-Carbon
Aluminum-Killed Steel 487

D.C. Wei
Microstructure-Fracture Correlations for Permanent Mold Gray
Iron 495

D.J. Diaz and S.E. Benson
Field Metallography Aids NDT of Evaluation of Indications
in Turbine Main Column Horizontal Plate Welds at Power
Plant 511

Advances in Metallographic Techniques
Chairpersons: W.J.D. Shaw and J.D. Braun

A.S. Holik and J.C. Grande
The Microcomputer in the Metallographic Laboratory 525

J.F. Henry
Field Metallography: The Applied Techniques of In-Place
Analysis 537

N.A. Shah, J.J. Moore, and J.H. Broadhurst
Detailed Macrosegregation Studies Using a Tandem Van de
Graaff Accelerator Facility 551

R.L. Bronnes and R.C. Sweet
Metallography of a Novel Stirling Engine Heat Receptor 571

D.J. Lloyd and M. Ryvola
Strain Inhomogeneity in Aluminum Alloys 577

1983 International Metallographic Exhibit 587

CONTRIBUTORS

Benson, S.E.
Lawrence Livermore National
 Laboratory
Livermore, California USA

Blau, P.J.
National Bureau of Standards
Washington, DC USA

Blum, M.E.
FMC Corporation
Santa Clara, California USA

Boyd, J.D.
CANMET
Ottawa, Ontario, Canada

Breisacher, P.
The Aerospace Corporation
El Segundo, California USA

Broadhurst, J.H.
University of Minnesota
Minneapolis, Minnesota USA

Broek, D.
FractuResearch
Columbus, Ohio USA

Bronnes, R.L.
Philips Laboratories
Briarcliff Manor, New York USA

Buchheit, R.D.
Battelle-Columbus Laboratories
Columbus, Ohio USA

Chattopadhyay, R.
EWAC Alloys, Ltd.
Powai, Bombay, India

Chittim, K.J.
University of Windsor
Windsor, Ontario, Canada

Cleary, H.J.
Bethlehem Steel Corporation
Bethlehem, Pennsylvania USA

Cooke, C.M.
MetCut-Materials Research Group
Wright-Patterson AFBase, Ohio USA

Dainty, R.V.
National Aeronautical Establishment
Ottawa, Ontario, Canada

DeVan, J.H.
Oak Ridge National Laboratory
Oak Ridge, Tennessee USA

Diaz, D.J.
Lawrence Livermore National
 Laboratory
Livermore, California USA

Eylon, D.
MetCut-Materials Research Group
Wright-Patterson AFBase, Ohio USA

Filer, E.W.
Cabot Corporation
Reading, Pennsylvania USA

Froes, F.H.
Air Force Wright Aeronautical
  Laboratories
Wright-Patterson AFBase, Ohio USA

Grande, J.C.
General Electric Corporation
Schenectady, New York USA

Gray, R.J.
Oak Ridge National Laboratory
Oak Ridge, Tennessee, USA

Hebsur, M.G.
University of Minnesota
Minneapolis, Minnesota USA

Henry, J.F.
Combustion Engineering, Inc.
Chattanooga, Tennessee USA

Hindum, H.
National Research Council
Montreal, Quebec, Canada

Hoeppner, D.W.
University of Toronto
Toronto, Ontario, Canada

Hofmann, H.
Max-Planck-Institut
Stuttgart, West Germany

Holbert, R.K., Jr.
Oak Ridge Y-12 Plant
Oak Ridge, Tennessee USA

Holik, A.S.
General Electric Corporation
Schenectady, New York USA

Hopple, G.
Lockheed Missiles & Space Co., Inc.
Sunnyvale, California USA

Keiser, J.R.
Oak Ridge National Laboratory
Oak Ridge, Tennessee USA

Kenny, P.J.
Ontario Research Foundation
Mississauga, Ontario, Canada

Koul, A.K.
National Aeronautical Establishment
Ottawa, Ontario, Canada

Lloyd, D.J.
Alcan International Ltd.
Kingston, Ontario, Canada

Louthan, M.R., Jr.
Virginia Polytechnic Institute
Blacksburg, Virginia USA

Mahadevan, P.
The Aerospace Corporation
El Segundo, California USA

Marchand, C.
National Aeronautical Establishment
Ottawa, Ontario, Canada

Miley, D.V.
E.G.&G. Idaho, Inc.
Idaho Falls, Idaho USA

Moore, J.J.
University of Minnesota
Minneapolis, Minnesota USA

Moore, V.
CANMET
Ottawa, Ontario, Canada

Northwood, D.O.
University of Windsor
Windsor, Ontario, Canada

Olsen, A.R.
Oak Ridge National Laboratory
Oak Ridge, Tennessee USA

Parsons, D.E.
CANMET
Ottawa, Ontario, Canada

Peggs, I.D.
Hanson Materials Engineering Ltd.
Edmonton, Alberta, Canada

Petzow, G.
Max-Planck-Institut
Stuttgart, West Germany

Rafiee, M.H.
C.E. Power Systems
Chattanooga, Tennessee USA

Rapp, R.A.
Ohio State University
Columbus, Ohio USA

Raynaud, G.M.
IREQ
Varennes, Quebec, Canada

Rice, R.C.
Battelle-Columbus Laboratories
Columbus, Ohio USA

Rondelli, G.
CNR
Cinisello B., Italy

Roper, H.
Sydney University
Sydney, NSW, Australia

Roth, T.A.
Kansas State University
Manhattan, Kansas USA

Rungta, R.
Battelle-Columbus Laboratories
Columbus, Ohio USA

Ryvola, M.
Alcan International Ltd.
Kingston, Ontario, Canada

Shah, N.A.
University of Minnesota
Minneapolis, Minnesota USA

Shaw, W.J.D.
University of Calgary
Calgary, Alberta, Canada

Shehata, M.T.
CANMET
Ottawa, Ontario, Canada

Sherman, I.
University of Toronto
Toronto, Ontario, Canada

Sinha, A.K.
University of Minnesota
Minneapolis, Minnesota USA

Sinigaglia, D.
CNR
Cinisello B., Italy

Skrastins, D.P.
Hanson Materials Engineering Ltd.
Edmonton, Alberta, Canada

Smith, F.
University of Toronto
Toronto, Ontario, Canada

Smolik, G.R.
E.G.&G. Idaho, Inc.
Idaho Falls, Idaho USA

Sutherland, J.K.
New Brunswick Electric Power
  Commission
Fredericton, New Brunswick, Canada

Sweet, R.C.
Philips Laboratories
Briarcliff Manor, New York USA

Thornley, J.C.
New Brunswick Research and
  Productivity Council
Fredericton, New Brunswick, Canada

Thrasher, T.H.
Oak Ridge Y-12 Plant
Oak Ridge, Tennessee USA

Tortorelli, P.F.
Oak Ridge National Laboratory
Oak Ridge, Tennessee USA

Vincentini, B.
CNR
Cinisello B., Italy

Wasielewski, R.C.
Virginia Polytechnic Institute
Blacksburg, Virginia USA

Wei, D.C.
Kelsey-Hayes Company
Ann Arbor, Michigan USA

Whittle, D.P.
Lawrence Berkeley Laboratory
Berkeley, California USA

Wu, D.C.
University of Toronto
Toronto, Ontario, Canada

Yanishevsky, M.
University of Toronto
Toronto, Ontario, Canada

PREFACE

Volume 12 of the *Microstructural Science* series records the technical presentations made at the 16th Annual IMS Technical Meeting of the International Metallographic Society (IMS) held at the Calgary Convention Center, Calgary, Alberta, Canada, July 25-28, 1983.

The papers presented provide information on a broad range of microstructural topics. Nearly half of the papers concerned microstructural aspects of corrosion, corrosion control and corrosion failure analysis. These papers describe various analytical approaches for studying corrosion and for diagnosing corrosion failure mechanisms. Other papers document the microstructural features of corrosion on a wide range of materials or coatings and in environments ranging from within the human body to outerspace. The remaining papers covered the following topics: microstructural-property relationships, microstructural-fracture relationships, and advances in metallographic techniques.

The annual IMS meeting also hosted the judging and premier showing of the International Metallographic Exhibit (sponsored jointly by IMS and the American Society for Metals). The exhibit maintained the high standards of technical and aesthetic quality for which it is internationally recognized. The exhibit awards are listed after the papers.

The success of the meeting was the result of the efforts of many persons, either directly or indirectly. The meeting was held under the able cognizance of the recent Past-President, George F. Vander Voort, and his Vice-President (now President), James E. Bennett, and the Treasurer, Richard D. Buchheit. The general chairmanship was performed by William E. White. The technical program chairman and vice chairman were Derek O. Northwood and Stuart A. Shiels, respectively.

The past eleven volumes of *Microstructural Science* were edited and typed for camera-ready copy by James L. McCall and Connie McCall, respectively. The IMS officers, board members and general membership extend their deep gratitude to Jim and Connie for their many years of devoted, outstanding service to IMS and the metallographic community.

For Volume 12, editing was primarily the responsibility of George F. Vander Voort. Mrs. Anne Louise Egan typed the camera-ready copy for our new publisher, the American Society for Metals. The success of this volume is also the result of participation of IMS members, their colleagues, and their organizations in presenting the results of their work at this meeting.

Derek O. Northwood, Technical Chairman
University of Windsor
Windsor, Ontario, Canada

William E. White, General Chairman
Petro-Canada, Inc.
Calgary, Alberta, Canada

George F. Vander Voort, Series Editor
Carpenter Technology Corporation
Reading, Pennsylvania USA

Metallography of Corrosion and Corrosion Related Failures

Chairpersons: I. LeMay, J.E. Bennett, M.R. Louthan, P.J. Kenny, D.W. Stevens, A. Kaul, and M.E. Blum

A GENERAL METALLOGRAPHIC APPROACH TO CORROSION FAILURE ANALYSIS

M. H. Rafiee*

ABSTRACT

A six-stage procedure for the management of a successful failure analysis is outlined and discussed briefly. While the outline considers all types of failures, particular attention is directed to corrosion-related failures. The described methodology will permit failure analysts to achieve greater success in their endeavors.

INTRODUCTION

A successful failure analysis requires a systematic and careful investigative approach as well as accurate and knowledgeable analytical work. In order to achieve this combination, a six stage method has been developed as a guideline for failure analysis. Table 1 presents the guide lines which could be used as a check list during the investigative processes.

Table 1. Stages in the Investigation of Failures

Stage I:	Site Inspection and History
Stage II:	Non-Destructive Examination
Stage III:	Destructive Examination
Stage IV:	Interpretation of Data
Stage V:	Conclusions
Stage VI:	Recommendations

---

* C. E. Power Systems, 911 W. Main Street, Chattanooga, TN 37402 USA.

Briefly described, stage I consists of site inspection and component history as detailed in Table 2. Items such as photographic documentation, eyewitness information as well as design, operating condition, and material specification belong to this category. Information on earlier failures at the same or related locations can often provide valuable clues. While the importance of this stage can hardly be over-emphasized, the objectives of this stage are among the hardest to achieve and, therefore, are often neglected.

Table 2. Stage I. Site Inspection and History

A. Site Inspection

1. Survey of Failure Site

    a) Condition of failed component
    b) Location of detached pieces (if any)
    c) Photographic documentation of site

2. Collection of Eyewitness Information

    a) Operating condition at time of failure
    b) Sequence of events leading to failure
    c) Apparent location of inception of failure

B. Component History

1. Service Environment

    a) Design environment
    b) Recorded departure from norm

2. Material Specifications

    a) Dimensions
    b) Composition
    c) Mechanical properties
    d) Hardness
    e) Heat treatment

3. Archives

    a) Sample of material never used in service
    b) Sample of material operating outside of damaging environment

Stage II deals with nondestructive examination encompassing various methods of nondestructive testing as well as visual characterization of the fractured surfaces and selection of the suitable areas for destructive examinations (Table 3). It is desirable in many cases to perform this type of inspection prior to removal of the sample, in order to assure that useful evidence is not destroyed.

Table 3. Stage II. Nondestructive Examination

A. Nondestructive Testing

   1. Visual Inspection
   2. Magnetic Particle Inspection
   3. Ultrasonic Testing
   4. Liquid Penetrant Inspection
   5. Eddy Current Testing
   6. Radiography
   7. Experimental Stress Analysis: Residual Stress Measurements

B. Macroscopic Fractography

   1. Selection and Presentation of Fracture Surfaces
   2. Visual Characterization of Surface Features

      a) Chevron marks
      b) Beach marks
      c) Surface texture
      d) Angle of fracture propagation
      e) Point of fracture initiation

Stage III is destructive examination of the selected areas for chemical composition, mechanical properties, and metallographic analysis (Table 4). This stage generally determines material integrity and its adherence to design specifications.

Table 4. Stage III. Destructive Examination

A. Chemical Analysis

　1. Determination of Material Composition
　2. Analysis of Surface Constituents and Deposits

B. Mechanical and Physical Properties Testing

　1. Mechanical Properties Testing

　　a) Hardness tests
　　b) Tensile tests
　　c) Impact tests
　　d) Others: fatigue, fracture toughness, creep tests

　2. Physical Properties Testing

　　a) Melting point
　　b) Conductivity
　　c) Coefficient of expansion

C. Examination of Selected Areas of Fracture Surfaces in SEM or TEM

　1. Characterization of Microscopic Surface Features

　　a) Dimple mode of rupture
　　b) Cleavage mode of rupture
　　c) Striations
　　d) Intergranular rupture

　2. Identification of Fine Microscopic Features

　　a) Carbide morphology
　　b) Dislocation patterns
　　c) Particle identification

D. Selection and Preparation of Metallographic Specimens

　1. Evaluation of Microstructure in Areas of Interest

a) Fracture path and fracture mode
b) Grain size
c) Microstructural constituents
d) Microhardness
e) Inclusion types and content

Stage IV consists of identification and classification of the types of damages present in areas chosen for metallographic analysis from stages II and III. The classification (Table 5) is made first in broad terms such as fatigue, corrosion, overheating, etc. and then divided into more specific sub-categories. For instance, the broad heading corrosion contains nine specific types of corrosion attack such as stress corrosion, corrosion fatigue, intergranular attack, and so on.

Table 5. Stage IV. Interpretation of Data
(Identify Nature of Observed Damage)

A. Fatigue

1. High Cycle
2. Low Cycle
3. Unidirectional Bending
4. Alternating Bending
5. Rotational Bending
6. Torsional Bending

B. Corrosion

1. General Corrosion
2. Pitting Attack
3. Corrosion Fatigue
4. Stress Corrosion
5. Intergranular Attack
6. Crevice Corrosion
7. Galvanic Corrosion
8. High-Temperature Corrosion
9. Fretting Corrosion

C. Overheating

1. Creep-Rupture Damage
2. Deformation
3. Thermal Fatigue
4. Grain Growth
5. Precipitation
6. Decomposition

D. Brittle Fracture

    1. Cleavage Mode
    2. Intergranular Mode

E. Tensile Failure

    1. Ductile Rupture
    2. Non-ductile Rupture

F. Hydrogen Effects

    1. Attack   3. Blistering
    2. Embrittlement   4. Cracks

G. Wear

    1. Adhesive Wear   4. Erosion-Corrosion
    2. Abrasive Wear   5. Surface Fatigue
    3. Erosion

Stage V identifies the primary reason/reasons for the failure (Table 6) among the types of damage noted in stage IV. The selection is based on the data gathered in previous stages, which enables the analyst to explain the reason for the failure rather than merely identifying the mode of rupture.

Table 6. Stage V. Conclusions
(Identify Cause of Failure)

A. Fatigue

    1. Poor Design   4. Damage in Handling
    2. Material   5. Service Related Damage
       Deficiencies   6. Abnormal Loading
    3. Processing   7. Fabrication Defects
       Deficiencies

B. Corrosion

   1. Wrong Material
   2. Unanticipated Environment
   3. Improper Processing

C. Overheating

   1. Wrong Material
   2. Excessive Service Temperature
   3. Disruption of Heat-Transfer
   4. Loss of Coolant

D. Brittle Fracture

   1. Processing Deficiencies
   2. Material Deficiencies
   3. Service Induced Embrittlement
   4. Mechanical Shock
   5. Thermal Shock

E. Tensile Failure

   1. Wrong Material
   2. Improper Processing
   3. Overload
   4. Wrong Design

F. Hydrogen Effects

   1. Wrong Material
   2. Unanticipated Environment
   3. Corrosion
   4. Surface Treatment (Pickling, Electroplating, etc.)
   5. Fabrication Procedures

G. Wear

   1. Abnormal Loading
   2. Lubricant Breakdown
   3. Wrong Material
   4. Improper Processing

Stage VI provides recommendations to prevent occurrence of similar damage in the future (Table 7).

Table 7. Stage VI. Recommendations

A. Improve Component Design
B. Improve Material
C. Alter Service Environment
D. Improve Quality Controls Governing Fabrication and Handling of the Component

SUMMARY

The experienced analyst can easily see not only the advantages but also the necessity for a complete set of data in order to achieve a meaningful analysis of the failure. However, those active in this field are also aware that in many cases they simply have to do with what is available. Economical, political, human, and time factors often prevent access to a great deal of information leaving the analyst with a poorly preserved piece of specimen and a great deal of pressure for some sort of answer. The guidelines presented in the six stages outlined (Table 2-7) will help the failure analyst organize their efforts for maximum efficiency in dealing with corrosion-related failures.

METALLOGRAPHY AND FAILURE ANALYSIS OF CORROSION RELATED FAILURES

Philip J. Kenny

ABSTRACT

Six failure analyses related to corrosion are presented. The materials involved are AISI 304, 316 and 321 stainless steels, admiralty brass, leaded red brass and silicon brass. The types of corrosion, either singly or in combination with each other, are stress-corrosion cracking, pitting, oxygen pitting, corrosion fatigue, hydrogen-sulphide attack, dezincification, bimetallic corrosion, oxygen-assisted chloride pitting, and bacteria-induced corrosion.

INTRODUCTION

This paper is a presentation of selected failures that illustrate the results of corrosion in its many forms. The underlying physical and chemical principles will not be discussed in order to present as many case studies as possible.

CASE 1. STRESS-CORROSION CRACKING OF AISI 316 STAINLESS STEEL

Several air heat exchangers failed in service in a pulp and paper operation. The tubes are made from AISI 316 stainless steel with an extruded aluminum fin mechanically bonded to the outside. Originally, the failures were blamed on poor tube to header welds. The units were sent back to

---

* Ontario Research Foundation, Sheridan Park Research Community, Mississauga, Ontario, Canada L5K 1B3.

 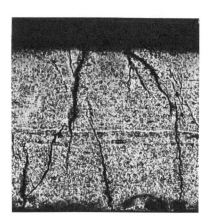

Figure 1. (Left) Branched transgranular cracks in AISI 316 stainless steel U bend (etched with water, HNO_3 and HCl, 50X).

Figure 2. (Right) Cracking in the U-bend tube originated at both the inner and outer surfaces (etched with water, HNO_3 and HCl, 20X).

the manufacturer for repair. Some of the units failed the hydrostatic test after they were repaired. Another laboratory reported the presence of stress-corrosion cracking. However, examination of the other half of the failed tubes by the Ontario Research Foundation (ORF) did not reveal any metallurgical defects. Further samples were obtained without finding any leaks. Finally a tube was found that had a leak at the "U" bend.

A metallographic section was cut through the leak. Microscopic examination revealed the presence of branched transgranular cracks, Figure 1. These cracks are characteristic of stress-corrosion cracking. The following facts were revealed by our investigation:

1) Cracking originated at both the inside and the outside surfaces of the tube, Figure 2.

2) Not all tubes leaked.

3) Tubes from all the failed units were mixed together during repairs.

4) The cracking appears to be highly localized in the tubes that failed.

5) The hardness of the tubes was at or slightly above the maximum of 90 HRB specified in SA249.

6) The tubes were not dried after hydrostatic testing.

The results of the investigation led to the following conclusions:

1) Only some of the tubes failed and these did so by stress-corrosion cracking.

2) The highly local nature of the cracking and the fact that it originated at both the water (inside) and the air (outside) surfaces tends to rule out operational irregularities as a cause of the failures.

3) The most probable primary cause of the stress-corrosion cracking is local high residual stresses indicated by the areas of high hardness in the tubes.

4) The corrosive environments, low halogens in the water and airborne corrodents found normally in a pulp and paper mill, were all that were required in the presence of high residual stresses in the tubes to initiate stress-corrosion cracking.

Our recommendations included:

1) Use a low-carbon grade of stainless steel such as 316L to facilitate formation of the tube without producing excessive residual stresses.

2) Segregate future failed units until it can be determined if the failure is related to operating pressure or some other unique cause.

CASE 2. STRESS-CORROSION CRACKING AND GALVANIC CORROSION OF ADMIRALTY BRASS

Not all case of stress-corrosion cracking result from the effect of chlorides on austenitic stainless steel. In this investigation, 3/4-inch admiralty brass tubes were failing in a heat exchanger. Only some of the tubes were failing. The heat exchanger cooled air by passing river water through the inside of the tubes.

Figure 3. (Left) Straight branched transgranular and intergranular cracks in failed admiralty brass heat exchanger tubes (unetched, 50X).

Figure 4. (Right) Intergranular cracks in failed admiralty brass heat exchanger tubes (unetched, 50X).

Subsequent examination of failed and unfailed tubes revealed the following:

1) The cracks were originating at the inside surface of all tubes, not just the failed ones.

2) Straight branched and unbranched transgranular cracks and general intergranular cracks were found, Figures 3 and 4.

3) Sand deposits were found in the tubes.

4) The pH of the river water varied from 7.5 to 8.3. In addition, sulphates are known to be present in the river water.

5) All the tubes that were examined had the same hardness, 105 VHN (100g) (59 HRB equivalent).

6) The wall thickness of all tubes ranged between 1.19-mm to 1.27-mm (0.047 to 0.050 inch).

The results of the failure analysis led to the conclusion that two and possibly three corrosion mechanisms were operative:

1) General intergranular corrosion at the inside surfaces of the tubes.

2) Transgranular stress-corrosion cracking, probably the result of sulphates under basic conditions. Some intergranular stress-corrosion cracks can form under neutral (6-8 pH) conditions.

3) Dezincification occurring as the result of galvanic corrosion under the deposits in the tubes.

Our recommendations were:

1) Use a closed-loop water system to eliminate sulphates, ammonia, etc.

2) Run trials on one unit with tubes of other alloys such as 80-20 copper-nickel or 70-30 copper-nickel to evaluate their performance prior to any large scale retubing operations.

CASE 3. CORROSION FATIGUE, STRESS-CORROSION CRACKING AND HYDROGEN-SULPHIDE ATTACK OF AISI 304 STAINLESS STEEL

Three separate corrosion mechanisms were involved in the failure of a 6-1/4-inch ID, 1/4-inch wall, AISI type 304 stainless steel pipe elbow. Examination of the inner surface of the elbow revealed that it was covered with a crazed pattern of cracks. The pattern was not uniform, being barely visible in some areas and quite deep in others. The area immediately around the leak was severely cracked.

Metallographic sections were prepared from the leak area and from a location 1-in. removed. The major cracks, including the one that penetrated the wall, tend to be wide-mouthed, tapering to a blunt tip, with corrosion products filling much of the crack space, Figure 5. This is characteristic of corrosion fatigue. The leakage crack is shown in Figure 6. There is no evidence that the weld, per se, contributed to the failure.

Figure 5. (Left) Wide-mouth, blunt-tipped crack in AISI 304 stainless steel pipe elbow (unetched, 12X).

Figure 6. (Right) Through-wall crack that produced leakage in AISI 304 stainless steel elbow (etched with water, HNO_3 and HCl, 10X).

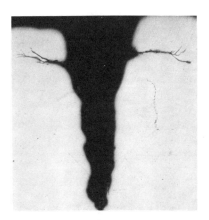 

Figure 7. (Left) Branched transgranular cracks that originated at large cracks in AISI 304 stainless steel elbow (unetched, 50X).

Figure 8. (Right) Intergranular crack network in AISI 304 stainless steel elbow (unetched, 50X).

The second type of cracking originated at some of
the major cracks, Figure 7. These cracks are branched and
transgranular which is characteristic of stress-corrosion
caused by chlorides.

The third crack mode, an intergranular network, is
most probably the result of hydrogen sulphide attack, Figure 8.
This type of corrosion, in combination with cyclic stress loads,
can develop into corrosion-fatigue; however, the corrosion-
fatigue cracks, Figure 5, are not necessarily associated with
the intergranular networks. This leads to the conclusion
that these two crack modes are independent of each other.

The 13-year service life of the elbow makes it
difficult, if not impossible, to determine the order of the
corrosion mechanisms or the length of time that it has
taken to reach the present state of degradation after the
initiation of corrosion. Based on the long service life
that the present material has given, it was recommended
that it be used again.

CASE 4. DEZINCIFICATION OF SILICON BRASS IN
        CONTACT WITH LEADED RED BRASS

After six years of service, three water valves on a
copper water line in a residential building were found to be
inoperative. These shut-off valves were closed to allow repair
work to be carried out in another part of the system. After
the repairs, these values could not be opened. They were
then removed for failure analysis.

Macroscopic examination of the valves after
disassembly revealed that all three failed at the key that
holds the spindle in the gate. In addition, the color near
the key has changed from yellow to red-brown. This strongly
suggests that dezincification has taken place.

A transverse metallographic section was prepared
from the spindle. Microscopic examination confirmed that
approximately two-thirds of the section had been affected by
dezincification, Figure 9. Higher magnification reveals the
striking difference between the sound metal, left, and the
dezincified material, right, Figure 10.

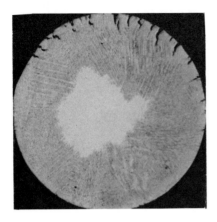 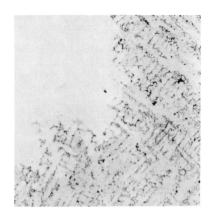

Figure 9. (Left) Dezincification of silicon brass spindle (unetched, 5.5X).

Figure 10. (Right) Interface between sound metal and dezincified region of silicon brass spindle (unetched, 50X).

The mode of failure was determined to be dezincification; however, the cause was not yet clear. The next step was to determine the composition of the components involved. The gate was found to be made from leaded red brass (85-5-5-5) while the spindle was made from silicon brass. The difference in alloy content of the two parts in contact in an electrolyte suggested a galvanic effect may have been operative. To determine this, the open-circuit potential in tap water was measured for both components. The gate was found to be -0.038V while the spindle was more anodic, i.e., more likely to corrode, at -0.075V. The potential between the spindle and the gate at 3-cm spacing was found to be 0.0522V with the spindle being anodic. From this we concluded that the valves failed by dezincification that was the result of bimetallic galvanic corrosion.

During our investigation we learned that it is common in the valve industry to use various alloys for components in the same valve. It may be advisable to review the design of such valves to see if there is an alloy that is suitable for all components in a given valve.

 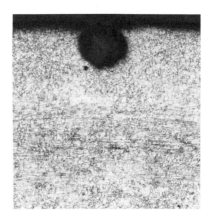

Figure 11. (Left) Typical pit appearance in AISI 321 stainless steel heat exchanger tube (etched with water, HNO_3 and HCl, 50X).

Figure 12. (Right) Typical pit appearance on the outside surface of an AISI 321 stainless steel blow-down tube (etched with water, HNO_3 and HCl, 50X).

CASE 5. OXYGEN-ASSISTED CHLORIDE PITTING OF
         AISI 321 STAINLESS STEEL

AISI type 321 stainless steel tubes - from both a heat exchanger, used to pre-heat make-up water, and from a blow-down unit in the same system - were submitted for examination. The exchanger tubes are finned while the blow-down tubes are not. The units failed after only 3-months service. The make-up water passes through the exchanger tubes and over the outside of the blow-down tubes. The make-up water is passed through a zeolite softener bed.

Selected tubes were sectioned longitudinally to allow the examination of the inside surfaces. Macroscopic examination revealed that the leaks were the result of localized pitting attack originating at the water side surfaces of the tubes.

Metallographic sections were prepared from both sets of tubes. Microscopic examination revealed that the pits have a small mouth with a large subsurface cavity. Figure 11 shows a typical pit at the inside surface of an exchanger tube while Figure 12 illustrates a similar pit in the outside surface of a blow-down tube.

This type of pitting is typical of chloride pitting of austenitic stainless steel. However, it should be noted that no pitting was found in other areas of the system where the chloride content of the process water is higher. This was attributed to the fact that they are down stream from a deaeration unit.

Our investigation led us to conclude that the pitting is the result of a synergistic effect of chlorine and oxygen in the make-up water. Since it is not possible to install a deaeration unit before the heat exchangers it was recommended that a molybdenum-bearing stainless such as 316L or 317L be used instead of 321.

CASE 6. BACTERIAL-INDUCED CORROSION OF AISI TYPE 304 STAINLESS STEEL TANKS

All failure analysis investigations are not simple and straightforward. This investigation involved two AISI 304L acid storage tanks and one AISI 304L spent solvent tank from a sewage treatment facility. After installation, these tanks were hydrostatically tested using sewage effluent. No leaks were found and the tanks were left with the effluent in them for 12 to 24 months when they were drained and filled with nitric acid in preparation for service. Three weeks later the two acid tanks were found to be leaking from the bottom only. The tanks were drained, rinsed and inspected. A black residue was still present on the tank bottoms. It was removed by wiping to expose numerous small pits at the weld toes and in areas remote from the welds.

Samples were cut from the tanks and sent to our lab for investigation. Macroscopic examination revealed that all of the pits originated at the inside of the tank, mainly near weld toes. However, no general corrosion of the heat-affected zone was found.

Metallographic sections were prepared through several of the pits and from noncorroded areas. The microstructure was found to be normal for 304L stainless steel, Figure 13. In addition, no signs of carbide precipitation or other abnormalities were found in the heat-affected zone, Figure 14.

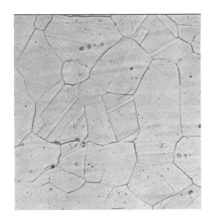 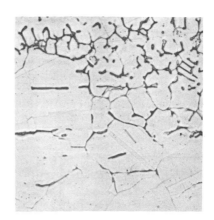

Figure 13. (Left) Normal microstructure of AISI 304L austenitic stainless steel (etched with oxalic acid solution, 500X).

Figure 14. (Right) Normal microstructure of the heat-affected zone in the AISI 304L stainless steel tank (etched with oxalic acid solution, 500X).

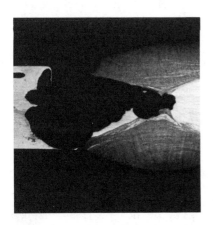 

Figure 15. (Left) Appearance of pits adjacent to welds in the AISI 304L stainless steel tank (etched with oxalic acid solution, 3.5X).

Figure 16. (Right) Appearance of pits in the bottom of the AISI 304L tank bottom (unetched, true size).

The pits were found to have a small mouth with a large subsurface cavity filled with a reddish brown slimy deposit, Figure 15. In addition, an oxalic acid test for sensitization proved negative as did a hot Murakami's etch for the presence of sigma phase.

The carbon content of the parent metal and the weld metal was found to conform to the applicable specifications.

At this point in the investigation, the failure of the spent solvent tank was reported. Examination of samples from this tank revealed that the pitting was located in a depressed area of the tank bottom near a suction hole, beneath a black residue. No welds were involved. A typical pit is illustrated in Figure 16. Weld spatter was found on the tank bottom. There does not appear to be a necessary cause and effect relationship between the weld spatter and the pits.

At this point, we endeavoured to find an explanation that would adequately explain the failure of all three tanks. To this end, analysis was performed on the three distinct corrosion products found in one of the pits. The results are listed below:

1) The brown iridescent film was found to be amorphous and could not be identified by diffraction; however, a small amount of metallic iron was found.

2) The dendrites were found to be metallic iron.

3) The white crystals were identified as calcite and vaterite, two forms of calcium carbonate. In addition, a small amount of metallic iron was found.

It is significant that only iron was found in the pit, since it would be expected that chromium and manganese would also be present if crevice or galvanic action was solely responsible for the corrosion. This fact, coupled with the reddish-brown slime deposit in the pit, points to the presence of an iron-type ion-concentrating bacteria. Corrosion due to this type of bacteria can take several forms such as:

1) The formation of oxygen concentration cells.

2) The deposits create anaerobic conditions which harbor sulphate-reducing microbes which actually cause the corrosion.

3) The production of ferric and manganic chlorides which cause severe local attack by penetrating the oxide film.

Microbial corrosion is normally located at or near welds or other small surface imperfections such as weld spatter or other debris which act as incubation sites.

The results of this investigation led us to conclude that the acid tanks failed by chloride-induced pitting that was initiated by microbial activity. Further, the spent solvent tank failed by a similar mechanism although the type of pitting appeared to be related to anaerobic conditions described in No. 2 above.

From these results, it appeared obvious that the use of the effluent for the hydrostatic test and the failure to remove it and clean and dry the tanks after, was the prime cause of the failure. However, a report by another investigator of the presence of grain boundary carbides in some areas led us to examine more metallographic sections and recheck the carbon content. This time we also found isolated areas of carbide precipitation and carbon levels in excess of 0.03% that were not necessarily related to welds. These areas would act as preferential corrosion sites if they occurred at the surface. The majority of the areas that we found had the highest concentration of carbides at the center of the plate and not at the surface.

Our final conclusion is that localized carbide segregation in the original plate served as preferential corrosion sites for the bacterial attack which developed chloride-induced pits, or anaerobic-induced pits that resulted in leaking of the tanks. Had the tanks been hydrostatically tested in a proper manner, it is possible, if not probable, that the pitting may not have occurred.

THE METALLOGRAPHY OF ELECTROCHEMICAL CORROSION TESTING

Ian D. Peggs* and David P. Skrastins*

ABSTRACT

Potentiodynamic anodic polarization is a very powerful and rapid means of evaluating corrosion characteristics of solid/fluid combinations. However, its acceptance by industry has been limited due to the superficial sophistication of the technique. To simplify an understanding of the method, we have related metallographic surface changes to the electrochemical corrosion curves for samples of Inconel 625, Monel 400, Nickel 200 and plain-carbon steels in various waters. General- and pitting-corrosion rates are compared.

INTRODUCTION

For maximum operating and planned shutdown efficiency, it is essential to know the corrosion rates of system materials. Ultimate reliability is attained by using test coupons <u>in situ</u> in real time but it is time consuming and cannot be used rapidly to assess the effects of changing chemistry, temperatures, etc.

Electrochemical probes have been developed and inserted into systems to directly convert measured corrosion currents into corrosion rate displays. Under optimum conditions, these probes provide a corrosion rate based on the assumption of uniform material loss from all surfaces, i.e., a general corrosion rate. Rapid measurements can be made and the

---

* Hanson Materials Engineering (Western) Ltd., Edmonton, Alberta, Canada.

effects of chemistry changes, etc. can be quickly evaluated. However, the most serious forms of corrosion are localized effects, such as pitting and crevice corrosion, and the simple electrochemical probe provides no information on such processes.

However, laboratory electrochemical equipment [1] has been available for some time which can evaluate the full corrosion characteristics of any metal-conducting fluid combination with respect to general corrosion rate, passivation, approximate pitting rate, and susceptibility to crevice corrosion. A portable unit is now available [2] which facilitates actual system testing in the field.

ELECTROCHEMICAL PRINCIPLES

For corrosion to occur, a fluid electrolyte is required to maintain contact between two electrodes: an anode at which metal oxidation occurs, and a cathode at which reduction occurs. In working systems the anode and cathode can be different areas on the same component.

The dissolution of the metal at the anode is accompanied by production of electrons -- an electric current:

$$M \rightarrow M^+ + e^-$$

Similarly, at the cathode some species in the fluid is reduced:

$$Z^+ + e^- \rightarrow Z$$

Both reactions proceed at the same rate in an equilibrium system which is corroding but the net corrosion current, the sum of both electrode currents of opposite signs, is zero. It is, however, possible to measure the corrosion potential, E_{corr}, relative to a reference electrode placed in a corroding system. Subsequently, if the potential is changed to anodically (increase) or cathodically (decrease) polarize the system, the current associated with the individual oxidation and reduction reactions can be measured. When the currents are characterized, corrosion rates can be calculated.

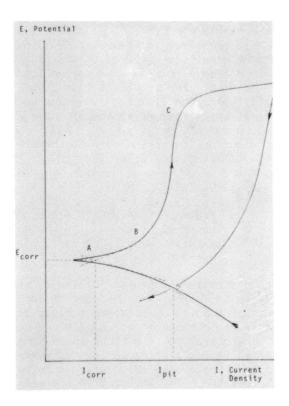

Figure 1. Idealized polarization curve.

The most useful potentiodynamic polarization cycle is to start with the test electrode cathodically polarized in the fluid of interest. The potential is slowly increased with respect to the reference electrode so that the sample passes through the equilibrium corrosion situation (E_{corr}) and then becomes anodically polarized, i.e., forced to corrode. At some level, the potential is reversed to the original value.

A plot is made of the potential and measured current density on the sample electrode to produce a curve similar to that which is idealized in Figure 1. Due to the large range of current densities measured, a log plot is used which obviates a change in sign when moving from anodic to cathodic activity. The resulting cusp in the curve clearly identifies the equilibrium corrosion potential, E_{corr}, and facilitates the definition of the equilibrium anodic corrosion current density, I_{corr} [3]. I_{corr} represents the uniform corrosion rate of the electrode material which is calculated from:

$$\text{rate} = 0.129\ I_{corr}\ M/D \quad (1)$$

where: M = equivalent weight of electrode material
               D = material density (g/cm^3), and
          I_{corr} = current density ($\mu A/cm^2$)

When the system is anodically polarized, the sample is forced to corrode and the current density increases until the build up of corrosion products forms a protective layer on the surface (B) and the sample is passivated. The corrosion rate stops increasing. As the potential increases a point is reached where the protective layer breaks down (C). At this point the potential is reversed. The amount of hysteresis shown by the curve is an indication of the susceptibility of the material to pitting and crevice corrosion. The current density at which the reverse scan intersects the initial cathodic scan or its extrapolation is proportional to the pitting rate assuming pitting is occurring [4]:

$$\text{Pitting rate} = 0.129\ I_{pit}\ M(D\ Ap) \quad (2)$$

where: Ap = area of pitting (cm^2)

In several instances, the area subject to pitting is found to be about 5% of the total electrode areas so as a first approximation this value could be used.

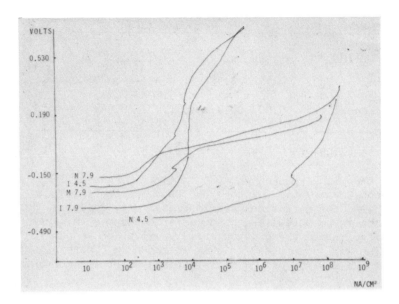

Figure 2. Anodic polarization curves for Inconel 625 (I), Nickel 200 (N) and Monel 400 (M) in chloride solutions of pH 4.5 and 7.9.

CASE EXAMPLES

An incinerator used for burning the last vestiges of vinyl chloride monomer, fabricated from Hastelloy C276 and plain-carbon steel, suffered severe crevice and pitting corrosion. The hot vinyl chloride gas is mixed with a water spray and the resulting HCl is recirculated to the water spray. The chlorides and varying pH fluids produce a very corrosive environment.

Inconel 625, Monel 400, and Nickel 200 were considered candidate materials to replace the initial plain-carbon steel components and were subjected to a non-reversing anodic potentiodynamic polarization analysis in two fluids of differing pH retrieved from the incinerator. Figure 2 shows the resulting curves and Table 1 shows the calculated general corrosion rates.

Table 1. General Corrosion Rates, mm/yr (MPY*)

Alloy	pH 4.5	pH 7.9
Nickel 200	0.100 (3.92)	0.002 (0.079)
Monel 400	- -	0.006 (0.235)
Inconel 625	0.002 (0.078)	0.001 (0.045)

* mils (0.001 inch)/year

The nickel curve (pH 4.5) initially shows a high current density (10^3 nA/cm^2) and as the potential increased the corrosion rate increased very rapidly. Small changes in potential can produce large changes in current density. Since there is no early passivation tendency, pitting will occur very quickly as shown in Figure 3a which shows the surface of the nickel sample after the corrosion cycle.

If the pH is increased to 7.9, the current density at equilibrium is reduced by almost two decades thus giving rise to a lower general corrosion rate. As the potential increased the rate of increase of current density decreased when passivation begins. However, the knee in the curve at 10^3 nA/cm^2 indicates that pitting has commenced and the current density continues to rise rapidly.

While the general corrosion rate is decreased, there is a severe pitting problem as shown by the surface condition after the test, Figure 3b. Small changes in potential again can result in large changes in current density making nickel very unstable in these solutions.

A very similar situation occurs with Monel 400 - a low general corrosion rate plus susceptibility to pitting, Figure 3c, at only slightly higher potentials. The Inconel curves show a very positive and extensive range of passivation with pitting probably just starting to occur as potentials reach 0.6 V.

The service corrosion conditions would probably equilibrate in the region of -0.150V and 10^3 nA/cm^2. Slight changes in interfacial conditions, and hence potential, would be unlikely to initiate pitting and accelerated corrosion processes. The Inconel is very stable in these solutions, Figures 3d and 3e.

Electrochemical Corrosion Testing / 31

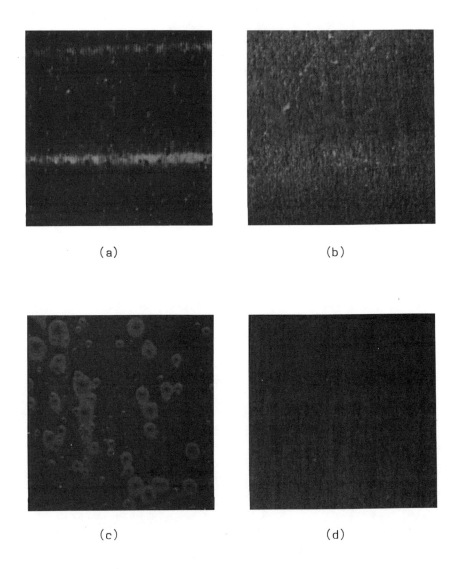

Figure 3. Surface condition of samples after polarization test. (a) Nickel, pH 4.5; (b) Nickel, pH 7.9; (c) Monel, pH 7.9,; (d) Inconel, pH 4.5; (e) Inconel, pH 7.9.

Figure 3.
(e)

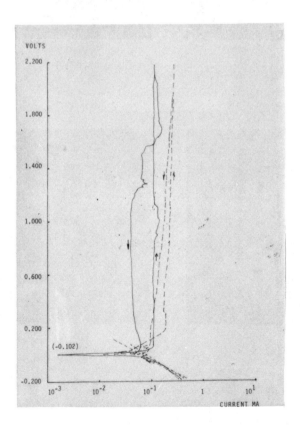

Figure 4. Reverse anodic polarization curve for piling in runoff water.

Electrochemical Corrosion Testing / 33

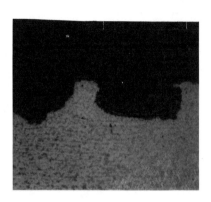 

Figure 5. (Left) Pitting corrosion in piling surface.

Figure 6. (Right) Waterside pitting on cooler.

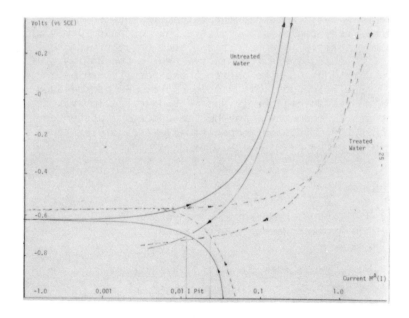

Figure 7. Corrosion curve for cooler material in treated and untreated feed water.

Some underground steel piling was periodically subjected to exposed-face wetting by ground run-off water containing 410-ppm chlorides and had suffered severe exfoliation. A reversible potentiodynamic polarization run, Figure 4, revealed the general corrosion rate to be 3 MPY (.076 mm/yr) which is not excessive but the pitting rate was 60 MPY. A section prepared normal to the surface of the piling, Figure 5, showed that pitting was the predominant corrosion mechanism.

A high pressure cooler showed extreme water side pitting, Figure 6, despite what was considered to be satisfactory feed-water inhibition. A reverse anodic polarization scan, Figure 7, with the cooler material in both untreated and treated feed-water, identified both general- and pitting-corrosion rates to be higher in the treated water. Water analysis showed that the treated water contained more chlorides, thus explaining the higher corrosion rates.

CONCLUSION

Potentiodynamic anodic polarization measurements can be performed very rapidly to evaluate the corrosion behavior of various materials in a specific system or determine the effects of changing system chemistry. While the analysis of results and methods used to obtain those results can be extremely sophisticated, very simple approaches can be used to determine reliable general- and pitting-corrosion rates of direct use to industry. The photomicrographs presented confirm that the corrosion mechanisms are as analyzed from the corrosion curves.

REFERENCES

1. Model 350 Corrosion Measurements System, EG & G Princeton Applied Research, Princeton, New Jersey.

2. Potentiodyne Analyzer, Petrolite Corporation, Houston, Texas.

3. R. Baboian, "Electrochemical Techniques for Corrosion," <u>NACE</u> (1977).

4. A. E. Woodson, "The Potentiodyne: A New Corrosion Monitoring Tool for the Lab and Field," <u>NACE Canadian Western Conference</u> (Feb. 1982).

FATIGUE AND CORROSION FATIGUE FAILURE SURFACES OF CONCRETE REINFORCEMENT

Harold Roper*

ABSTRACT

Some corrosion processes in the presence of chlorides, for steel embedded in concrete, are described and illustrated with the aid of scanning electron microscope EDXA data. Observations made of failure surfaces of reinforcements removed from the concrete beams after being subjected to sinusoidal load fluctuations at 6.7 Hz in air, 3 percent sodium chloride solution, or natural sea water are described. Reinforcement types studied included: hot-rolled mild steel bar, hot-rolled alloyed high strength bar, cold-worked high strength bar, galvanized bar of all these three types, nickel-clad bar and epoxy-coated bar.

INTRODUCTION

In a properly designed, constructed, and used structure, there should be no problem of steel corrosion in concrete during the design life. Such protection is attributable to the highly alkaline chemical environment in which the steel resides. The integrity and protective quality of the oxide film formed on reinforcing steel due to the alkaline condition may be reduced under unfavorable circumstances, in particular due to carbonation of the concrete and depassivation by chloride ions. The first portion of this paper presents information, using metallographic techniques, on the depassivation of reinforcing steel in the presence of Cl^- ions.

---

* Associate Professor, Sydney University, School of Civil and Mining Engineering, Sydney, NSW 2006 Australia.

Although there are no reports of failures in primary elements of structural concrete members due to fatigue, there are a number of reasons for increasing concern. Adoption of strength design procedures and use of higher strength materials require these members to perform under higher stress levels. New uses are being made of concrete, such as in offshore structures and railroad ties, where the full design loading may be repeated for a very large number of cycles, and occasional overloadings are expected. The second portion of this paper describes some metallographic features of failed reinforcing bars after being subjected to sinusoidal load fluctuations at 6.7 Hz in air, 3 percent sodium chloride solution, or natural sea water. For details of loading procedures, fatigue life and other aspects of the test program reference can be made to a series of papers by Roper [1-3].

PROCEDURE AND RESULTS

Depassivation by Cl^-

In order to study some aspects of depassivation by Cl^-, a series of corroded reinforcing bars were removed from a structure, whose history was known, and were sampled and examined by scanning electron microscopy (SEM), energy-dispersive x-ray analysis (EDXA), and electron spectroscopy for chemical analysis (ESCA). After preparation of the steel specimens using chloride-free polishing lubricants, a search was made for evidence of the presence of chlorides on the surface of the corroding steel. Figure 1 shows a scanning electron image of corrosion product present at an embayment in the steel. This figure also shows an x-ray scan image revealing the presence of chlorine. It should be noted that the chlorides are not evenly distributed, but are concentrated to a significant extent in the oxidation product close to the steel interface. Banding is obvious in both photographs.

In Figure 2, a similar pair of SEM photographs show corrosion product developing within the steel ahead of the advancing corrosion front. In the x-ray scan, tiny blebs of chloride-rich material can be seen. At a much higher magnification an eruption associated with such a bleb is shown in Figure 3. This photograph utilized secondary electrons which

Concrete Reinforcement Failures / 39

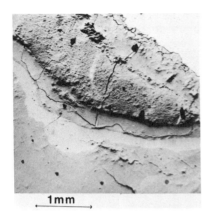 

Figure 1. (Left) Electron image showing unoxidized steel in upper right hand corner (mottled light gray), and banded oxidation products. (Right) X-ray image showing Cl concentration and the oxidation bands.

 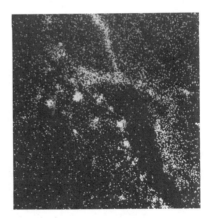

Figure 2. <u>Left</u> - Electron image showing bright metal (left), oxidation products (right) and the commencement of oxidation development on the surface of the metal. <u>Right</u> - X-ray image showing Cl concentration in the oxidation product and in spots where the metal is actively corroding.

accentuate the surface topography of the area. Scans using EDXA showed two predominant elements at such corrosion points, iron and chlorine (Figure 4). ESCA was used to determine that in such areas atomic percentages were up to 22.05% Cl, compared with 34.14% Fe and (by difference) 43.81% O. These figures indicate that in this local spot, about half the iron is present as iron chloride, a deliquescent salt at which the local pH will be about 3.5, as opposed to over 11.5 in the concrete. Under such conditions corrosion of the reinforcement will continue to occur despite the alkalinity of the surrounding concrete.

Observations from Light and Scanning
Electron Microscope Fractographs

Figure 5 shows a typical fatigue failure surface of a 24-mm diameter reinforcing bar which failed after 5.985×10^6 cycles of flexural loading of a reinforced concrete beam at a stress range of 214MPa tested in air. The fatigue crack initiation site, indicated by arrow A, is located at the intersection of the bottom longitudinal lug (bottom center), and a transverse lug (light profile right of bottom center). All fatigue fractographs of in-air beam tests of all hot-rolled and cold-worked bars, showed crack initiation sites at such intersections, indicating the sensitivity of the crack initiation mechanism, under this test condition, to geometric stress raisers.

The boundary of region B, the stable crack growth zone, and D, the rupture zone, is not completely concave to A, the stable fatigue crack preferring to propagate across the bar at a greater rate in proximity to the surfaces, designated by the letter C, than away from the bar surface. This is evidence of the presence of mild stress concentrations existing close to the bar surface, which accelerated crack propagation.

Transgranular cleavage and intergranular tearing were observed at the crack initiation site by scanning electron microscopy (Figure 6), whereas 8-mm from the initiation site, ductile fatigue striations were observed (Figure 7).

Figure 8 shows a fracture surface of a 24-mm diameter reinforcing bar which failed after 7.462×10^6 cycles of flexural loading of a reinforced concrete beam at a stress range of 149MPa tested in contact with sea water. Corrosion products, which could not be removed from the fracture

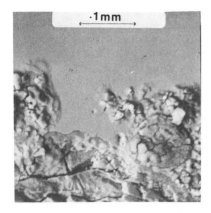 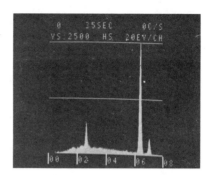

Figure 3. (Left) Secondary electron image of one of the points of high chloride concentration in Figure 2. The oxidation "eruption" is clearly seen to be related to the presence of Cl ions (Approx. X 2000).

Figure 4. (Right) Spectral lines obtained by EDXA showing the presence of Fe (two peaks on right) and Cl (peak to the left) in the area of an oxidation "eruption" (Figure 3).

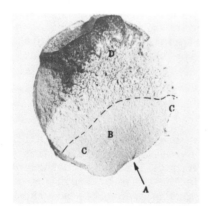 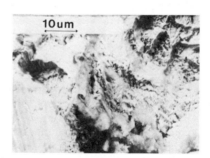

Figure 5. (Left) Fracture surface of reinforcement from beam tested in air.

Figure 6. (Right) Transgranular cleavage (lower left) and intergranular tearing (upper right), crack initiation site of Figure 5.

surfaces by ultrasonic treatment, appear as dark mottled areas (Figure 8). At least two crack initiation sites, A1 and A2, are apparent. It is possible that more than one crack has initiated from site A2. Unlike the earlier described specimen, which had only one smooth stable crack growth surface, this specimen has three such surfaces, B1, B2, and B3. Multiple crack initiation sites indicate an increased potential for crack initiation to occur.

It can be seen from Figure 8 that, unlike the former specimen, neither initiation site A1 nor A2 are located at the intersection of a longitudinal and a transverse lug. This, and the fact that there are more than one initiation site, shows that the corrosive environment reduces the dependence of the crack initiation stage on pre-existing stress concentrations. Corrosion pits at the crack initiation site A1 of Figure 8 are shown in Figure 9.

Failure surfaces of nickel- and epoxy-coated bars subjected to fatigue in the presence of chloride ions are similar in many respects to that shown in Figure 8. Corrosion occurs on the ribs and lugs of epoxy-coated bars where the perforated epoxy material separates the high and low pH zones. Intense corrosion of the steel surface of nickel-coated bars occurred even when the bar was removed from the concrete shortly after the conclusion of testing. This phenomenon is in contrast to the behaviour of zinc-dipped bars, which exhibited only a mild development of zinc salts on the whole bar surface, while no corrosion of the steel on the fracture surface was observed.

CONCLUSIONS

Explanation of the behaviour of the reinforcements in load tests have been aided considerably by metallographic studies of the type outlined. Decreases in fatigue life under corrosion fatigue conditions are explicable on the basis of rapid development of pits from which cracks can propagate, and the increase in propagation rate of the cracks. The longevity of zinc-coated bars in particular is related to the delay in crack initiation and propagation under cathodic protection conditions.

Concrete Reinforcement Failures / 43

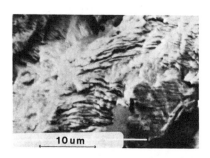 

Figure 7. (Left) Ductile fatigue striations, crack propagation zone of Figure 5.

Figure 8. (Right) Fracture surface of reinforcement from beam tested in sea water.

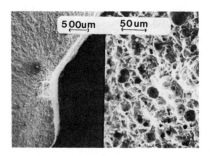

Figure 9. Crack initiation site A1 and corrosion pits, reinforcement from sea water test.

ACKNOWLEDGEMENTS

Funds supporting this work were in part obtained from the International Lead Zinc Research Organization and Argonne National Laboratories as portion of their OTEC research. The Electron Microscope Unit staff of the University of Sydney are thanked for their cooperation in these studies.

REFERENCES

1. H. Roper and G. Hetherington, "Fatigue Tests of Reinforced Concrete Beams in Air, Chloride Solution and Sea Water," American Concrete Institute Symposium on Fatigue of Concrete Structures, SP-75, pp. 307-330 (1982).

2. H. Roper, "Reinforcement for Concrete Structures Subject to Fatigue," International Association for Bridge and Structural Engineering, Proceedings of the Colloquium Fatigue of Steel and Concrete Structures, Lausanne, pp. 239-245 (1982).

3. H. Roper, "Investigations of Corrosion, Fatigue and Corrosion Fatigue of Concrete Reinforcement," National Association of Corrosion Engineers Symposium, Corrosion 83, Paper 169, Anaheim, pp. 1-12 (March 1983).

METALLURGICAL INVESTIGATION OF A TURBINE BLADE AND A
VANE FAILURE FROM TWO MARINE ENGINES

A. K. Koul*, R. V. Dainty*,
and C. Marchand*

ABSTRACT

The paper describes the circumstances surrounding the in-service failure of a cast Ni-base superalloy second stage turbine blade and a cast and coated Co-base superalloy first stage air-cooled vane in two turbine engines used for marine application. An overview of a systematic approach, analyzing the nature of degeneration and failure of the failed components, utilizing conventional metallurgical techniques, is presented.

The topographical features of the turbine blade fracture surface revealed a fatigue-induced crack growth pattern, where crack initiation had taken place in the blade trailing edge. An estimate of the crack-growth rate for the stage II fatigue fracture region coupled with the metallographic results helped to identify the final mode of the turbine blade failure. A detailed metallographic and fractographic examination of the air-cooled vane revealed that coating erosion in conjunction with severe hot-corrosion was responsible for crack initiation in the leading edge area. The role of large amounts of sub-surface porosity in the vane and the salt-rich marine environment in contributing to crack propagation has been discussed.

---

* Structures and Materials Laboratory, National Aeronautical Establishment, National Research Council, Ottawa, Ontario, Canada.

INTRODUCTION

It is the purpose of this paper to present a logical approach that systematically analyzes the nature of degeneration and failure of turbine engine vanes and blades utilizing conventional metallurgical techniques. This objective is achieved by referring to two separate marine engine failures, one involving a turbine blade disintegration and another one concerning a vane failure.

Examination of the second stage turbine rotor blades of the first engine (no. 1), revealed that the majority of the blade tips were ripped off. One exception to this was a blade with most of its airfoil section removed that exhibited a transverse fracture located some 15-20 mm from the blade root, Figure 1. The engine had operated at a maximum service temperature of 815 C and since its installation it had operated for 7000 h with 3000 start-ups. Apparently the engine failure occurred shortly after the ship's power was increased from 2,000 HP to 12,000 HP.

The other engine (no. 2) had a cracked first stage air-cooled turbine engine vane detected during a routine hot-section inspection. The vane had developed a well defined crack in the airfoil section, Figure 2. The airfoil had also suffered a considerable amount of erosion along the leading edge area and general pitting of the concave surface. At the time of the vane removal, the engine (no. 2) had logged 5500 h of running time, including 1100 start-ups. However, the complete service history of the vane could not be ascertained because the engine (no. 2) had been purchased as an overhauled engine.

Details of the service-induced degeneration in these failed components are described in the next few sections.

EXPERIMENTAL METHODS AND APPROACH

In any failure analysis, a close examination of the fracture topography is the first step in identifying the mode of fracture. Scanning electron microscopy (SEM) in conjunction with optical microscopy is generally used to that end.

Figure 1. Fractured second stage Alloy 713LC turbine blade sample. Arrow M indicates the metallographic sampling location.

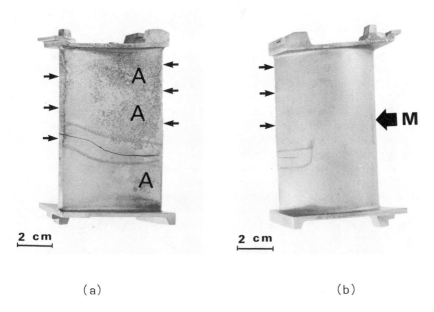

(a) (b)

Figure 2. Cracked first steps MAR-M302 Turbine engine vane in the as-received condition. (a) Concave airfoil surface; (b) Convex airfoil surface. Metallographic sampling location indicated by arrow M.

Optical examination is particularly important in determining the gross fracture features and in locating the presence or absence of any gross surface corrosion products. Scanning electron microscopy, on the other hand, can reveal the precise mode of failure, e.g., transgranular, intergranular or a mixed mode type fracture [1], including other details of crack propagation, e.g., creep, fatigue or a mixed mode of propagation [2,3].

Prior to any detailed SEM examination of the fractured surface of a high temperature turbine engine component, the use of a surface cleaning treatment is generally necessary to remove some of the adherent oxidation and/or corrosion products formed during service. Techniques such as ultrasonic cleaning and/or repeated replica cleaning are usually applied to these surfaces to obtain the desired results. In instances where a component is cracked (e.g., the vane sample in Figure 2), extreme care should be taken while opening the crack in the laboratory. This is because it is important to retain the original cracked surface features to obtain any meaningful fractographic results.

To assess the microstructural damage incurred by the hot turbine engine components during service, metallographic samples were taken from locations adjacent to the cracked or fractured areas, such as indicated by the arrows M in Figures 1 and 2. Samples from the cooler sections of these components, e.g., the root section of a turbine blade, are very useful for comparison purposes. A direct microstructural comparison with a service exposed component from the same row of the failed engine or with a new component (where available) is even better for failure assessment purposes. Metallographic samples are first examined in the unetched condition to study the distribution of primary carbides, sulphocarbides, inclusions and various forms of casting defects. Coated components are also best suited for an unetched microstructural examination, since coating degeneration effects, including the soundness of the coating-substrate interface, can readily be identified by this technique.

Metallographic examination in the etched condition provides useful information pertaining to service-induced microstructural degradation, e.g., creep voiding and triple-point cracking, topologically closed-packed (TCP) phase precipitation, grain boundary carbide precipitation and the degree of oxide penetration. Also, engine overtemperature effects (a feature not uncommon in service exposed hot turbine engine components) can sometimes be revealed in the etched microstructures. The presence of abnormal temperature gradients along or across the airfoil section may be determined from changes in carbide, γ' particle size and the matrix grain size in Ni-base superalloys [4] and from changes in matrix carbide particle size and grain size distribution in Co-base alloys. Because of the high magnification/resolution required, transmission electron microscopy, using carbon replica or thin foil techniques, is normally used for this purpose. An alternative method of analyzing any significant loss in strength due to the presence of abnormal temperature gradients involves room temperature hardness testing (HV30* data) at varying sampling locations. Microhardness testing may be necessary in the case of very thin components. A significant decrease in the hardness value relative to new material properties would be indicative of strength degradation.

In instances where minor phase precipitates or their degeneration (e.g., MC degeneration of $M_{23}C_6$, sigma and primary M_2SC precipitates) could have contributed towards overall failure, x-ray analysis of the component phases at varying sampling locations would be required to quantify the service induced degradation [5]. Electrolytic extractions are generally carried out to separate the minor phase content from the bulk samples. The Debye-Scherer x-ray diffraction patterns of these extractions are used to identify the minor phases. The relative diffraction line intensity measurement of different phases provides useful information pertaining to the extent of the minor phase changes that might have occurred during service. Similar techniques are employed to identify the principal phases present within the corrosive deposits, e.g., $NiCr_2O_4$, Na_2SO_4, Co_3O_4, etc., and these reaction products are invariably given a ranking in order of their relative abundance within the deposits.

RESULTS AND DISCUSSION

Millings from the blade and the vane samples were chemically analyzed to establish their respective alloy identity. The chemical analyses of both samples are shown in Table 1. It is evident that the second stage turbine blade composition is well within the Alloy 713LC specification,

---

* HV30 -- Vicker's hardness testing at 30 Kg load.

Table 1. Elemental Distribution of the Blade and Vane Material (wt. %)

Alloy Distribution	C	Mn	Si	Cr	Ni	Co	Mo	W	Nb	Ti	Al	Fe	Ta	Others
Turbine Blade Sample	0.024	-	-	12.0	Bal.	-	4.0	-	2.0	0.7	6.0	-	-	B 0.01 Zr 0.10
Alloy 713LC Spec.	0.05	-	-	12.0	Bal.	-	4.5	-	2.0	0.6	5.9	-	-	B 0.01 Zr 0.10
Turbine Vane Sample	0.80	0.10	0.15	21.0	0.8	Bal.	-	9.8	<0.1	-	-	-	8.4	B 0.005 Zr 0.14
MAR-M302 Spec.	0.85	0.10	0.20	21.5	-	Bal.	-	10.0	-	-	-	-	9.0	B 0.005 Zr 0.15

whereas the first stage turbine vane composition corresponds to MAR-M302 specification. MAR-M302 is a cast Co-base superalloy typically used in very high temperature (1200 C) corrosive environments, e.g., turbine vanes, and Alloy 713LC is a cast high strength Ni-base superalloy generally used in corrosive environments at high temperatures (750-900 C), e.g., turbine blade applications [7]. The fractographic and the metallographic analyses of the failed samples is described in the following sub-sections.

Second Stage Turbine Blade (Engine No. 1)

a. Fractographic Analysis

Optical examination of the fracture surface, Figure 3, revealed a relatively rough fracture surface that was generally covered with a dark oxide layer that probably resulted from exposure to the hot gases at elevated engine temperatures. The fracture surface exhibited a lightly shaded region, area A in Figure 3, located at the trailing edge of the airfoil section. Optical and SEM examination revealed that a large portion of the fracture surface in area A had been mechanically damaged and consequently the oxide layer had generally been removed. This damage was probably due to impacting of the mating fracture surfaces in this area during crack propagation.

Marine Engine Failures / 51

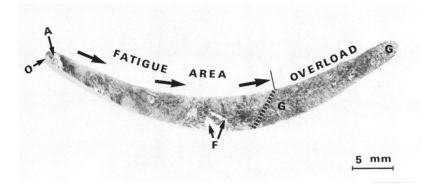

Figure 3. Turbine blade fracture surface showing different fracture zones.

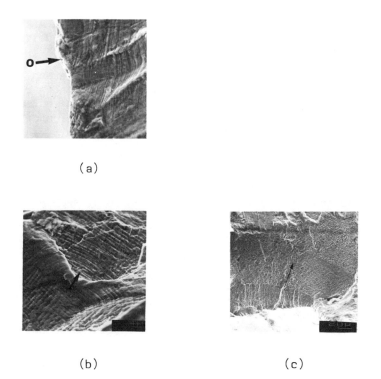

Figure 4. Fatigue crack initiation and propagation in the turbine blade sample. Arrow 'O' indicates initiation region and small arrows depict crack growth direction. (a) Initiation region; (b) Fatigue striations 0.1-mm from crack initiation region; (c) Crack growth bands 1-mm from crack initiation region.

Figure 5. Crystallographic facet in the initiation area.

Alternatively, this area might have been impacted and wiped by flying debris subsequent to final separation of the blade during engine disintegration. Although much of this area was damaged, examination in the SEM revealed several areas of well-defined fatigue striations and microscopic crack growth bands, Figure 4. A specific fatigue crack nucleation site could not be readily identified, but the orientation of the crack growth bands and striations in area A, Figure 3, suggested that fatigue crack initiation occurred at the trailing edge tip area as indicated by arrow O, Figures 1, 3, 4a and 5.

Also observed in the initiation region of area A, Figure 3, was a triangular or wedge-shaped protrusion approximately 1-mm in length, Figure 5. The smooth mirror-like surfaces of this protrusion or facet revealed crystallographic or cleavage-like features. Similar features have been observed previously in a wide variety of Ni-based superalloys, where the predominant mechanisms of crack initiation below 0.6 T_m* is extensive Stage I cracking along persistent slip bands [3,8-10]. This is true for both wrought and cast alloys. At temperatures greater than 0.5-0.6 T_m, fatigue crack initiation usually

---

* T_m - Melting temperature in degrees K.

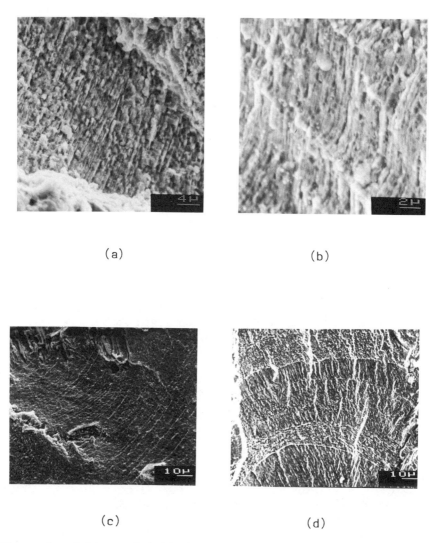

(a) (b)

(c) (d)

Figure 6. Fatigue striations and crack growth bands along varying fracture locations in the turbine blade. (a) 2-mm from crack initiation; (b) 13-mm from crack initiation region; (c) 18-mm from initiation region; (d) 26-mm from initiation.

occurs intergranularly and crack propagation can be either transgranular or intergranular [11,12]. Faceted intergranular crack nucleation generally contains a grain boundary at the far end of the facet [13], and a boundary of this nature was not observed in the present investigation, Figure 5. It is therefore concluded that the crack initiation in the Alloy 713LC blade occurred transgranularly along the slip bands. The proportions of fracture surface produced by Stages I (initiation), II (propagation) and III (overload fracture) during crack growth varies with material and applied stress-strain level. In terms of fraction of life spent in different stages, in high-cycle, low-stress fatigue (e.g., turbine blades), Stage I dominates, whilst in low-cycle, high-strain fatigue (e.g., turbine discs), Stage II plays a significant role [14].

Subsequent to localized crack growth in the crack initiation region, the fatigue crack propagated in a general chordwise direction across the airfoil section for approximately 28-mm as shown in Figure 3. Although a large portion of this fatigue area consisted of Stage II crack propagation, exhibiting crack growth bands and fatigue striations, Figure 6, some crystallographic facets were also observed in this region, arrows F, Figure 3. Figures 6(a) and 6(b) show two regions of single cycle striations located some 2-mm and 13-mm, respectively, from the fatigue crack initiation site. The average striation spacing in these two areas was similar at approximately 0.25 micrometer indicating about 4000 cycles per millimeter of crack growth in these areas. The remaining fracture surface, area G, Figure 3, revealed topography typical of rapid overload failure for this type of material. This topography consisted of ductile dimples, cleavage and relatively large areas of dendritic fracture that contained numerous script-type carbides, Figure 7.

The presence of a few well-defined areas of single cycle striations (Figures 6(a) and 6(b)) beyond the faceted Stage I initiation region also suggests that relatively few cycles may have been involved in the crack growth from 2-mm to final fracture some 28-mm from the crack initiation site. Based on an average crack growth rate (da/dN) of 0.25 micrometer/cycle (4000 cycles/mm) from 2-mm to 28-mm, it appears that an estimated 104,000 cycles may have accounted for the final 26-mm of fatigue crack growth. Based on this assumption, it is possible that these relatively few load cycles may have been applied during the last several minutes of engine operation when the engine was operating at maximum speed of ~10,000 rpm.

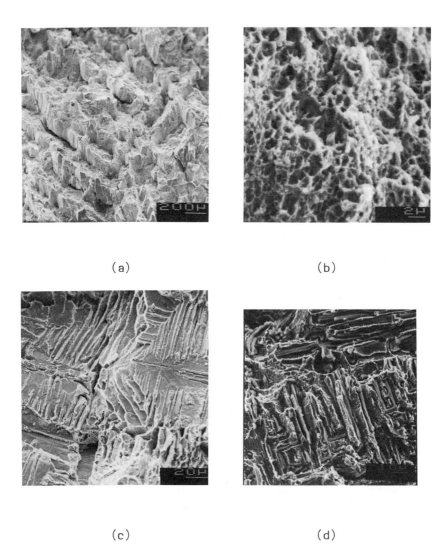

Figure 7. Turbine blade overload fracture surface features.
(a) General features; (b) Ductile dimples; (c) Dendritic fracture area; (d) Script-type MC carbide containing areas.

b. Metallographic Analysis

The metallographic sample (Figure 1) in the unetched condition revealed a reasonable population of primary MC carbide and an occasional sulphocarbide in the blade airfoil, Figures 8(a) and 8(b). Script-type primary MC carbide morphology was

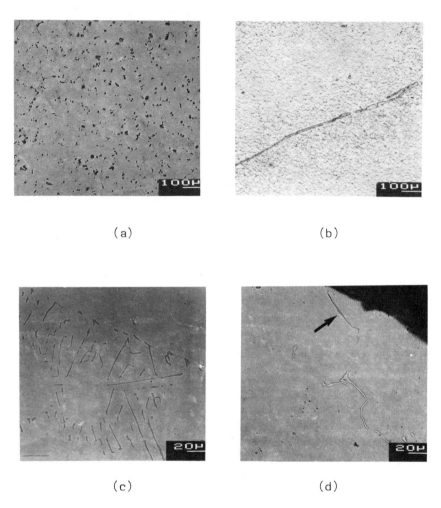

Figure 8. Unetched microstructural features observed in the turbine blade sample. (a) Primary MC carbide; (b) Sulpho-carbides; (c) Script-type primary MC carbides; (d) Subsurface script-type carbides.

evident in the thicker blade sections, Figure 8(c), and the presence of an occasional script-type carbide very close to the blade surface is indicated by an arrow in Figure 8(d). The etched microstructure revealed coarse grained dendritic structures with intense γ' mottling in the background, Figure 9. No creep cavities were observed at the grain boundaries thus indicating the absence of any secondary creep damage.

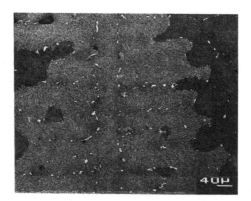

Figure 9. Etched microstructural features of the turbine blade sample.

Evidence of any creep cavitation may be obtained sometimes by examining the intergranular facets at higher magnification using SEM. Providing the ductile dimples do not overly obscure the intergranular facets, the presence of creep cavities can be verified by this technique [15]. Creep voids were not observed during optical examination of the Alloy 713LC blade; it can therefore be concluded that creep was not responsible for the present service failure. The metallographic sample was also prepared to locate any regions containing undesirable sigma-phase (following the techniques described elsewhere [5]) and its presence was not evident in the blade sample examined.

Room temperature hardness testing of the metallographic sample from the failed blade showed a hardness reading of 383 HV30. A second stage turbine blade from the same engine that had suffered a partial tip rub during engine disintegration was hardness tested along different airfoil locations for comparison purposes. The hardness data as a function of the distance of sampling location from the root section is plotted in Figure 10. The HV30 data varied between 383-393 in both blades, thus indicating no significant variation in strength due to an abnormal temperature gradient along the blade airfoil. To precisely elucidate any service-induced high temperature strength degeneration effects through γ' precipitate size and morphology changes [16], the γ' particle sizes of both failed

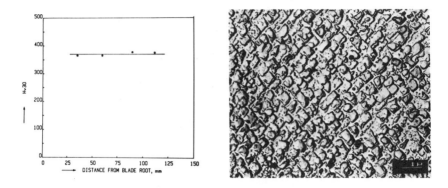

Figure 10. (Left) Hardness profile of another turbine blade from the same row in the same engine varying as a function of sampling location along the blade airfoil.

Figure 11. (Right) A general view of the primary and secondary γ' morphology observed in the turbine blade sample.

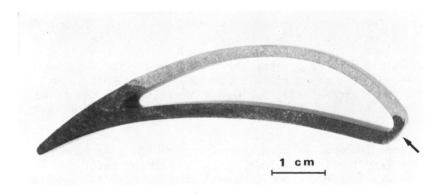

Figure 12. The cracked turbine vane fracture surface (lightly shaded area is the lab induced overload region).

blade and the reference blade were determined along different airfoil and root sections. All samples revealed 660 nm cuboidal primary γ' particles and spherical secondary γ' particles 110 nm in diameter, uniformly distributed throughout the matrix, Figure 11. The uniformity of the γ' particle sizes and HV 30 data clearly suggests that the blade was not subjected to any over-temperature effects during service.

First Stage Turbine Vane (Engine No. 2)

a. Visual Examination and Fractographic Analysis

As shown in Figures 2 and 12, optical examination indicated that the crack had penetrated through the vane wall thickness across the entire concave side of the airfoil section and extended approximately 20-mm from the trailing edge on the convex side of the vane. As indicated in Figure 13, a slight amount of axial deformation of the vane, in the form of bowing of the airfoil section, was also evident. It must be pointed out that the problem of airfoil bowing in turbine vanes is not perfectly understood [17]. In particular, the mechanical properties governing the bow resistance of a given material and the specific engine conditions that produce bowing have not been fully elucidated [17]. Optical examination also revealed a significant amount of surface erosion in the leading and trailing edge areas (small arrows, Figures 2, 12 and 14) and a lesser amount on the concave surface adjacent to the trailing edge, areas A, Figures 2(a) and 14(b). No significant degree of erosion was evident on the convex airfoil surface of the vane.

Both optical and SEM examination of the fracture surfaces, Figures 12 and 15, showed them to be so severely corroded that virtually no evidence of conventional fracture topography was observed. It should be noted that repeated replica cleaning of the fracture surfaces did not remove any appreciable amount of this adherent corrosion layer. As indicated in Figure 15, examination in the SEM of the areas immediately adjacent to the laboratory-induced overload region were also corroded to the extent that no meaningful fractographic information could be obtained from these areas. Generally, the primary fracture areas adjacent to the overload fracture are very helpful in determining the precise mode of failure, e.g., creep, high-cycle fatigue (HCF), or corrosion in cracked components exposed to corrosive environments. In the present case, however, the fracture surface corrosion was so severe that a specific crack initiation site could not be positively established.

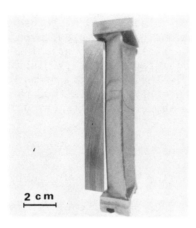

Figure 13. Axial deformation, i.e., bowing, in the turbine airfoil section.

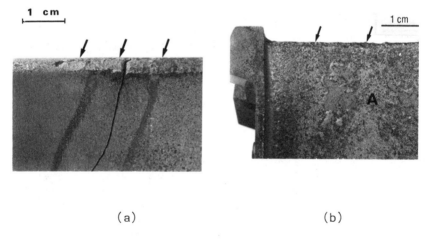

(a)   (b)

Figure 14. Turbine vane erosion in the leading and the trailing edge areas. (a) Leading edge; (b) Trailing edge. Arrows indicating severe corrosion.

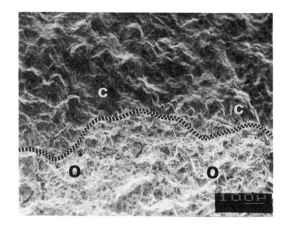

Figure 15. The corroded and overload fracture features in the cracked MAR-M302 turbine vane. (a) Transition of corroded region 'C' and the laboratory induced overload region 'O'; (b) Corroded area adjacent to overload region; (c) Laboratory induced overload area with cleavage and ductile dimples.

Apart from looking for the well established types of service induced degeneration, e.g., creep, HCF, or corrosion, a failure analyst should always make routine checks for other types of damage utilizing conventional fractographic techniques. Overtemperature damage (e.g., re-solution of the alloy constituents and incipient melting) of a vane can occur during a momentary temperature overshoot associated with a hot start. The fracture in the case of incipient melting is typically intergranular, revealing clean-sharp grain facets, with molten-solidified debris present at some triple point and/or grain boundary locations [18]. The presence of incipient melting areas should also be verified through conventional metallographic techniques. Turbine engine vanes and blades are also known to crack under thermal fatigue (a form of low-cycle fatigue) conditions [17]. It is believed that oxidation sites, particularly at grain boundaries, provide potential areas for thermal fatigue crack nucleation. Thermal fatigue crack propagation may be either intergranular or transgranular and this depends upon the exact service conditions and the inherent characteristics of the material. Fractographic evidence in the form of striations should, however, be provided to support any low-cycle fatigue arguments [19]. None of these features were observed during fractographic examination of the MAR-M302 vane.

b. Metallographic and EDX Analysis

Figure 16(a) shows that the MAR-M302 vane was coated to provide protection against oxidation and hot corrosion [20]. It was interesting to note that the coating on the convex side of the vane had a uniform thickness of $\sim$0.11-mm over the surface, Figure 16(b), whereas the coating on the concave side had virtually disappeared in places, Figure 16(c). This form of degradation of coatings is generally attributed to the abrasion and erosion caused by carbon particles and ingested particulate material that remove protective oxides at airfoil leading edges and render the components vulnerable to premature failure [21,22]. Maximum erosion of blades and vanes generally occurs at a point near the midspan, where the highest temperatures occur. More frequent start-ups generate larger quantities of carbon particles thus increasing the degradation effects. High temperature corrosion product formation and penetration on the concave side of the specimen was abundant throughout the cracked vane, Figure 17(a). Coating degradation and corrosion penetration of this nature is well known to provide potential crack nucleation sites under normal service conditions [22].

(a)

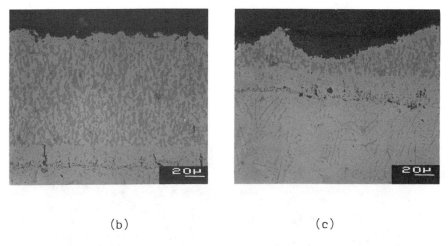

(b) (c)

Figure 16. Coating features of the vane along the airfoil section. (a) Convex side showing intact coating and shrinkage porosity; (b) Coating-matrix interface along convex side; (c) Vane coating and coating-matrix interface along concave side.

Figure 17. Coating degeneration and corrosion attack in MAR-M302 vane along the concave airfoil side. (a) Fine cracking; (b) Subsurface shrinkage porosity; (c) Corrosion product penetration in the subsurface porosity.

A number of areas in the vane airfoil sample contained a large amount of shrinkage porosity, Figure 16(a), and some of these pores could well be classified as subsurface shrinkage cavities, Figure 17(b). Along the concave surface, the corrosion product penetration had extended in places right into these subsurface cavities, Figure 17(c), thus rendering them potentially hazardous critical crack nucleation sites. Generally, castings containing unfavorable shrinkage porosity are rejected and the rate of rejection depends largely on the quality control procedures and the inspection standards set by both the manufacturer and the user.

The coating and the diffusion zones between the coating and the matrix were carefully examined to identify any areas containing inherent coating defects or service induced microstructural degradation. With intermetallic coatings in general, corrosion failures based mainly on coating-substrate diffusion reactions are difficult to single out, e.g., Co-Al coating on a Co-base superalloy [23]. Besides, a variety of defects, e.g., intermetallic inclusions, thickness variations, shallow edge cracks, subsurface holes, imbedded grit, internal porosity, hairline fissures, interfacial cracks, composition variations, etc., are invariably introduced during the application of the coating itself. The only defects that are introduced during use are cracks that result from thermal expansion mismatch, excessive mechanical strain or impact [22]. The service environments or thermal cycling generally serve to increase the number and/or severity of these inherent coating defects. Apart from the coating thickness variations on the convex and concave sides of the MAR-M302 vane, the diffusion interface appeared reasonably sound. The microstructures in the coating-substrate diffusion zones were similar on both sides of the airfoil section, Figures 16(b) and 16(c), which suggests that the coating was properly bonded originally on both sides of the vane. These observations further confirm that the severe coating degradation on the concave side of the vane is a direct consequence of the external environmental factors rather than a reflection of any internal coating defects.

Energy-dispersive x-ray analysis (EDXA) of the overload fracture, the corroded surface, and the coating was carried out to reveal any differences in the surface elemental distribution. A strong Al peak of the coating, Figure 18(a), clearly indicates that an aluminide coating was used for corrosion protection.

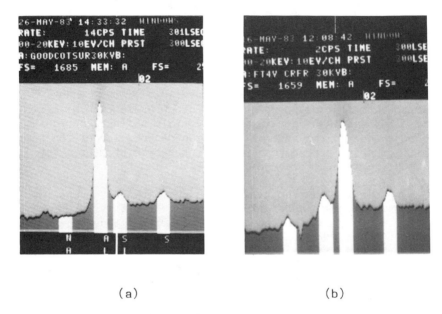

Figure 18. EDX analysis of the coating and corroded area of MAR-M302 vane. (a) Al and S peaks observed in the coating; (b) Corroded fracture region showing Na, S, Al and Si peaks.

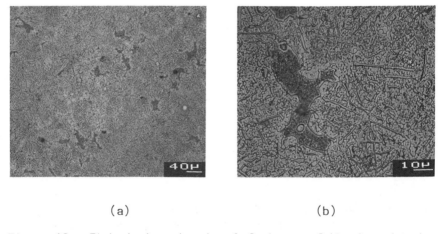

Figure 19. Etched microstructural features of the investment cast MAR-M302 vane. (a) Eutectic carbide pools and script-type carbides; (b) Coarse peppery secondary carbide in the background.

The presence of a small S peak on the coating surface is also noteworthy. Both overload and corroded fracture surfaces showed very strong Co and Cr peaks and relatively lower intensity peaks of Ta, W and Si. The corroded surface, however, revealed additional peaks of S and Na, Figure 18(b), and these two elements were not present in the overload region. The presence of S on the coating surface and the Na and S pick-up from the corroded fracture region suggests that Na/S participated in the corrosion reactions. It is well established that an accelerated form of high-temperature corrosion is associated with the presence of NaCl (ingested by marine engines from NaCl-rich air) and S impurity in fuel which react above 600 C to form Na_2SO_4 [20,24]. Condensed Na_2SO_4 disrupts and dissolves the protective Cr-oxide films, and this is followed by sulfide formation, chromium depletion in the substrate, and accelerated oxidation of the Cr-depleted regions. This type of corrosion is generally termed "sulfidation" or hot corrosion. The morphological features of the corrosion products vary with the environmental and temperature conditions, including the chemical composition of the materials used [25]. Sulfidation attack is known to occur at S levels as low as 0.004 wt. % (0.05 wt. % S is present in the normal fuel), and Na levels of the order of 1-15 ppm can also trigger the corrosion [26,27]. It is therefore evident that a very small quantity of NaCl in the intake can cause hot corrosion problems under normal service conditions in marine environments.

The etched vane sample revealed cast microstructural features, Figure 19, with interdendritic dark islands (eutectic carbide) and a needle-like constituent (script-type primary MC carbides) uniformly distributed throughout the microstructure. These are features typical of a MAR-M302 investment casting [28]. The microstructure also showed a coarse, peppery constituent abundant in the background, Figure 19(b), which corresponds to the secondary carbide precipitated upon aging of the MAR-M302 vane. The secondary carbide is, however, much finer in the original heat-treated material [28], thus indicating that the coarsening effects were imparted during normal high temperature service exposure.

The vane airfoil was thoroughly examined for any evidence of incipient melting and chilling effects, a feature not uncommon to vane airfoil surfaces [29]. It appears that the vane was not subjected to any over-temperature and rapid cooling conditions during service. Also, creep cavities in the etched microstructure were not observed in the airfoil section. These observations are indicative of a normal engine operation, where the marine engine in question was not subjected to any overload or over-temperature effects.

FAILURE ASSESSMENT

Second Stage Turbine Blade (Engine No. 1)

The fractographic examination indicated that the Alloy 713LC turbine blade failed as a result of high cycle fatigue (HCF). Although the lack of continuous, well defined striations along the entire length of the fatigue fracture precluded a detailed quantitative assessment of the fatigue-crack growth rate, the fractographic evidence does permit some general conclusions. The presence of the mirror-like crystallographic facets in the Stage I initiation region suggests that relatively low crack growth rates, probably at near threshold stress intensity levels, were involved in the initial 2-mm of fatigue fracture [3,8-10]. Although somewhat speculative, it is possible that the initial 2-mm of crack growth occurred over a long period of engine operation involving millions of load cycles [14]. Although no obvious surface or metallographic irregularities were observed in the trailing edge initiation region, it is possible that a relatively small subsurface discontinuity, such as a pore, a minute crack, or a slip band, possibly associated with a near surface carbide [30], could have contributed to fatigue crack initiation. The effects of the subsurface script-type MC carbide and sulpho-carbides on the in-service fatigue properties have not been fully categorized in the literature to date. Obviously, more research is needed to take a careful look at their role during initiation stages under HCF fatigue conditions at normal service temperatures.

The estimate of cycles involved in crack propagation from 2-mm to 28-mm, that may have occurred within the final few minutes of engine operation, have already been pointed out. Apparently, the ship's power was increased from 2,000 HP to 12,000 HP a few minutes prior to engine failure. Although no specific operational conditions were reported that could account for so relatively few cycles, it is possible that increased loads, due to the increase in engine speed, in conjunction with an incipient fatigue crack, may have created sufficiently high levels of cyclic stresses to cause relatively rapid crack growth over this final 26-mm of fatigue fracture.

First Stage Turbine Vane (Engine No. 2)

Metallographic and fractographic examination indicated that hot corrosion was responsible for cracking in the vane leading edge area. It was also evident that hot corrosion cracking occurred once the coating had completely eroded away in the general vicinity of the leading edge area. Since precise service life details of the overhauled engine remained uncertain, it was unclear whether coating degeneration was premature or simply a consequence of normal service deterioration. Upon examining the rest of the first stage vanes from the engine, severe leading edge erosion was observed in a large majority of them. However, the subject vane was the only one with a visible crack. In view of the presence of subsurface shrinkage porosity in the cracked vane, it is suggested that a small hot corrosion crack could have reached a critical length much earlier than anticipated. In addition, the engine was continuously exposed to NaCl-rich atmospheres that could have contributed to accelerated sulfidation [20,22-27]. Therefore, a combination of coating erosion and subsurface porosity under sulfidation conditions describes the vane degeneration and cracking.

REFERENCES

1. M. F. Ashby, "Fracture Mechanism Maps," Cambridge University, (Sept. 1978).

2. A. J. Perry, J. Mat. Sci., Vol. 9, p. 1016 (1974).

3. Fatigue Mechanisms, ASTM STP 675, J. T. Fong (ed.) (1978).

4. The Nimonic Alloys, Edward Arnold, W. Betteridge and J. Heslop, (eds.), 2nd ed. (1974).

5. A. K. Koul and W. Wallace, Met. Trans., Vol. 14A, p. 183 (1983).

6. C. J. Spengler and P. P. Singh, Corrosion/Erosion of Coal Conversion System Materials, Berkeley, California, U.S.A., p. 746 (Jan. 1979).

7. Aerospace Structural Metals Handbook, Mechanical Properties Data Centre, Belfour Stulen Inc., Vol. 4 and 5 (1983).

8. C. H. Wells and C. P. Sullivan, ASM Quarterly Transactions, Vol. 60, p. 217 (1967).

9. D. M. Moon and S. P. Sabol, "Fatigue at Elevated Temperatures," ASTM STP 520, p. 438 (1972).

10. H. T. Merrick, Met. Trans., Vol. 5, p. 891 (1974).

11. C. J. McMahon and L. F. Coffin, Jr., Met. Trans., Vol. 1, p. 3443 (1970).

12. L. F. Coffin, Jr., Met. Trans., Vol. 5, p. 1053 (1974).

13. N. Gell and G. R. Leverant, "Mechanisms of High Temperature Fatigue," ASTM STP 520, p. 37 (1973).

14. B. Tompkins, "Creep and Fatigue in High Temperature Alloys," Applied Science Publishers Ltd., London, p. 111 (1981).

15. A. Venugopal Reddy, Pract. Met., Vol. 19, p. 659 (1982).

16. A. K. Koul and W. Wallace, Met. Trans., Vol. 13A, p. 673 (1982).

17. Source Book on Materials for Elevated Temperature Application, E. F. Bradley, (ed.), ASM, Metals Park, Ohio, p. 309 (1979).

18. A. K. Koul and W. Wallace, private communication with Westinghouse Canada Inc., May (1983).

19. M. F. Day and G. B. Thomas, High Temperature Alloys for Gas Turbines, Leige, Belgium, p. 641 (Sept. 25-27, 1978).

20. C. T. Sims, Superalloys, W. C. Hagel and C. T. Sims, (eds.), Vol. 1, Chapter 5, John Wiley and Sons Inc., New York (1972).

21. L. D. Graham, et al., ASTM STP 521, p. 105 (1967).

22. High Temperature Oxidation Resistant Coatings, National Academy of Sciences - National Academy of Eng., Washington, ISBN 0-309-01769-6, p. 105 (1970).

23. R. Pichoir, <u>High Temperature Alloys for Gas Turbines</u>, Liege, Belgium, p. 191 (Sept. 25-27, 1978).

24. M. A. DeCrescente and N. S. Bronstein, <u>Corrosion</u>, Vol. 24, p. 127 (1968).

25. J. F. Stringer, <u>Gas Turbine Materials in Marine Environments</u>, MCIC 75-27 (1975).

26. H. T. Quigg, R. M. Shirmer, and L. Bagnetto, "Final Report to Naval Air System Command on Contract N00019-70-C-0293," Phillips Petroleum Co. R and D Report 5903-71 (Jan. 1971).

27. P. Huber, <u>High Temperature Alloys for Gas Turbines</u>, Liege, Belgium, p. 251 (Sept. 25-27, 1978).

28. <u>Metals Handbook, Atlas of Microstructures</u>, Vol. 7, 8th Edition, ASM, p. 196 (1980).

29. W. Wiebe and A. K. Koul, unpublished NAE-NRC Research (April 1981).

30. W. Wallace, et al., NAE-NRC Report LTR-ST-514 (Feb. 1972).

INVESTIGATION INTO PRECIPITATION PHENOMENA FOLLOWING
HEAT TREATMENT OF ELI FERRITIC STAINLESS STEELS
AND THEIR INFLUENCE ON MECHANICAL
AND CORROSION PROPERTIES

Gianni Rondelli*, Bruno Vicentini*,
and Dany Sinigaglia*

ABSTRACT

The effects of heat treatments in the range
650 C - 900 C on an extra-low interstitial (ELI)
ferritic steel (25%Cr-4%Mo-4%Ni-0.5%Nb) have been
evaluated. Embrittlement of the alloy occurs more
rapidly at 800-850 C. Susceptibility to attack by
nitric acid is maximum after heat treatment at
700-750 C; even just 5 minutes heat treatment is
sufficient to impair the corrosion resistance of
the alloy.

The above phenomena are discussed in relation
to the precipitation processes of intermetallic
phases revealed by metallographic investigation and
x-ray diffraction. The presence of sigma, chi and
Z phases was demonstrated in relation to temperature
conditions and to duration of heat treatment. The
alloy is immune to intergranular corrosion due to
chromium depleted zones under all heat treatment
conditions.

INTRODUCTION

High chromium-molybdenum ELI ferritic stainless steels,
besides possessing excellent resistance to stress corrosion,
also exhibit optimum resistance to localized corrosion.

---

* Instituto per la Tecnologia dei Materiali Metallici non
  Tradizionali del CNR. 20092 Cinisello B. Italy.

Hence, they find potential application under conditions of high chloride concentration [1]. These steels were developed in order to overcome the toughness limitation and susceptibility to intergranular corrosion exhibited by conventional ferritic stainless steels [2,3,4]. This was achieved by lowering the interstitial content and by adding stabilizing elements (Ti,Nb) in suitably studied quantities [4,5]. They are currently manufactured by means of argon-oxygen decarburization (AOD) and vacuum-oxygen decarburization (VOD) techniques whereby low interstitial contents are achieved (C+N <400 ppm) [2,3] at competitive prices. However, even these new alloys are not immune to embrittlement, due to intermetallic phase precipitation [5,6], after exposure to temperatures ranging from 650-900 C.

Particular attention has been paid to the precipitation of hard, fragile sigma phase with tetragonal structure and of hard chi phase with complex cubic structure [7]. Streicher [8] demonstrated, in a 29%Cr-4%Mo-2%Ni steel heat treated for 100 hours at 800 C, the formation of sigma phase as isolated particles at the grain boundary as well as chi phase precipitation, both at the grain boundary and in the ferritic grain. Kiesheyer [9] revealed, in a 24%Cr-4%Mo steel heated for 5000 hours at 800 C, the presence of coarse sigma phase at the grain boundary as well as a finer intergranular and intragranular chi phase; the sigma phase is much richer in chromium than the matrix and is also richer in molybdenum, while the chi phase is particularly rich in molybdenum.

The sigma and chi phases were revealed by x-ray diffraction on Fe-Cr alloys heated for 100 hours in the range 760-930 C [10]. Nichol [6] revealed the sigma phase in a 29%Cr-4%Mo steel heated for 3 hours at 760 C, while after 100 hours heat treatment [11] at 700 C he found the simultaneous presence of sigma and chi phases. He also revealed the presence of at least two phases, one of which was identified as sigma phase, in a 29%Cr-4%Mo-2%Ni steel after heat treatment for 1000 hours at 600 C [12].

Intermetallic phase precipitation in Cr-Mo steels is also strongly influenced by the other elements contained in the steel. Kiesheyer [13] found that the presence of nickel accelerated the process of intermetallic phase precipitation in steels containing 28%Cr-2%Mo and 20%Cr-5%Mo. According to

Dundas [14], stabilization with titanium reduces the precipitation time of sigma phase, while niobium increases it. Kiesheyer [13] states that an excess of titanium over the quantity required for correct stabilization of the steel is particularly damaging as it accelerates formation of intermetallic phases.

The ELI ferritic stainless steels can loose their corrosion resistance in strongly oxidizing environments, such as nitric acid, if submitted to heat treatments in the same temperature range which caused precipitation of intermetallic phases. This appears to depend on the composition of the steel, as well as on the temperature and duration of heat treatment [14] and is attributed to intermetallic phase precipitation processes [15]. However, heat treatments at 800-900 C, which cause complete embrittlement of the alloy, do not cause an equivalent lowering of corrosion resistance in boiling nitric acid [16].

As a further contribution to the knowledge of these problems, the authors considered it interesting to evaluate the behavior of a steel containing 25%Cr-4%Mo-4%Ni-0.5%Nb. This paper is concerned with the effects on toughness and corrosion resistance of the intermetallic phases precipitated. Investigation was also made into the identification of the precipitated phases.

EXPERIMENTAL PROCEDURE

Material

The chemical composition (weight percent) of ITM 43 experimental steel was: 24.8 Cr, 4.05 Mo, 3.93 Ni, 0.12 Si, 0.013 P, 0.002 S, 0.074 C, 0.0117 N and 0.51 Nb. This steel was melted in a vacuum induction furnace (VIM) and cast into a square water-cooled mold to obtain an ingot weighing 7 kg. The ingot was then hot rolled from 1200 C to 850 C (75% reduction in thickness) and cold rolled (75% further reduction) until a 3-mm thick strip was obtained. Heat treatment at 1035 C for 5 min followed by water quenching was then carried out to provide recrystallization of the ferritic grain free from any precipitated second phases. The ferritic grain size, measured in accordance with the ASTM E112 procedure, is 5.5.

Heat Treatments

Some specimens of the annealed ITM 43 were submitted to a second heat treatment at: a) 600 C for 1 hour/water quench; b) 850 C for 1 hour/water quench; and, c) 1250 C for 1 hour/water quench. All heat treatments were performed in air, except that at 1250 C for 1 hour, which was carried out in a vacuum sealed tube [17]. Other specimens were heat treated in the temperature range of 650-900 C for time intervals of 4 min, 20 min, 1 h, 2.5 h, and 5 h followed by water quenching.

Impact Tests

Charpy tests were performed in accordance with ASTM E23. The quarter-size V-notch samples were cut perpendicular to the rolling direction and notched in the short-transverse direction.

Streicher Test

Intergranular corrosion resistance was evaluated using a modification of the sulphuric acid - copper sulphate test in accordance to the Streicher procedure [18]. Those specimens with weight loss below 1 g/dm^2, and without cracks and grain dropping, were considered immune to intergranular attack.

Huey Tests

In order to evaluate the corrosion resistance in strong oxidizing acid media, tests were carried out in boiling 65% nitric acid solution according to ASTM A262-75 (Practice C).

Metallography

Systematic metallographic investigation was carried out to reveal the different phases present after the heat treatments performed in the range 650-900 C for times ranging between 5 min and 5 h. The following reagents were used:

a) Groesbeck's reagent (4 g KOH, 4 g KMnO4, 100 ml H_2O, heated to boiling; colors the sigma phase brown and reveals the other intermetallic phases by their different coloring); and,

b) Two stage attack [(1) 10% sulphuric acid, 6 V for 5 s; and, (2) 30% NaOH, 2 V for 10 s; colors the intermetallic phases orange-red and reveals the presence of any austenite].

X-Ray Diffraction

X-ray diffraction was performed on extractions from samples heat treated in the temperature range 650-900 C for 5 h. Extraction was performed by dissolving the specimens in 30% aqueous solution of $FeCl_3 \cdot 6H_2O$ at 60-70 C. The extracts were filtered through 0.45-micrometer cellulose acetate paper. Diffractions were performed on a Siemens D-500 diffractometer using Cu radiation and a diffracted beam monochromator. The x-ray tube operated at 40 kv and 30 mA. The patterns were obtained at a scan rate of 0.5 degrees per minute.

RESULTS AND DISCUSSION

Metallography

Figure 1 shows the micrographs of specimens heated for 5 h in the range 650-900 C. The specimens heat treated between 750 and 850 C contained two phases at the grain boundary, one of which is colored brown (sigma phase). The other (chi phase), whose coloring depends on the etching conditions, is also present inside the grain [13]. The sigma phase occurs in the form of coarse isolated particles in the specimens heat treated for 5 h at 900 C and more likely at the triple line junctions. The specimens heated for 5 h at 700 C and 650 C contain microprecipitation along the grain boundaries and inside the grain. It was not possible to determine, through optical metallography, the nature of the precipitated phases under these conditions.

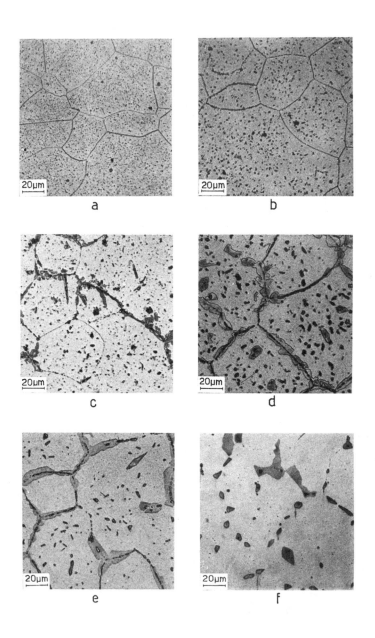

Figure 1. Microstructures of ITM 43 after 5 hours at:
a) 650 C; b) 700 C; c) 750 C; d) 800 C: e) 850 C;
f) 900 C (Groesbeck reagent).

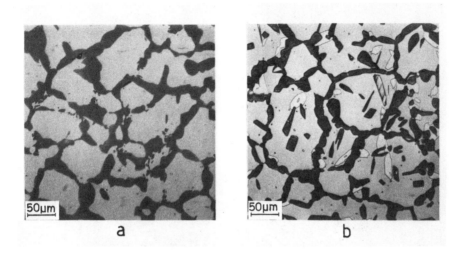

Figure 2. Microstructure of 25Cr-3.5Mo-3.5Ni-Ti experimental steel after 100 h at 850 C: a) Groesbeck reagent; b) two-stage attack.

Figure 2 shows the microstructure of an experimental steel of analogous composition (25%Cr-3.5%Mo-3.5%Ni-Ti) heat treated for 100 h at 850 C. The Groesbeck reagent (Figure 2a) reveals the presence of just the sigma phase, which was also confirmed by x-ray diffraction. The two-stage etch attack (Figure 2b) also reveals the presence of another light-colored phase. Microprobe analysis gives the following composition of the brown colored phase (Figure 2a): Cr = 33.8%, Mo = 7.7%, Ni = 3.3%. The light-colored phase revealed by the two-stage attack (Figure 2b) has the following chemical composition: Cr = 20% (avg.), Mo = 2.2%, Ni = 6.5% (avg.). Microprobe analysis results confirm that the brown colored particles are sigma phase [9]; moreover, it shows that the light colored phase is austenite, which was also found by Kiesheyer [13]. Steel ITM 43 heated for 100 h at 850 C also gave the same results regarding both metallographic examination and microprobe analysis.

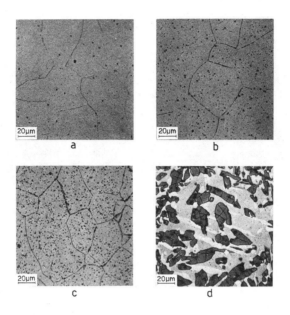

Figure 3. Microstructures of ITM 43 heat treated at 800 C for: a) 5 min; b) 20 min; c) 1 h; d) 100 h. Etchant: a), b), c): Groesbeck reagent; d): two-stage attack.

Figure 3 shows the microstructures of ITM 43 samples heat treated at 800 C for different times. After only 5 min, a continuous network of very fine precipitates can be seen along the grain boundaries. After one hour, the sigma phase and a second phase have precipitated inside the grains; after 5 hours (Figure 1-d), a network of finer chi phase particles surrounded by coarser sigma phase particles can be seen; the chi phase also occurs inside the grain. After 100 h, the extent of precipitation was even greater and the presence of austenite was also noticed.

X-Ray Diffraction

X-ray diffraction patterns gave the following results. For heat treatments carried out at 650 C and 700 C just the Z phase occurred; at 750 C there is the simultaneous presence of Z, sigma and chi phases; at 850 C and 900 C just sigma and chi phases were revealed. These phases were identified

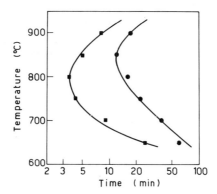

Figure 4. Temperature-time embrittlement curves for ITM 43. Charpy V-notch quarter size samples tested at room temperature. Impact energy: ■ 25 joules; ● 17.5 joules; annealed samples 27.5 joules.

according to ASTM powder diffraction files 11.339, 7.356 and 6.074 respectively. However, it should be pointed out that the extracts obtained on specimens heated at lower temperatures show a low diffraction intensity; therefore, the presence of other phases in the extract cannot be excluded.

Furthermore, x-ray diffraction has never revealed the presence of austenite, even on extracts obtained from specimens in which the presence of austenite was revealed by other means. This was probably because the austenite was dissolved during matrix dissolution process.

Impact Tests

From the impact tests, we obtained curves of impact energy vs. heat treatment time at fixed temperatures (650-900 C). From these, temperature-time embrittlement curves were constructed by plotting a constant impact energy curve (Figure 4). A nose can be seen corresponding to the maximum rate of embrittlement and centered around 800-850 C with minimum embrittlement times of 5-15 min. The rate of embrittlement decreases in the order: 850 to 800 C > 900 C > 750 C > 700 C > 650 C.

From the comparison of the optical micrographs and the results of the impact tests, it can be said that the embrittlement seems to be independent from the type of the intermetallic phases precipitated, but it depends upon their morphology, distribution, and amount. The behavior of the specimens heat treated at 900 C can be explained by the different morphologies of the precipitates obtained at that temperature.

Streicher Tests

The modified Strauss test [18] was carried out on specimens heated for one h at 600 C, 850 C and 1250 C and on the annealed material. The test was passed in all cases. It should be noted that the samples heated for 1 h at 850 C fractured during the bending test; however, this was due to the precipitated embrittling phases and not to intergranular corrosion.

The test was also carried out on specimens heated for 1 h at 700 C, 1 h at 800 C, 5 h at 850 C and 100 h at 850 C. These specimens passed the modified Strauss test because their weight losses were always appreciably lower than 1 g/dm^2 and neither intergranular attack nor grain dropping was observed by optical examination.

Hence, it can be concluded that for each heat treatment condition in the temperature range under investigation, the ITM 43 steel is not susceptible to intergranular corrosion due to the presence of chromium-depleted zones [4,19].

Huey Test

As is known, the Huey test strongly attacks all the less resistant zones besides the chromium-depleted zones, such as intermetallic phases, chromium carbides and titanium carbides [16]. The test was used to evaluate the effects of different heat treatments on the corrosion resistance of the steel in strongly oxidizing environments.

As can be seen in Table 1, the specimens heated in the range 850-900 C for times between 20 min and 5 h passed the test even if they contained appreciable quantities of sigma and chi phases. Figure 5d shows an SEM micrograph of the specimen heat treated for 5 h at 900 C at the end of the second cycle; the presence of a phase not attacked by nitric acid can be seen. Comparison of the optical and SEM micrographs suggested it is sigma phase. Strong attack can be seen in the zones in which optical metallography revealed the presence of chi phase. However, as the chi phase is present in the form of coarse, discontinuous particles, there was no onset of intergranular corrosion with grain dropping. These results are in agreement with those found by Kiesheyer [16]. Specimens heated for 5 h at 850 C showed a similar behavior (Figure 5c).

Table 1. Corrosion Rate of ITM 43 ($g/m^2 h$) in the Huey Test

Heat Treatment Temperature C	Cycle	Heat Treatment Time			
		5 min	20 min	1 hour	5 hours
650	1	5.2	35.3	54.1	55.6
	2	21.8	64.7	72.6	62.4
	3	25.0	-	-	59.1
700	1	42.1	46.8	34.5	35.4
	2	54.7	52.8	35.3	38.5
	3	50.2	50.8	32.8	36.0
750	1	14.3	20.1	7.5	7.6
	2	35.1	35.4	39.4	35.7
	3	33.5	31.0	31.0	29.0
800	1	3.5	1.2	.23	1.4
	2	22.5	16.5	1.6	1.3
	3	22.1	21.2	6.5	1.21
	4	17.6	22.4	9.8	1.50
	5	18.8	23.1	15.6	2.11
850	1	.17	.17	.14	.10
	2	1.38	.10	.14	.12
	3	5.18	.20	.26	.15
	4	10.60	.26	.32	.21
	5	10.01	.36	.35	.24
900	1	.12	.11	.08	.10
	2	.12	.22	.11	.08
	3	.18	.17	.17	.12
	4	.20	.22	.25	.13
	5	.20	.20	.26	.15

ANNEALED MATERIAL					
Cycle	1	2	3	4	5
Corrosion Rate	.07	.07	.16	.15	.16

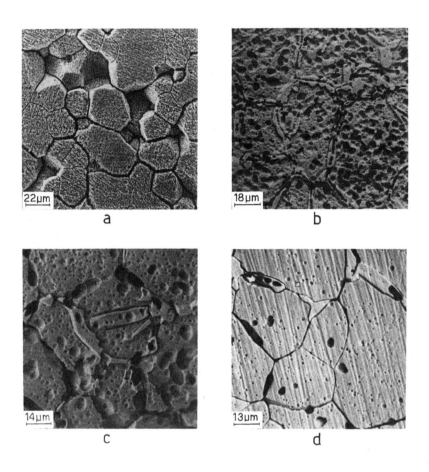

Figure 5. SEM micrographs of ITM 43 samples heat treated at different temperatures: a) 5 min at 700 C, tested 15 h in boiling HNO_3; b) 5 h at 800 C, after second cycle of the Huey test; c) 5 h at 850 C, after the fourth cycle of the Huey test; d) 5 h at 900 C, after the second cycle of the Huey test.

Specimens heat treated for 5 h at 800 C showed a higher rate of corrosion; however, the corrosion morphology for these specimens was also similar to that of the specimens heated at the higher temperature (Figure 5b).

In those specimens heated in the range 650-750 C, it can be seen that heating for 5 min was sufficient to cause high rates of corrosion. Figure 5a shows a SEM micrograph of a specimen heated for 5 min at 700 C, after being immersed just 15 h in boiling nitric acid; strong intergranular attack and grain dropping can be seen. The strong attack (see Table 1) of the specimens heat treated at 650-750 C can be ascribed either to the presence of chi phase in continuous form along the grain boundary or to another phase especially susceptible to attack in nitric acid. As x-ray diffraction showed the presence of Z phase on specimens treated in this temperature range, it can be assumed that this is the phase which determines the loss in corrosion resistance in nitric acid (at least for the ITM 43 niobium-stabilized steel).

The high corrosion rate exhibited by the specimens treated for 5 min at 850 C (Table 1) can be attributed to the presence of chi phase in continuous form along the grain boundary. Longer heat treatment times at this temperature (20 min) modify the morphology of the chi phase in a spot distribution, so that the specimen is resistant to nitric acid attack. In this connection, it should be noted that the specimens treated at 800 C show corrosion rates decreasing on increasing the length of time of the heat treatment; an hypothesis similar to the previous one can be put forward.

Embrittlement and loss of corrosion resistance to boiling nitric acid are both caused by heat treatment of the steel in the same temperature range (650-900 C). However, embrittlement depends on the quantity and distribution of the precipitated second phases; hence, it occurs with maximum rate in the range 800-850 C. The attack by nitric acid, instead, is due to the presence of intergranular precipitates of specific phases in a continuous form and it occurs at a maximum rate in the range 700-750 C.

CONCLUSIONS

Heat treatments in the range 650-900 C on the ELI ferritic stainless steel containing 25%Cr-4%Mo-4%Ni-0.5%Nb caused the precipitation of intermetallic phases which, on one hand, determine embrittlement; and, on the other hand, loss of corrosion resistance in boiling nitric acid.

The amount of precipitation and the nature of the precipitated phases depend on the temperature and duration of the heat treatments. In the range 750-900 C, metallographic examination and x-ray diffraction revealed the presence of sigma and chi phases. The Z phase was identified at lower temperatures (650-700 C); this phase was also present at 750 C. Conditions were also found in which austenite is formed. The steel is not susceptible to intergranular corrosion due to chromium-depleted zones in any of heat treatment conditions.

Embrittlement of the material occurs at a maximum rate in the range 800-850 C. Loss of corrosion resistance in nitric acid is, instead, maximum after heat treatments at 700-750 C. Embrittlement is correlated to the quantity and distribution of the phases precipitated; corrosion in boiling nitric acid is ascribed to precipitation of well defined phases preferably along the grain boundary.

ACKNOWLEDGEMENTS

The authors wish to express their thanks to Mrs. W. Cucciaro for her assistance in the metallographic investigation and to Mr. E. Signorelli for the microprobe analysis.

REFERENCES

1. Advanced Stainless Steels for Sea-Water Application. Proceedings of the Symposium, Editor, H. Morrow III, Piacenza (Italy) (February 28, 1980).

2. R. A. Lula, Metal Progress, Vol. 110, pp. 24-29 (1976).

3. R. M. Davison and R. F. Steigerwalt, Metal Progress, Vol. 115, pp. 40-48 (1979).

4. J. J. Demo, "Structure, Constitution and General Characteristic of Wrought Ferritic Stainless Steels," ASTM STP 619, pp. 1-65 (1977).

5. R. F. Steigerwald et al., Stainless Steel '77, London, pp. 52-57 (September 1977).

6. T. J. Nichol, Metallurgical Transactions, Vol. 8A, pp. 229-237 (1977).

7. R. N. Wright, "Toughness of Ferritic Stainless Steels," ASTM STP 706, pp. 2-33 (1980).

8. M. A. Streicher, Corrosion, Vol. 30, pp. 115-125 (1974).

9. H. Kiesheyer and H. Brandis, Z. Metallkunde, Vol. 67, pp. 258-263 (1976).

10. J. C. McMullin, S. F. Reiter and D. G. Ebeling, Transaction of ASM, Vol. 46, pp. 799-806 (1954).

11. G. Aggen, H. E. Deverell and T. J. Nichol, "Optimization of Processing, Properties and Service Performance,"

13. H. Kiesheyer and H. Brandis, Z. Werkstofftech, Vol. 8, pp. 69-77 (1977).

14. H. J. Dundas, Internal report Climax Molybdenum, R.P. 33-75-03 (May 1976).

15. M. A. Streicher, Stainless Steel '77, London, pp. 1-34 (September 1977).

16. H. Kiesheyer, G. Lennartz and H. Brandis, Werkstoffe u. Korrosion, Vol. 27, pp. 416-424 (1976); H. Kiesheyer and H. Brandis, Corrosion-81, paper No. 126, Toronto (April 6-10, 1981).

17. B. Vicentini et al., Werkstoffe u. Korrosion, Vol. 33, pp. 132-143 (1982).

18. M. A. Streicher, "Intergranular Corrosion of Stainless Alloys," ASTM STP 656, pp. 3-84 (1978).

19. R. J. Hodges, Corrosion, Vol. 27, pp. 119-127 (1971); M. A. Streicher, Corrosion, Vol. 29, pp. 337-360 (1973).

NICKEL-BERYLLIUM ALLOYS RESISTANCE TO
SULFIDE STRESS CRACKING

E. W. Filer*

ABSTRACT

This study demonstrates that wrought (BERYLCO alloy 440)** and cast (BERYLCO alloy 42C) nickel-beryllium alloys are suitable for use in environments containing hydrogen sulfide (sour gas). Testing was performed according to NACE Standard TM-01-77, covering ambient temperature testing of metals for resistance to cracking failure when subjected to tensile stresses in a low pH aqueous environment containing hydrogen sulfide.

The wrought Ni-2Be alloy, BERYLCO alloy 440AT, passed the 720 h specification when tested at 135 ksi (930 MPa), a stress equivalent to 75% of the yield strength (0.2% offset). Even when tested at 180 ksi, 100% of the yield strength, the life of the material exceeded 250 h.

The cast Ni-2Be-12Cr alloy, BERYLCO alloy 42C, passed the 720 h specification when tested at 120 ksi (827 MPa), 100% of the yield strength (0.2% offset).

The properties, method of testing, and results of failure analyses are discussed for these nickel-beryllium alloys which are well suited for use in sour gas applications.

INTRODUCTION

Failures due to sulfide stress cracking have long been recognized as a problem for metals exposed to oil field environments containing hydrogen sulfide. Both laboratory and

---

* Cabot Corporation, Reading Technology, Reading, PA 19603 USA.
** BERYLCO is a registered trademark of Cabot Corporation.

field data have shown that concentrations of a few parts per million of hydrogen sulfide may be sufficient to lead to failure. However, laboratory and field experiences have allowed engineers to select materials having minimum susceptibility to sulfide stress cracking [1].

Both wrought and cast nickel-beryllium alloys were evaluated for use as a petroleum industry material. The wrought 440 AT alloy was evaluated because of its outstanding formability, high temperature mechanical properties, and fatigue and corrosion resistance. With this alloy it is possible to produce complex shapes having good mechanical strength since it can be formed before age hardening. After a short heat treatment, the alloy achieves a minimum ultimate tensile strength of 215,000 psi (1480 MPa). The cast 42C alloy was evaluated because it is a highly castable, air-melting nickel-base alloy which is precipitation-hardenable to high strength and hardness. This alloy is "stainless" in nature and exhibits resistance to wear, corrosion, and oxidation comparable or superior to stainless steels and superalloys, but with better castability. The beryllium addition to the nickel lowers the melting point by more than 100 C and increases the fluidity of the liquid metal. This provides for superior castability at a lower temperature than that of the competitive nickel-, iron- or cobalt-base alloys.

This evaluation was done by determining the tensile stresses at ambient temperature at which each alloy will be resistant to sulfide stress cracking failure in low pH aqueous environment containing hydrogen sulfide. This test was run in accordance with NACE Standard TM-01-77 [2] as used in the petroleum industry.

MATERIALS

The commercial nickel-beryllium alloys of this investigation are designated BERYLCO alloys 440 and 42C. The 440 alloy is a wrought alloy with a nominal composition of 1.95% Be, 0.50% Ti and the balance Ni. The 42C alloy is a casting alloy with a nominal composition of 2.75% Be, 12% Cr and the balance Ni. The composition of the actual alloy samples are given in Table 1. Both alloys were tested in the "AT" condition (i.e., solution heat treated, quenched, and precipitation hardened, see Table 2) with the properties shown in Table 3. The ultimate tensile strength of the 440 alloy can be further increased to values approaching 300,000 psi (2070 MPa) depending upon the processing procedures.

Table 1. Chemical Analyses of Nickel-Beryllium Alloys Tested for Resistance to Sulfide Stress Cracking (Wt. %)

	Wrought 440 AT	Cast 42C
Be	2.02	2.51
Cr	0.001	11.72
Fe	0.003	0.13
Si	0.002	0.012
Al	0.003	0.017
Co	<0.010	0.015
Mn	<0.001	0.002
Mg	0.002	0.002
Cu	0.002	0.002
C	ND	0.015
Ti	0.38	<0.010
Ni	Bal	Bal

Table 2. Heat Treatments for Nickel-Beryllium Alloys Subjected to Sulfide Stress Cracking Tests

BERYLCO Alloy	Heat Treatment	
Wrought 440	Solution Heat Treated:	1900 F, 1/2 h - Water Quench
	Aged:	900 F, 2 h
Cast 42C	Solution Heat Treated:	1900 F, 1/2 h - Water Quench
	Aged:	900 F, 3 h

Table 3. Mechanical Properties of Nickel-Beryllium Alloys Tested for Resistance to Sulfide Stress Cracking

BERYLCO Alloy	Ultimate Tensile Strength		Yield Strength (0.2% Offset)		% Elong. in 1"
	(ksi)	(MPa)	(ksi)	(MPa)	
440 AT	243.9	1682	186.1	1293	14.2
42C	171.2	1180	116.9	806	15.5

TEST PROCEDURE

NACE Standard TM-01-77 was developed to facilitate conformity in testing so that data may be accumulated from different sources and compared on an equal basis. However, evaluation of the data still requires judgement on several difficult points which must remain a matter for the individual user's decision. This method is conceded to be a severe, accelerated test for sulfide stress cracking. This makes the evaluation of the data extremely difficult since, in testing the reproducibility of this test method among different laboratories, several undesirable side effects were noted: 1) The environment may cause failure by hydrogen blistering as well as sulfide stress cracking; the blistering form of failure, however, may be detected by metallographic observation. 2) The test environment may cause corrosion for some alloys which normally do not corrode in actual field service and thereby induce sulfide stress cracking failures in alloys which ordinarily do not fail by sulfide stress cracking. Other factors which may strongly influence sulfide stress cracking are pH, temperature, H_2S concentration, corrosion potential, stress level, manufacturing considerations, bimetallic effects, etc.

Stressed specimens are immersed in acidified sodium chloride solution saturated with hydrogen sulfide at ambient temperature and atmospheric pressure. Applied stresses at convenient increments of the yield strength can be used to obtain sulfide stress cracking data. Time-to-failure at a fixed stress is an important parameter for experimental correlation purposes. A 30-day (720 h) test period is

Nickel-Beryllium Alloys / 93

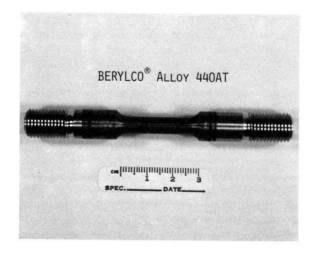

(a) Tested at 135 ksi for 720 h

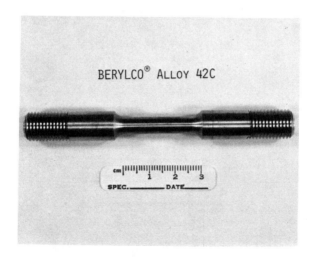

(b) Tested at 120 ksi for 720 h

Figure 1. Photographs of specimens of wrought and cast Ni-Be alloys after sulfide stress cracking tensile tests according to NACE Standard TM-01-77 showing no corrosion.

considered sufficient to reveal failures in materials susceptible to sulfide stress cracking. These materials are tested either to failure or 30 days (720 h), whichever occurs first.

The test specimen is 0.250-inches (6.4-mm) in diameter with a one-inch (25.4-mm) gage length (Figure 1). It is very important to avoid overheating and cold working of the specimen during machining. The specimens are finished to a surface toughness of 32-microinch (0.81 micrometers) or better. The sample is loaded in the test cell (Figure 2) and the calibrated proofing ring places the desired stress on the specimen. The chamber is then filled with the acidified sodium chloride solution that is saturated with hydrogen sulfide. The testing is done at ambient temperature and atmospheric pressure.

Figure 2. Apparatus used in running sulfide stress cracking tensile tests according to NACE Standard TM-01-77 (Courtesy of Cortest, Inc., Eastlake, Ohio).

Since the two alloys have different mechanical properties, they were tested at 50 to 100% of the actual yield strength (0.2% offset). The mechanical properties are given in Table 3.

TEST RESULTS AND DISCUSSION

The wrought Ni-2Be alloy, BERYLCO 440AT, was tested at 100, 75 and 50% of the yield strength (0.2% offset), i.e., 180, 135 and 90 ksi (1240, 930, 620 MPa), respectively, in a low pH aqueous environment containing hydrogen sulfide in accordance with NACE standard TM-01-77. The material passed the 720 h requirement when tested at the 135 and 90 ksi (930 and 620 MPa) tensile loads. These specimens did not corrode (Figure 1). The specimens tested at 180 ksi (1240 MPa) failed at 249.8 and 345.7 h, Table 4.

Table 4. Results of Sulfide Stress Cracking Tests on Wrought BERYLCO Alloy 440AT Per NACE Standard TM-01-77

Specimen	Percent of Yield Strength (0.2% Offset)	Tensile Stress (ksi)	(MPa)	Hours to Failure
1	50	90	620	Passed*
2	50	90	620	Passed
3	75	135	930	Passed
7	75	135	930	Passed
8	75	135	930	Passed
5	100	180	1240	249.8
6	100	180	1240	345.7

* Test stop at 720 h. Specimen did not fail.

The specimens tested at 180 ksi (1240 MPa) were examined metallographically. The structures are typical of solution heat treated and aged 440 alloy specimens. The large grain size is due to the 1900 F (1040 C) solution heat treating temperature with little or no prior cold work. The material had been hot worked prior to this time. The fractures are intergranular. The cross sections of the specimens that

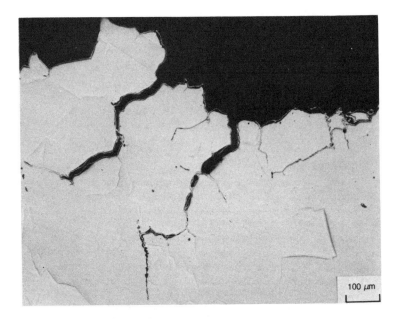

(a) Fractured end

(b) Edge (top) and fractured end (right)

Figure 3. Photomicrographs of 440AT alloy tested at 180,000 psi to failure at 249.8 h (As polished, 100X).

Figure 4. Photomicrograph taken normal to the fracture (left) and edge (top) of the 440AT specimen tested at 180,000 psi to failure at 345.7 h (As polished, 100X).

failed at 249.8 and 345.7 h are shown in Figures 3 and 4. No internal voids or separations were found. Examination of the fractured surface on the scanning electron microscope showed no abnormalities. There had been an oxide layer formed on the fractured surface from standing in the solution, which made it difficult to study the fracture mechanism.

Cast Ni-2.5Be-12Cr alloy, BERYLCO 42C, is not as strong as cast Ni-2.5Be alloy; 170 vs. 240 ksi (1170 vs. 1655 MPa) ultimate tensile strength. However, since it was the opinion of experts in the petroleum industry that the chromium would improve the corrosion resistance, it was decided to test this alloy. The cast 42C alloy was tested at various loads up to 100% of the yield strength (0.2% offset). All tests were stopped after 720 h with no failures (Table 5). No metallographic studies were made since there were no failures. The specimens showed no corrosion (Figure 1).

Table 5. Results of Sulfide Stress Cracking Tests on Cast BERYLCO Alloy 42C in the Solution Heat Treated and Aged Condition Per NACE Standard TM-01-77

Specimen	Percent of Yield Strength (0.2% Offset)	Tensile Stress (ksi)	Tensile Stress (MPa)	Hours to Failure
E-1	50	60	414	Passed*
E-2	60	72	496	Passed
E-3	70	84	579	Passed
D-3	80	96	662	Passed
D-4	90	108	745	Passed
D-1	100	120	827	Passed

* Test stopped at 720 h. Specimen did not fail.

Data on competitive stainless steels, specialty alloys and alloy steels [3-8] commonly used in the oil fields were compared with the data on the nickel-beryllium alloys (Table 6). The 440AT alloy passed at a higher test stress than the competitive alloys listed (Table 6) when tested according to NACE standard TM-01-77. This test stress of 135,000 psi (75% of yield strength) is below the threshold stress for 440AT.

The cast 42C alloy had no failures when tested according to NACE standard TM-01-77 at test stresses up to 120,000 psi (100% of the yield strength). This alloy was tested at approximately the same stress levels as A286 with no failures.

CONCLUSIONS

The results of these studies indicate that nickel-beryllium alloys are resistant to sulfide stress cracking at ambient temperature in a low pH aqueous environment containing hydrogen sulfide at high stress levels.

Table 6. Comparison of Results of Sulfide Stress Cracking Tests Per NACE Standard TM-01-77

Alloy	Yield Strength (0.2% Offset) (ksi)	(MPa)	Test Limit Stress* (ksi)	(MPa)	Ratio Test Limit Stress to Yield Strength (Percent)
Stainless Steels					
AISI 304	33.0	228	32.4	223	NF**
AISI 316	34.4	237	33.7	232	NF
AISI 321	39.3	271	38.5	265	NF
AISI 410	111.9	772	11.1	77	10
CUSTOM 450#	82.0	565	42.6	294	52
17-4 PH	110.4	761	21.1	145	20
A-286	128.6	887	126.0	869	NF
Specialty Alloys					
HASTELLOY# C-276	50.4	347	49.4	341	NF
INCONEL# X-750	120.0	827	117.6	811	NF
MONEL# 400	90.0	621	88.2	608	NF
MONEL K-500	98.6	680	96.6	666	NF
Alloy Steels					
N-80	94.1	649	62.3	430	66
AISI C-4130	92.5	638	80.0	552	86
AISI C-4135	120.2	829	76.9	530	64
AISI C-4135 Mod.	117.5	810	75.2	518	64
Nickel-Beryllium Alloys					
BERYLCO 42C	116.9	806	120.0	827	NF
BERYLCO 440	186.1	1293	135.0	930	73

* Test Limit Stress - The maximum tensile stress at which no failures occurred in 720 h.
** No failure in 720 h at 98% of the yield strength.
\# HASTELLOY is a registered trademark of Cabot Corporation.
\# INCONEL and MONEL are registered trademarks of the Inco family of companies.
\# CUSTOM 450 is a registered trademark of Carpenter Technology Corporation.

The wrought BERYLCO alloy 440AT can be used in this environment up to a stress level of 135 ksi (930 MPa). This material is easy to work or form in the solution heat treated and quenched condition; subsequently a short heat treatment allows the material to reach the desired properties.

The BERYLCO alloy 42C can be cast at a lower temperature than competitive nickel-, iron- or cobalt-base alloys and has greater fluidity providing superior castability. With a solution heat treatment followed by an age hardening treatment, the alloy can be used up to 120 ksi (827 MPa) in this environment.

Based on testing according to NACE standard TM-01-77, the 440AT alloy can be used at higher stress levels than the stainless steels, specialty alloys and alloy steels used in the sour gas fields. The cast 42C alloy has not failed at any strength levels tested.

REFERENCES

1. J. B. Greer, "Factors Affecting the Sulfide Stress Cracking Performance of High Strength Steels," H_2S Corrosion in Oil and Gas Production - A Compilation of Classic Papers, eds. R. N. Tuttle and R. D. Kane, NACE, Houston, Texas, pp. 181-192 (1981).

2. NACE Standard TM-01-77, National Association of Corrosion Engineers (July 1977).

3. J. B. Greer, "Results of Interlaboratory Sulfide Stress Cracking Using the NACE T-1F-9 Proposed Test Method," H_2S Corrosion in Oil and Gas Production - A Compilation of Classic Papers, eds. R. N. Tuttle and R. D. Kane, NACE, Houston, Texas, pp. 290-296 (1981).

4. D. S. Burns, "Laboratory Test for Evaluating Alloys for H_2S Service," H_2S Corrosion in Oil and Gas Production - A Compilation of Classic Papers, eds. R. N. Tuttle and R. D. Kane, NACE, Houston, Texas, pp. 275-282 (1981).

5. R. D. Kane, et al., "Factors Influencing the Embrittlement of Cold Worked High Alloy Materials in H_2 Environments," H_2S Corrosion in Oil and Gas Production - A Compilation of Classic Papers, eds. R. N. Tuttle and R. D. Kane, NACE, Houston, Texas, pp. 401-412 (1981).

6. Private communication from Allen F. Denzine; Cortest, Inc., Eastlake, Ohio.

7. J. A. Straatmann, P. J. Grobner, and D. L. Sponseller, "Results of Sulfide Stress Cracking Tests in Different Laboratories on SAE 4135 Steel Modified with 0.75% Mo and 0.035% Cb," H_2S Corrosion in Oil and Gas Production - A Compilation of Classic Papers, ed., R. N. Tuttle and R. D. Kane, NACE, Houston, Texas, pp. 579-592 (1981).

8. R. R. Gaugh, "Sulfide Stress Cracking of Precipitation Hardening Stainless Steels," H_2S Corrosion in Oil and Gas Production - A Compilation of Classic Papers, eds. R. N. Tuttle and R. D. Kane, NACE, Houston, Texas, pp. 333-338 (1981).

THE MICROSTRUCTURE AND CORROSION RESISTANCE OF 55% Al-Zn COATINGS ON SHEET STEEL

H. J. Cleary*

ABSTRACT

Metallographic, scanning Auger microprobe (SAM), and scanning transmission electron microscopy (STEM) techniques were employed to characterize 55% Al-Zn coatings applied to sheet steel by the hot-dip method. The effect of cooling rate in the coating solidification temperature range was of particular interest. Increased cooling rates produced: (a) smaller dendrite arm spacings in the coating microstructures, and (b) improved corrosion performance both in a laboratory acidified salt spray test and in long term atmospheric exposure tests in an industrial environment. The relationships between coating morphology and corrosion mechanisms are considered.

INTRODUCTION

The development of 55% Al-Zn coated steel sheet** has been described in the literature [1,2]. In current practice, the coating is applied by continuous hot-dip lines and about 1.6% Si is added to the coating bath to control growth of the alloy layer that forms at the coating/steel interface.

---

* Bethlehem Steel Corporation, Homer Research Laboratories, Bethlehem, PA 18016 USA.
** A sheet steel having a nominal coating composition of 55% Al-43.4% Zn-1.6% Si, by weight, and sold under the trademark GALVALUME, a trademark of Bethlehem Steel Corporation.

Corrosion resistance of the coated-sheet product has been studied with emphasis on: (a) the effect of hot-dip coating line conditions on coating microstructures; and, (2) the relation between coating microstructural variations and corrosion resistance. The discussion in this paper will be limited to effects of the cooling rate during solidification of the coating as the steel strip exits the coating pot. The cooling rate has been found to be of major importance to the corrosion resistance of the 55% Al-Zn coating.

EFFECT OF COOLING RATE ON MICROSTRUCTURE

Conventional metallographic studies of coated sheet cross sections reveal network-like cored dendrite microstructures (Figure 1). These structures consist of approximately 80% (by volume) Al-rich dendrites and about 20% interdendritic Zn-rich material. The coarseness of the network-like microstructures varies with cooling rate, with faster cooling producing finer networks.

Quantitative measurement of these morphological variations is facilitated by viewing the coating in the plane of the sheet; more specifically, by locating and viewing properly oriented dendrites as in Figure 2. Here, the Al-rich phase is oriented with the (111) plane parallel to the sheet surface as indicated by x-ray diffraction analysis, and there is a six-fold symmetry in the directions of the primary dendrite arms. The smaller secondary dendrite arms branching off each of the six larger arms at regular intervals provide quantitative characterization of the coarseness of the microstructure, and these intervals are referred to as the dendrite arm spacings (DAS). These primary and secondary arms represent intersections of {100} planes with the plane of the sheet surface [3]. This is true because the easy growth directions for Al alloys are <100>, and growth from a nucleus occurs by orthogonal branching to form filled-in {100} planes.

Several of these symmetrical dendritic patterns can usually be found on the surface of 12.7x12.7 mm sheet specimens. The DAS measurements are obtained by applying a light metallographic polish in the surface plane of the coating, followed by chemical etching, photographing at 200X, and averaging the number of secondary dendrite arms on both sides of each of the six primary arms over a length of 25.4 mm on the photograph.

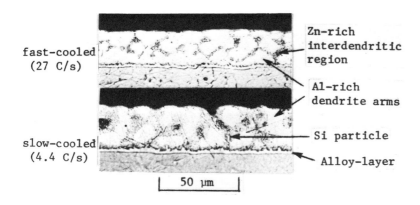

Figure 1. Random cross sections of 55% Al-Zn coatings. Amyl-nital etch.

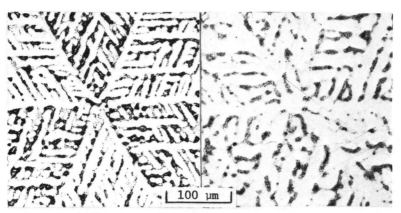

Small dendrite arm spacing (DAS = 12.7 μm)
Fast-cooled (27 C/s).

Large dendrite arm spacing (DAS = 24.4 μm)
Slow-cooled (4.4 C/s).

Figure 2. Spangle center microstructure in a view normal to the sheet surface after polishing to a depth of about 2.5 micrometer. Amyl-nital etch.

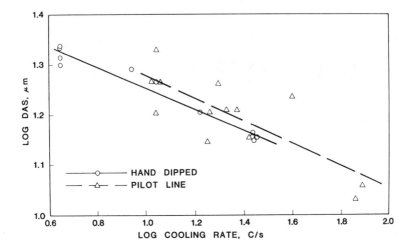

Figure 3. Effect of cooling rate on dendrite arm spacing of 55% Al coatings.

To determine effects of cooling rate on DAS, a series of hand-dipped coated specimens was prepared in the laboratory using sheet preheated in hydrogen at 538 C, a coating bath temperature of 610 C, and controlled cooling conditions ranging from 4.4 to 27 C/s. The sheet and bath temperatures are close to those for currently recommended commercial practice. A second series of coatings produced on a pilot line under varied cooling conditions was also examined. The cooling rate for these coatings was determined by dividing the non-equilibrium freezing range [4] by visually observed freezing times. Figure 3 is a log-log plot of DAS value versus cooling rate for these two sets of specimens. The magnitude of the DAS diminishes with faster cooling rates. A least squares regression line with a slope of -0.21 and a correlation coefficient of 0.98 represents the hand-dipped specimens. For the pilot line materials, the regression line has a similar slope of -0.22, although the correlation coefficient, 0.78, is somewhat lower, probably because of the less precise method of determining the cooling rate. Relationships between DAS and average cooling rate (CR) have the form:

$$DAS = k(CR)^{-n}$$

where k is a constant and the exponent n is in the range of 0.33 - 0.50 for secondary spacings [5]. The reason for the differences between these values and the 0.21 determined here is not known but may be related to differences between bulk and thin-film solidification.

In the region of the coating adjacent to the alloy layer (Figure 4), a dark-etching, thin Zn-rich layer is visible at the alloy-layer/overlay interface of a slow-cooled specimen but is much less evident in the fast-cooled material. Even with careful metallographic preparation of specimens, this difference in appearance is subtle. Scanning Auger microprobe (SAM) point analyses (one micrometer between points) were performed along lines traversing the alloy-layer/overlay interface starting from a point in the base steel and ending at a point in the overlay well above the Zn-rich layer. Care was taken to avoid line termination in a Zn-rich phase in the overlay. Typical quantitative measurements for the two cooling conditions (Figure 5) show a maximum in the Zn concentration in both materials in the region of the alloy-layer/overlay interface, but the maximum is clearly greater in the slow-cooled specimen. Similar results were obtained for traverses at five additional locations on each specimen, and the average maximum was found to be 23% (atomic percent) Zn for the slow-cooled specimen and 1% Zn for the fast-cooled specimen. The layer of interest is from 1.5 to 3 micrometers thick.

Analysis of the polished coating surfaces using the Leitz Texture Analyzing System (TAS) revealed that the volume fraction and size of interdendritic Zn-rich areas (Figure 6) and the density of Si particles (Figure 7) can be used to ascertain the cooling rate history of 55% Al-Zn coated steel sheet.

Scanning transmission electron microscopy (STEM) was used to study the fine structures of thin foils of slow- and fast-cooled coatings. Dislocation lines within the Al-rich solid solution (Figure 8a) were typically found in coatings cooled at 9 C/s and faster. A more complex structure was found in Al-rich areas of the slowly cooled coating (4.4 C/s), as seen in Figure 8b. The mottled appearance indicates a non-homogeneous structure of Zn-rich (dark) particles within

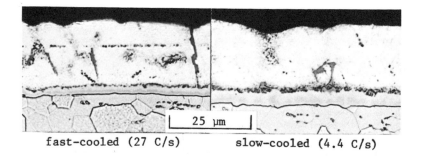

fast-cooled (27 C/s)　　slow-cooled (4.4 C/s)

Figure 4. Optical micrographs of polished and etched (amyl-nital) cross sections of 55% Al-Zn coatings.

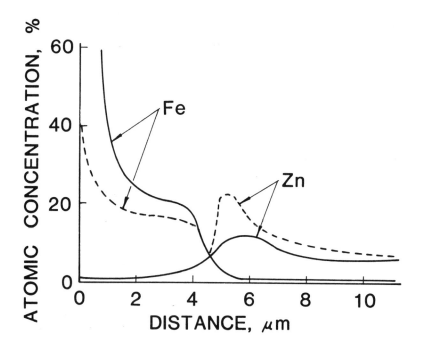

Figure 5. AES-measured atomic concentrations across steel/alloy-layer/overlay interface. Solid curve for fast-cooled (27 C/s). Dashed curve for slow-cooled (4.4 C/s).

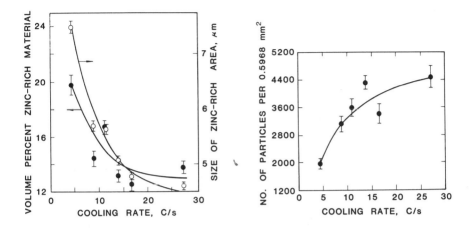

Figure 6. (Left) Effect of cooling rate on volume fraction and size of zinc-rich areas in microstructure. TAS data for 100 fields. 95% confidence levels.

Figure 7. (Right) Effect of cooling rate on the number of silicon particles per unit area. TAS data for 100 fields. 95% confidence level.

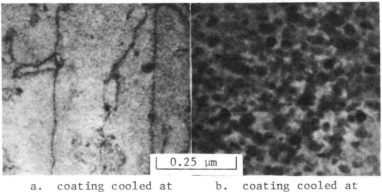

a. coating cooled at 9 C/s.

b. coating cooled at 4.4 C/s.

Figure 8. STEM images of Al-rich regions, bright field.

an Al-rich (light) matrix rather than a solid solution. These fine structures in the Al-rich regions give an indication of the cooling rate history of the coating. A similar comparison of Zn-rich interdendritic areas representative of different cooling rates was not possible because these areas were preferentially dissolved and lost in the electrochemical thinning operation.

LABORATORY AND ATMOSPHERIC CORROSION TESTS

In addition to the commonly used salt spray test, an acidified salt spray test was employed during the development of the commercial product [6]. The test is designed to simulate a severe environment such as a seashore or industrial environment where the atmosphere has acidified chloride characteristics. In this test, slowly cooled coatings were observed to corrode by a flaking mechanism in which patches of the coating overlay separated from the sheet along the interface between the overlay and the alloy layer. When sufficient accelerated cooling (about 11 C/s or more) was applied to the sheet as it exited the coating pot, this tendency to flaking was eliminated. A plausible explanation for the corrosion flaking phenomenon is the greater degree of Zn enrichment of the near-alloy layer region of slow-cooled coatings as compared to fast-cooled coatings, although the coarseness of the dendritic structure may also play a role. On the basis of these findings, accelerated cooling (>11 C/s) has been applied to all commercially produced 55% Al-Zn coated sheet.

Replicate 10x15 cm panels of 5 early commercial and 2 pilot line coated sheet materials were exposed to the industrial atmosphere at Bethlehem, PA for a period of about 7.5 years. A direct relationship (least squares line) between DAS and coating weight loss was found (Figure 9). The adverse effect of a "coarse" microstructure (i.e., large DAS) on corrosion performance is evident, with a 55% increase in DAS resulting in a weight loss increase of about 2 times. Data for shorter exposure times (not shown) indicate that the harmful effect of high DAS values accelerates as the exposure time increases. No significant amounts of red rust or rust staining were present on these specimens.

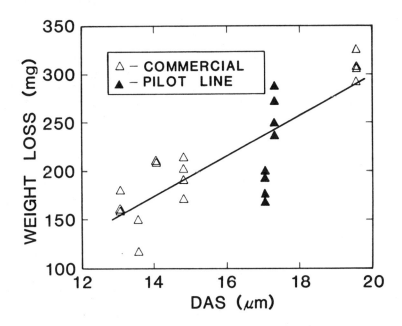

Figure 9. Effect of dendrite arm spacing on weight loss after 7.5 years of exposure at Bethlehem, PA.

CORROSION MECHANISMS

The first stage in the corrosion of the 55% Al-Zn coating almost always involves corrosion of the interdendritic Zn-rich channels. This has been observed for both atmospheric and acidified salt spray tests. As corrosion progresses further in acidified salt spray, coating flaking occurs in slow-cooled material, with separation of the coating overlay at the alloy-layer/overlay interface. This phenomenon has not been observed in the atmosphere to date, but it is conceivable in aggressive chloride-containing industrial environments for slow-cooled materials at long exposure times.

Rapid cooling is believed to improve corrosion resistance by decreasing the DAS values and by lowering the Zn concentration in the interface region between the alloy layer and the coating overlay. Finer dendritic microstructures are comprised

of narrower and more circuitous interdendritic paths that are
believed to retain corrosion products better than coarser
structures and provide more resistance to penetration by
corroding species. The lowering of the Zn concentration at
the alloy-layer/overlay interface is expected to diminish the
tendency of the interfacial material to corrode preferentially
and act as a plane of weakness in the coating microstructure.
Rapid cooling may also improve corrosion resistance by lowering
the volume of Zn-rich material in the coating, as indicated by
Figure 6.

CONCLUSIONS

Key findings in this study include:

1. The secondary dendrite arm spacing (DAS) of 55% Al-Zn
   coatings decreases with increasing cooling rate (CR)
   according to the equation:

   $DAS = k(CR)^{-n}$ where $n = 0.21$

2. A 55% increase in DAS produced a twofold increase in
   coating weight loss after 7.5 years of exposure in
   the industrial atmosphere at Bethlehem.

3. Fast cooling eliminates corrosion flaking in an
   acidified salt spray test employed in the laboratory
   to simulate a severe atmospheric environment.

4. Metallographic and SAM investigations revealed a
   greater concentration of Zn adjacent to the alloy
   layer in the slow-cooled coatings that are susceptible
   to flaking in acidified salt spray.

5. The Texture Analyzing System can be used to obtain the
   volume fraction and size of interdendritic Zn-rich
   areas in the microstructure as well as the number of
   Si particles per unit area, all of which reflect a
   specimen's cooling history at particular bath and strip
   temperatures. The fine structure of Al-rich dendritic
   material was found by STEM to exhibit a marked change
   when the cooling rate varies from 4.4 C/s to 9 C/s,
   which is also an indication of cooling history.

ACKNOWLEDGEMENT

The author wishes to thank the Bethlehem Steel Corporation for permission to publish these results; H. E. Townsend and L. Allegra for support and assistance during the course of the work; Professor J. B. Vander Sande, of M.I.T., for the STEM examination; and, R. G. Hart for the Auger studies.

REFERENCES

1. J. C. Zoccola, et al, ASTM STP 646, S. K. Coburn, Ed., American Society for Testing and Materials, pp. 165-184 (1978).

2. A. R. Borzillo and J. B. Horton, U.S. Patent Nos. 3,343,930, September 26, 1967 and 3,393,089, July 16, 1968.

3. T. F. Bower and M. C. Flemings, Transactions of the Metallurgical Society of AIME, Vol. 239, p. 1620 (1967).

4. A. A. Presnyakov, Yu. A. Gorban, and V. V. Chervyakova, Zhurnal Fizicheskoi Khimii, Vol. 35, No. 6, p. 1289 (1961).

5. M. C. Flemings, Solidification Processing, McGraw-Hill, p. 148 (1974).

6. H. J. Cleary, J. B. Horton, and G. F. Melloy, U.S. Patent No. 3,782,909, January 1, 1974.

FRACTOGRAPHIC OBSERVATION OF CORROSION FATIGUE AND FRETTING FRACTURE SURFACES

David W. Hoeppner*
and
Irina Sherman*

ABSTRACT

Fretting fatigue and corrosion fatigue are common causes of failure of various aircraft components. Micrographic study of fretting fatigue and corrosion fatigue surfaces were carried out on fracture surfaces in the present study that were generated under various applied stress and frequency conditions. The fretting fatigue fracture surfaces were from AMS 6415 steel specimens. It was observed that in the case of fretting fatigue the area damaged by the fretting process becomes the location of the origin of the crack due to the elevation of strain. The corrosion fatigue fracture surfaces were from 1-1/2 dogbone 7075-T76 A1 alloy specimens. In the case of corrosion fatigue crack nucleation of the joints it was observed to be facilitated by the fretting action between hole bore/faying surface corner and the faying surface itself, rather than by the influence of the corrosive environment. The corrosive environment had a greater effect on the rate of the crack growth at the lower frequency and corrosion products were observed on the surface of the crack.

---

* Department of Mechanical Engineering, University of Toronto, Toronto, Ontario, Canada M5S 1A4.

INTRODUCTION

Fretting is a form of wear involving small relative motion or "slip", usually not greater than 25 μm [1], between two contacting surfaces. This motion is usually due to vibrations or to the presence of a cyclic stress and, therefore, is oscillatory in nature.

Fretting fatigue is a failure mode that occurs at the interface of two components clamped together under a normal pressure, with at least one of the components subjected to a cyclic strain [2]. The phenomena occurring at the interface are generally accepted to be similar to those that occur under "pure" fretting. However, the existence of the repeated stress can propagate cracks (fatigue) from the "fretting damage" thus leading to the term fretting fatigue.

The situation which led to some of the work described herein was the in-field discovery of a crack in a propeller/hub-propeller shaft flange bolted interface on a turboprop engine, diagnosed to be fretting fatigue related.

The other part of this work deals with the investigation of the corrosion fatigue surfaces which were from 1-1/2 dogbone specimens. These specimens were initially designed to test the integrity of Hi-Lok fasteners under simulated operating conditions. However, part of the current research relates to an extensive series of corrosion fatigue studies underway at the University of Toronto.

Corrosion fatigue is the term used to describe the interaction of cyclic stresses and aggressive environments [3]. Failure by fatigue and degradation by corrosion continue to be major considerations in aircraft design.

EXPERIMENTAL PROCEDURE

The test rig used for the fretting fatigue studies consisted of a fretting apparatus mounted on an axial-load fatigue test rig. A coupon specimen was mounted vertically between the cross-head and the main plate. Axial-fatigue load was applied via a hydraulic actuator with closed loop control. The specimen was clamped between a fretting pad and a reaction pad in the transverse direction. Clamping load was applied via a wrench-tightened screw. The rest of the fretting apparatus was a fixture which held the clamping load train. This fixture allowed the contacting surfaces to align freely when they were brought together, yet provided rigid support to the fretting pad when all bolted joints were tightened.

The upper link between the cross-head and the specimen was of adjustable length so that the overall stiffness of this link could be adjusted. This permitted the total displacement of the specimen, and hence, the slip amplitude, to be adjusted accordingly. Calibration of slip amplitude versus length of the link was experimentally carried out for different axial loads.

The test rig used for corrosion fatigue studies consisted of a corrosion fatigue salt spray cabinet mounted on a closed loop servo-controlled electrohydraulic fatigue system. The specimen was mounted vertically in the chamber which provided a uniform acidified salt spray environment.

Fretting Fatigue Specimens

Coupon specimens and fretting pads were cut from a turbine flange forging of AMS 6415 steel. Dimensions of the specimen and pad are shown in Figure 1.

The specimens were ground to give a surface finish of 3 micro-inch RMS in the longitudinal direction. The pads were lapped to give a surface finish of 12 micro-inch RMS. Both were degreased with a chlorinated solvent prior to testing.

The fretting fatigue tests were performed under the following conditions:

```
        Normal Stress    -   40 MPa
        Frequency        -   20 Hz
        Slip Amplitude   -   0.001 inch
        Temperature      -   22 C
        "Lab Air"
```

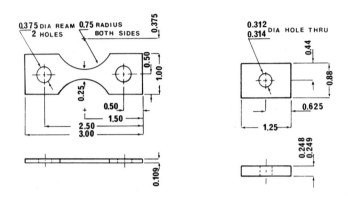

Figure 1. Dimensions of the specimen and pad used for fretting fatigue.

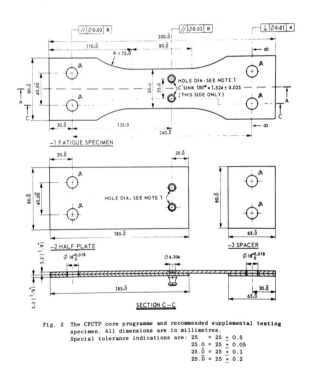

Figure 2. The 7075-T76 aluminum alloy specimen used for corrosion fatigue.

Corrosion Fatigue Specimens

The 7075-T76 aluminum alloy specimen used for the corrosion fatigue programme was a 1-1/2 dogbone containing a single row of two countersunk Hi-Lok fasteners shown in Figure 2. The specimen was developed by the Laboratorioum für Betreibsfestigkeit (LBF), West Germany. It simulates the load transfer and secondary bending characteristics of the runouts of stiffeners attached to the outer skin of aircraft structures.

Corrosion fatigue tests were performed under the following conditions:

 Frequency — 2 Hz, 0.5 Hz, 0.05 Hz
 S_{max} (maximum stress) — 210 MPa and 144 MPA
 R (stress ratio, S_{min}/S_{max}) — 0.1

Salt spray environment conditions were:

 Solution: 5% aqueous NaCl acidified with H_2SO_4 to a pH of 4 $\pm$ 0.1

 Temperature: Solution and humidified air at 295 $\pm$ 2 K

The corrosion protection system for the specimens was as follows: A chromate conversion coating was applied to all surfaces of the specimen. An inhibited epoxy polyamide primer was then applied to all surfaces except the countersunk fastener holes. The assembly of the fatigue specimen (part #1) and half plate (part #2) with Hi-Lok fasteners and collars was followed by application of primer to the fastener areas. An aliphatic polyurethane top coat was applied to all exterior surfaces.

EXPERIMENTAL RESULTS

Results from four fretting fatigue tests are presented herein. The four specimens were tested at three different axial-stress levels. The test run-out was defined at 4×10^7 cycles, but all tests were completed before run-out. The resultant stress-life data are shown in Table 1.

Table 1. Fretting Fatigue Test Results

Specimen	Axial Stress (MPa)	Fretting Fatigue Life (Cycles)
TA3C	700	41,800
RA3C	600	66,400
TA3B	400	371,200
TA1C	400	472,000

Four corrosion fatigue specimens were analyzed for this study and the data are presented in Table 2. All tests were continued to specimen fracture.

Table 2. Corrosion Fatigue Test Results

Specimen	Maximum Stress (MPa)	Frequency (Hz)	Corrosion Fatigue Life (Cycles)
638	210	0.5	14670
634	144	0.5	54040
552	210	2.0	8700
550	210	0.05	7490

MICROSCOPIC OBSERVATIONS

Fretting Fatigue

Subsequent to testing the specimens were examined under a scanning electron microscope (SEM). Figures 3 through 6 are the micrographs of the fretting fracture surfaces. Figure 3 shows the origin of the crack. There is evidence of the debris which are likely the product of the fretting. Figure 4 shows the secondary cracking in the area of the low crack growth rate. Oxide particles at the origin of the crack are shown in Figure 5. Figure 6 is the micrograph of the surface of the crack with inclusions which were analyzed using an energy dispersive x-ray analyzer (EDXA). Inclusions consist of Mn and S and the results of the analysis are shown in Figure 7 and 8. Figure 9 shows the striations in the area of the fast crack growth rate, i.e., close to the final fracture.

 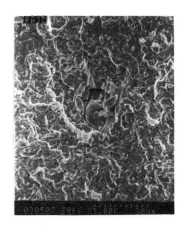

Figure 3. (Left) Fretting fracture surface. Origin of the crack.

Figure 4. (Right) Fretting fracture surface. Secondary cracking in the area of the low crack growth rate.

 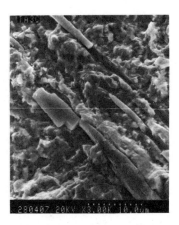

Figure 5. (Left) Fretting fracture surface. Oxide particles at the origin of the crack.

Figure 6. (Right) Fretting fracture surface. Inclusions which were analyzed using an energy dispersive x-ray analyzer.

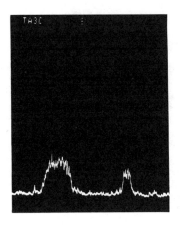 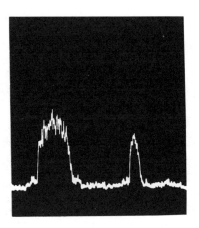

Figure 7. (Left) Distribution of Mn along the line which is shown in Figure 6.

Figure 8. (Right) Distribution of S along the line which is shown in Figure 6.

Corrosion Fatigue

Multi-point fatigue crack nucleation was evident in all of the primary origins of the specimens fatigued in salt spray. Figure 10 is the micrograph of the area close to the primary origin of the crack in specimen 638. The fracture surface exhibits striations and secondary cracking. Figures 11 and 12 are the micrographs of specimen 634 fatigued at S_{max} = 144 MPa. Cleavage facets indicate very little ductility in the area close to the origin. Secondary cracking and some poorly defined striations are shown in Figure 13. Fractographic examination of the two specimens fatigued in salt spray at higher frequency (2 Hz) and lower frequency (0.05 Hz) showed no macroscopic effects. Origin sites were similar in all cases. At lower frequency there is greater evidence of environmental effects. Figure 14 shows the random crack growth paths, secondary cracking, and some intergranular cracking for specimen 550 fatigued at the lowest frequency (0.05 Hz).

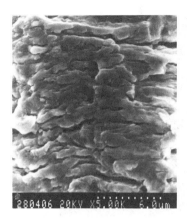 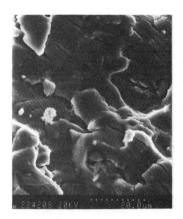

Figure 9. (Left) Fretting fracture surface. Striations in the area of the fast crack growth rate.

Figure 10. (Right) Corrosion fracture surface. Area close to the primary origin of the crack in specimen 638.

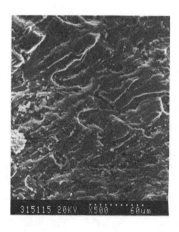 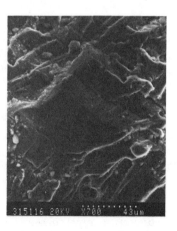

Figure 11. (Left) Corrosion fracture surface. Specimen 634 S_{max} = 144 MPa. Cleavage facets.

Figure 12. (Right) Corrosion fracture surface. Specimen 634. Cleavage facets in the area close to the origin.

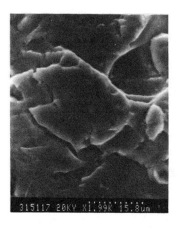 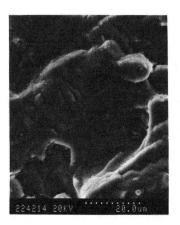

Figure 13. (Left) Corrosion fracture surface. Secondary cracking and some poorly defined striations.

Figure 14. (Right) Corrosion fracture surface. Intergranular cracking for specimen 550.

DISCUSSION AND CONCLUDING REMARKS

Fretting fatigue is usually distinguished from corrosion fatigue by the fact that mechanical factors tend to be involved and may dominate the former phenomenon.

In the present fractographic examination of fretting fatigue surfaces, evidence of oxide particles in the origin of the crack suggested that the nucleation of the crack is due to the stress concentration created in the contact area arising from the frictional stresses between surfaces. In the case of corrosion fatigue, crack nucleation appeared to be similar. This may be attributed to fretting as well due to the configuration of the specimen. Crack propagation was influenced by the corrosive environment and evidence of this influence was shown by the character of the striations and the presence of corrosion products along the crack propagation path. The latter observations suggest that the 1-1/2 dogbone specimen may not be valid for evaluating corrosion fatigue behavior since many of the crack origins appeared to be connected with fretting.

Fractographic techniques are a valuable aid in delineating the characteristics of fatigue fractures produced under conditions where fretting and corrosion act synergistically with the cyclic load.

REFERENCES

1. R. B. Waterhouse, Fretting Corrosion, Pergamon Press, pp. 232-244 (1975).

2. R. K. Reeves and D. W. Hoeppner, "The Effect of Fretting on Fatigue," Wear, Vol. 40, pp. 395-397 (1976).

3. Corrosion Fatigue, AGARD Conference Proceedings No. 316 (1981).

4. R. J. Wanhill and J. J. DeLuccia, "An AGARD Coordinated Corrosion Fatigue Cooperative Testing Programme," AGARD Report No. 695 (February 1982).

5. D. W. Hoeppner and G. L. Goss, "A Fretting Fatigue Damage Threshold Concept," Wear, Vol. 27, pp. 61-70 (1972).

6. A. R. MacDonald, Corrosion Fatigue of 7075-T76 Aluminum 1-1/2 Dogbone Specimens, MASc. Thesis, University of Toronto (1983).

CORROSION FATIGUE BEHAVIOR OF Ti-6Al-4V
IN SIMULATED BODY ENVIRONMENTS

Marko Yanishevsky*
and
David W. Hoeppner*

ABSTRACT

Ti-6Al-4V is currently the most popular implant material in use. This research presents an anlysis of the corrosion-fatigue crack-growth (CFCG) behavior of Ti-6Al-4V in three environments: ambient laboratory air (LA), Ringer's Injection (RI) and 10% by volume Newborn Calf Serum in Ringer's Injection (CS).

The experiments were designed to observe the CFCG behavior of the alloy at constant stress intensity levels. To overcome microstructural variability from specimen to specimen, portions of each specimen were tested in the three above-mentioned environments. The CFCG was found to be statistically the same in RI and in LA. In comparison to the LA and RI baseline data, the CS CFCG rate behaved differently decreasing to a level at least tenfold below the baseline data.

INTRODUCTION

The aim of implant design is to provide long term implant performance, preferably exceeding the lifespan of the recipient. Unfortunately, the vast assortment of load bearing prosthetic devices such as hip, knee, and shoulder implants are still prone to failure.

---

* Department of Mechanical Engineering, University of Toronto, Toronto, Ontario, Canada M5S 1A4.

Tests performed in sacrificial animals and compiled patient histories have identified the following causes of implant failure:

a) metallurgical defects and geometric stress concentrations [1-3]

b) poor surgical technique employed in implant bonding leading to implant loosening [3-9]

c) infection and hematoma of recipient [8,10,11]

d) fatigue failures associated with patient walking (1-3 million cycles per year) [1,5,6,9,12-16]

e) loads are often too high or too low and in both cases degradation occurs along the bone-implant interface [3,6]

f) body chemistry may change temporarily or permanently due to surgical ingress and presence of implant [11]

g) significant corrosion and metal ion release are found to arise from implants in the body environment [8,9,16,17].

The thrust of in vitro testing has been to investigate three major failure processes: fatigue [2,3,11,13,18,19], fretting [14,15] and corrosion [8,9,14,16,17,20]. Conclusions drawn from rotating bend fatigue samples of various implant metals exposed to physiological solutions and tested at a frequency of 1 Hz indicate reduced fatigue lives as compared to the results obtained in laboratory air. Wheeler and James [2], using single-edge notched specimens made of stainless steel, a tension-tension square load wave, and a frequency of 0.85 Hz, found that FCG increases in physiological solutions as compared to laboratory air, particularly for "low and medium" values of the alternating stress-intensity range.

Recently, Brown et al [14-16] carried out research in corrosion and fretting interactions comparing the behavior (in physiological solutions with and without serum protein additions) of a plate and screw specimen made of either stainless steel or titanium. The serum proteins were found to reduce stainless steel degradation processes, while for titanium the degradation rate was found to increase.

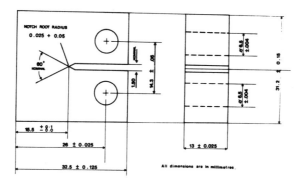

Figure 1. Dimensions and tolerances of the Ti-6Al-4V compact tension specimens.

EXPERIMENTAL EXPOSITION

Material Characterization

The specimens of compact tension (CT) configuration, shown in Figure 1, were supplied by Lockheed California Company and Ladish Company in 1973. Energy-dispersive x-ray qualitative element identification confirmed the material to be Ti-6Al-4V. From 44 hardness indentations for this alloy, the average hardness measured was 65.0 $\pm$ 1.0 in terms of the Rockwell A scale.

Microstructural Analysis

The microstructures found on the specimen surfaces can be categorized into three classes. The first microstructure is characterized by dark alpha titanium (equiaxed HCP structure) with some transformed beta regions (acicular HCP structure) with white beta concentrated at the boundaries. The second microstructure contains a higher percentage of equiaxed alpha and little transformed beta. Again beta is concentrated at the colony boundaries. The last microstructure contains about a 50-50 blend of equiaxed alpha and acicular alpha colonies in a basket weave formation. The primary alpha appears elongated indicating working of the material.

Since each specimen contained complex combinations of these three microstructures, classification of the specimens according to microstructure proved futile.

Environments

The CFCG tests were conducted in each of the following environments:

(1) LAB AIR (LA)
    Temperature: 22.5 ± 0.5 C
    Relative Humidity: 50 ± 20%

(2) RINGER'S INJECTION (RI)
    Temperature: 37.4 ± 0.5 C (Body)
    Composition per 100 ml:
    NaCl USP 86 mg
    CaCl USP 33 mg
    HCl USP 30 mg
    approximate pH = 6.0

(3) 10% by Volume Newborn Calf Serum in Ringer's Injection (CS)
    Temperature: 37.4 ± 0.5 C (Body)
    Composition of Newborn Calf Serum:
    Range of proteins 5.0 - 6.5 gm %
    content: - Albumin   2.69 gm %
               - Globulin α = 1.38 gm %
                          β = 1.15 gm %
                          γ = 0.69 gm %
    approximate pH = 6.5
    Sodium azide was added to this solution at 0.5 g/litre to inhibit bacterial growth in the serum

This last environment is an attempt to simulate the extracellular body fluids present in the implant vicinity long after the trauma periods of surgery and recovery.

Test Hypotheses

The research undertaken in this study examines the validity of the following two hypotheses:

(1) The test environment alters the FCG behavior of Ti-6Al-4V. The highest FCG rates should occur for CS, followed by RI, and finally LA environments.

(2) The FCG response to CS is affected by varying the test frequency.

Test Parameters and Approach

To help alleviate potential specimen-to-specimen variability effects, the following test matrix was developed, as shown in Table 1. Each specimen was divided into three portions, such that for each specimen testing could be performed in each of the three environments.

A sinusoidal load wave was employed with an R ratio $P_{min}/P_{max} = 0.1$. A frequency of 1-Hz was chosen to simulate a fast walk and the maximum stress intensity was set at 15 $MPa.m^{1/2}$ to provide a test schedule of 1-3 days per specimen.

Testing was performed in accordance to ASTM standard E647 for fatigue crack growth testing, with the stress intensity kept constant via computer control.

Table 1. Test Matrix

Specimen	Environment			Test Frequency (Hz)	Test Stress Intensity Range ($MPa.m^{1/2}$)
	Lab Air	Ringer's Injection	Calf Serum Solution		
FFT 43	-	-	ABC	1 & 10	11.0
FT 16	A	C	B	1	13.6
FT 22	A	B	C	1	13.6
FT 32	A	B	C	1	13.6
FT 42	B	C	A	1	13.6
FFT 45	A	B	C	1	13.6

Legend

A - Testing performed on the specimen in the range $0.46 \leq a/W \leq 0.55$
B - Testing performed on the specimen in the range $0.55 \leq a/W \leq 0.66$
C - Testing performed on the specimen in the range $0.66 \leq a/W \leq 0.80$ where a = crack length, W = specimen width

In the constant stress intensity range load test format adopted, the specimen is cyclically loaded according to a load shedding schedule which allows the crack to grow at constant load for a 0.125-mm interval after which the load is shed to return the stress intensity back to the initial level and testing is resumed. After the crack grows 0.250-mm, the number of cycles are totalled and the maximum stress intensity and the stress intensity range are determined based on the average crack length for the given interval. The FCG rate is determined for this interval.

Using the above technique, a statistical analysis of the determined FCG rates can be employed to determine more conclusively if the test environment affects the FCG behavior of the material.

DATA AND RESULTS

The obtained data has been organized into frequency distributions of the FCG rates. These are provided in Figures 2-4 for LA, RI and CS respectively. As well, the effect of test frequency on FCG rates for the CS environment is presented in Figure 5.

DISCUSSION

Statistical analysis of the raw data for the three environments indicates that there is no significant difference between the means and the variances for the FCG data generated in LA and RI environments (Table 2). Statistical t and F tests, comparing CS to each of LA and RI environments, in both cases indicate no significant difference in the means; however, the variances were found to be significantly different. Analysis of refined data where transient effects are eliminated, indicates that there is only a significant difference between the mean of the FCG data generated in RI and that obtained in CS. Further refinement of the data consisted in the elimination of the results from specimen FT 32 for all environments. This was done on the basis that this specimen did not reach the reduced range of FCG rate in CS shown in Figure 6. Analysis of the remaining data indicates that there is no statistically significant difference between the results in LA and RI, whereas both these environments exhibit a significant difference when compared to the CS environment.

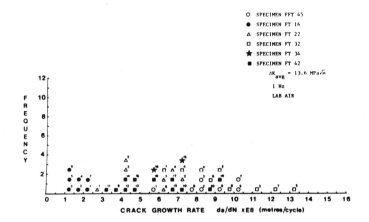

Figure 2. Frequency distribution of the fatigue crack growth data generated in LA.

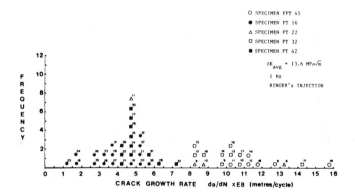

Figure 3. Frequency distribution of the fatigue crack growth data generated in RI.

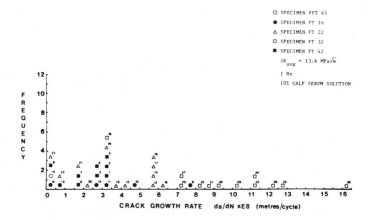

Figure 4. Frequency distribution of the fatigue crack growth data generated in CS.

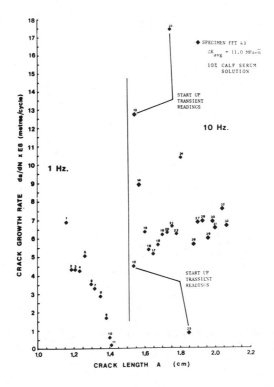

Figure 5. Fatigue crack growth rate versus crack length data in CS indicating the decreasing FCG rate trend for testing at 1-Hz. At 10-Hz no such reduction is evident.

Table 2. Comparison of Environmental Effects
for the Consolidated Data

State of Data		LA-RI	Table Value	LA-CS	Table Value	RI-CS	Table Value
RAW	t	0.73	1.671	0.66	1.671	1.19	1.671
	F	1.36	1.68	2.87	1.69	2.11	1.69
REFINED	t	0.60	1.671	1.26	1.671	1.74	1.671
	F	1.26	1.69	1.64	1.74	1.31	1.74
REFINED WITHOUT FT32 Data	t	0.45	1.671	3.46	1.684	3.14	1.684
	F	1.49	1.84	1.93	1.92	2.87	1.92

Values are based on: 10% level of significance (two-sided test)
5% level of significance (one-sided F test)

In Figure 5, a trend in FCG reduction is noted for testing at 1-Hz in the CS environment. If the test frequency is changed to 10-Hz, an increase in FCG rate occurs. Two further attempts were made to restore FCG at 1-Hz. Testing for over 50,000 cycles in both cases did not restore growth. However, once the test frequency was again increased to 10-Hz, FCG restarted freely.

From the above observation, it is evident that the mechanisms responsible for the diminished FCG behavior found while testing at 1-Hz in 10% calf serum solution are time-dependent, being active only at the lower test frequency.

Observation of the crack tip region during testing in the liquid environments divulged a type of sucking and blowing action by the crack. Debris in solution were noted to be attracted to the crack tip for both RI and CS environments. Over an extended period of time a sludge was found to begin formation at the crack tip for the CS environment. As this sludge enlarged, the generated FCG data was found to diminish. Using load versus crack-opening displacement (COD) data collected during the experimental process (Figure 7) a rough estimate of crack closure forces can be made by assuming

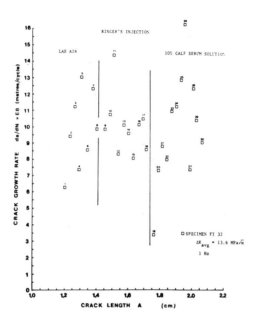

Figure 6. Fatigue crack growth rate versus crack length data for the three environmental conditions. Note the lack of a diminishing FCG rate trend in CS for this specimen.

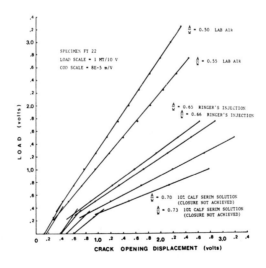

Figure 7. Load versus COD curves for various crack lengths and the three test environments for specimen FT 22.

that the closure load is defined by the intersection of the two different slopes evident for each group of load-COD measurements. It is evident that the closure load is more than twice as large for CS, in Figure 8, than for the other cases. Protein-adsorption on to the failure surface appears to be responsible for the observed sequence of events for the CS environment.

Fractographic Observations

Photographs of the failed specimen surfaces reveal several features (Figure 9). In specimen FFT 43, where the first microstructural class dominates, the failure surfaces appear quite smooth. Little evidence of crystallographic influence is seen and crack front curvature is minimal. In specimen FT 16, primarily of the second microstructural class, evidence of a very rough serrated surface is found. The bands indicate three distinct areas of crack growth in the different environments. Uneven crack growth is also evident, occurring across the crack front. The third specimen, FT 32, predominately of the last microstructural class, shows evidence of a much less serrated rough surface. Bands are again apparent indicating the three different environmentally influenced crack growth regimes. As well, preferred orientation is suspected by the aligned appearance of the microstructural bands across the fracture surfaces.

Further analysis of the specimen surfaces using a scanning electron microscope reveals similar features for all specimens. These features include:

(a) striations indicating cyclic FCG (Figure 10)

(b) cleavage-facets occuring along the various grain orientations (Figure 11)

(c) secondary cracking with some secondary cracks growing intergranularly and arresting at grain boundaries (Figure 12)

(d) intergranular cracking with striation formation (Figure 13)

(e) evidence that during crack growth, cracks tend to choose paths of least resistance (Figure 14)

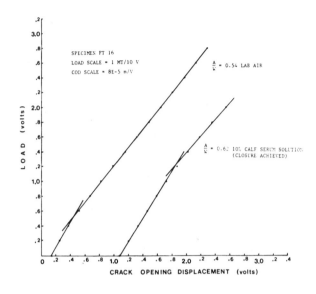

Figure 8. Load versus COD curves for two crack lengths in LA and CS for specimen FT 16.

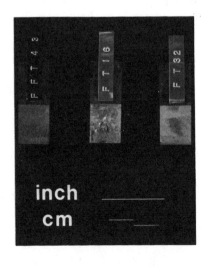
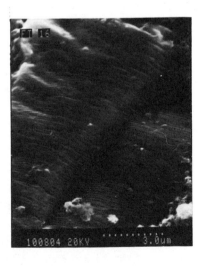

Figure 9. (Left) Failure surfaces of specimens FFT 43, FT 16 and FT 32.

Figure 10. (Right) High magnification of well defined striations. (LA at 1-Hz and ΔK_{avg} = 13.6 MPa.m$^{1/2}$).

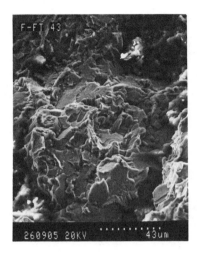 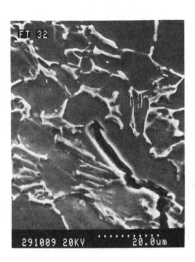

Figure 11. (Left) Cleavage-facets are found to occur along the different crystallographic orientations of the grains. (CS at 10-Hz and ΔK_{avg} = 11.0 MPa.m$^{1/2}$).

Figure 12. (Right) A growing secondary crack appears to have been arrested at the equiaxed alpha grain boundary. (CS at 1-Hz and ΔK_{avg} = 13.6 MPa.m$^{1/2}$).

There is no evidence of failure mechanisms being affected by either microstructure or different environments since ductile striation forming, intergranular secondary cracking, quasi-cleavage and cleavage-facets were found to occur in all three microstructural groups and all three environmental conditions.

CONCLUSIONS

Results of this study permit the following conclusions to be drawn.

(1) The FCG rates generated in LA and RI environments do not differ in terms of statistical significance for Ti-6Al-4V.

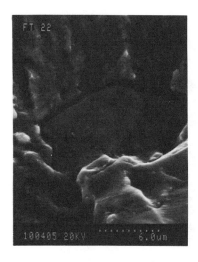 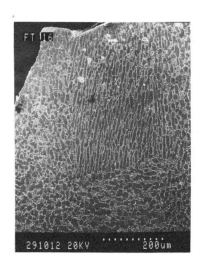

Figure 13. (Left) Intergranular cracking and striation formations on the failure surface. The grain appears to have separated from the rest of the structure. (LA at 1-Hz and $\Delta K_{avg} = 13.6$ MPa.m$^{1/2}$).

Figure 14. (Right) The crack avoids the area of the elongated equiaxed alpha microstructure following the path of least resistance. (CS at 1-Hz and $\Delta K_{avg} = 13.6$ MPa.m$^{1/2}$).

(2) Serum proteins influence the FCG behavior of Ti-6Al-4V, promoting a reduction in crack driving forces which results in decreased FCG rates. This behavior is time dependent requiring an incubation period before the phenomenon takes effect. As well, a frequency dependence exists, with the diminished FCG rates occurring only for the 1-Hz test frequency and not the 10-Hz case.

ACKNOWLEDGEMENT

Portions of this paper are extracted from the M.A.Sc. thesis of M. Yanishevsky, "Corrosion Fatigue Behavior of Ti-6Al-4V in Simulated Body Environments," University of Toronto, 1983.

REFERENCES

1. E.Y.S. Chao and M. B. Coventry, "Fracture of the Femoral Component after Total Hip Replacement," Journal of Bone and Joint Surgery, Vol. 63-A, No. 7, pp. 1078-1094 (Sept. 1981).

2. K. R. Wheeler and L. A. James, "Fatigue Behavior of Type 316 Stainless Steel Under Simulated Body Conditions," Journal of Biomedical Material Research, Vol. 5, pp. 267-281 (1971).

3. R. A. Blackwell and R. M. Pillar, "Fatigue Testing of Materials for Surgical Implants," Closed Loop, MTS Publication, Vol. 6, No. 2, pp. 17-22 (Sept. 1976).

4. H. S. Dobbs, "Survivorship of Total Hip Replacements," Journal of Bone and Joint Surgery, Vol. 62-B, No. 2, pp. 168-173 (May 1980).

5. T. A. Gruen et al., "Modes of Failure of Cemented Stem-Type Femoral Components. A Radiographic Analysis of Loosening," Clinical Orthopaedics and Related Research, Vol. 141, pp. 17-27 (1979).

6. J. Charnley, "Fracture of Femoral Prostheses in Total Hip Replacement," Clinical Orthopaedics and Related Research, Vol. 111, pp. 105-120 (Sept. 1975).

7. P. Ducheyne et al., "Performance Analysis of Total Hip Prostheses: Some Particular Metallurgical Observations," Journal of Biomedical Materials Research, Vol. 14, pp. 31-40 (1980).

8. H. S. Dobbs and J. T. Scales, "Fracture and Corrosion in Stainless Steel Total Hip Replacement Stems," Corrosion and Degradation of Implant Materials, ASTM STP 684, pp. 245-258 (1979).

9. S. A. Brown and K. Merritt, "The Effects of Serum Proteins on Corrosion Rates in Vitro," 2nd. European Conference on Biomaterials, Gotenborg, Sweden (August 1981).

10. H.A.C. Jacob and A. H. Huggler, "An Investigation into Biomechanical Causes of Prosthesis Stem Loosening Within the Proximal End of the Human Femur," Journal of Biomechanics, Vol. 13, No. 2, pp. 159-173 (1980).

11. A. C. Fraker and A. W. Ruff, "Metallic Surgical Implants: State of the Art," Journal of Metals, pp. 22-28 (May 1977).

12. J. R. Cahoon and H. W. Paxton, "Metallurgical Analyses of Failed Orthopaedic Implants," Journal of Biomedical Material Research, Vol. 2, pp. 1-22 (1968).

13. L. E. Sloter and H. R. Piehler, "Corrosion-Fatigue Performance of Stainless Steel Hip Nails - Jewett Type," Corrosion and Degradation of Implant Materials, ASTM STP 684, pp. 173-192 (1979).

14. S. A. Brown and J. P. Simpson, "Crevice and Fretting Corrosion of Stainless Steel Plates and Screws," Journal of Biomedical Materials Research, Vol. 15, pp. 867-878 (1981).

15. S. A. Brown and K. Merritt, "Fretting Corrosion in Saline and Serum," Journal of Biomedical Materials Research, Vol. 15, pp. 479-488 (1981).

16. S. A. Brown and K. Merritt, "Electrochemical Corrosion in Saline and Serum," Journal of Biomedical Materials Research, Vol. 14, pp. 173-175 (1980).

17. R. J. Solar et al., "Titanium Release from Implants: A Proposed Mechanism," Corrosion and Degradation of Implant Materials, ASTM STP 684, pp. 161-172 (1979).

18. M. A. Imam et al., "Corrosion Fatigue of 316L Stainless Steel, Co-Cr-Mo Alloy and ELI Ti-6Al-4V," Corrosion and Degradation of Implant Materials, ASTM STP 684, pp. 128-143 (1979).

19. V. J. Colangelo, "Corrosion Fatigue in Surgical Implants," Journal of Basic Engineering, pp. 581-586 (1969).

20. B. C. Syrett and E E. Davis, "Crevice Corrosion of Implant Alloys - A Comparison of In-Vitro and In-Vivo Studies," Corrosion and Degradation of Implant Materials, ASTM STP 684, pp. 229-244 (1979).

ON-SITE METALLOGRAPHIC INVESTIGATIONS
USING THE REPLICA TECHNIQUE

M. E. Blum*

ABSTRACT

Nondestructive metallography is an important technique for the on-site investigation of corrosion-related problems. However, in certain situations, although the surface can be prepared for metallographic examination, access with a portable microscope is not always possible. In these cases, the problem can be overcome by taking replicas of the polished and etched surface and then observing them with a microscope at a more convenient location. This paper briefly reviews the preparation techniques involved, and provides practical examples drawn from field investigations.

INTRODUCTION

The investigation of corrosion-related problems can often be greatly facilitated by the use of nondestructive metallography [1,2,3]. However, because of accessibility problems, it is not always feasible to examine the areas of concern directly with a portable microscope. In these cases, it is possible to overcome the problem by taking replicas of the polished and etched surfaces and then studying them under more convenient conditions [4]. This can be done in the field with a portable microscope or in the laboratory using a metallograph. Often, immediate and precise information can

---

* FMC Corp., Central Engineering Laboratories,
  1185 Coleman Ave., Box 580, Santa Clara, CA 95052 USA.

Figure 1. Portable electropolisher for nondestructive metallographic preparation.

Figure 2. Portable microscope with integral camera.

Figure 3. Reflective replica showing circular area prepared by electropolishing.

be obtained which enables construction or operating personnel to make rapid decisions, which can save considerable time and money.

In this paper, a brief review is given of the electrochemical approach to on-site metallographic preparation with the complementary use of replicas and a portable microscope. This is followed by examples which illustrate how the replica technique played a crucial role in resolving encountered problems during the construction and operation of petrochemical and process plants.

METALLOGRAPHIC PREPARATION

Metallographic preparation was carried out using a portable electropolisher of the type shown in Figure 1. This polisher consists of a power pack, an electrolyte reservoir, and a pump for circulating the electrolyte to the polishing head. The head contains a cathode, and the circuit is completed by making the workpiece the anode. When this system is in operation, depressing the polishing head opens a valve and the electrolyte is able to circulate freely.

Before proceeding with the electropolishing step, the surface was prepared by hand to a 220-grit finish. The electrolyte used had the following composition: 78 ml perchloric acid, 90 ml distilled water, 730 ml ethyl alcohol, and 100 ml butylcellusolve. After polishing was completed, the surfaces were electrolytically etched with aqueous 10% oxalic acid using an external etcher working off the power pack.

MICROSCOPE

Figure 2 shows the portable microscope that was used in this work. This particular model has the advantage of accepting high quality planachromat objectives and provides good resolution at 400X. In addition, it has an integral polaroid camera so that photomicrography can be carried out on site.

REPLICAS

The reflective replicas used are shown in Figure 3. These are essentially 20 x 30 mm acetate plates with an aluminum backing to reflect the light. The acetate is softened with an ethyl acetate solvent and then firmly pressed against the polished and deeply-etched surface. The acetate hardens in about two minutes, taking up an accurate profile of the surface to be examined.

APPLICATION EXAMPLES

Case I: Sorting Stainless Steel Components
         During the Erection of a Urea Plant

In the production of urea at elevated temperatures and pressures, the preferred material for many of the pressure vessels and associated piping is AISI type 316L austenitic stainless steel. The low carbon content of this steel is essential if chromium carbide formation during welding or stress relieving is to be avoided. These carbides, which tend to form in the 450-850 C temperature range, precipitate at the austenite grain boundaries and promote rapid intergranular corrosion in a steel which normally has excellent corrosion resistance. Consequently, during the construction of a plant using this technology, great care was taken to ensure that stainless steel samples representing the key items of equipment was subjected to suitable screening corrosion tests.

However, in order to keep the project on schedule, a series of type 316L high pressure pipe fittings that had been manufactured in Europe were shipped to the construction site in South Africa while their qualifying corrosion tests were still being completed. The corrosion tests were to ASTM A262, Practice C, which calls for five, forty-eight hour boiling periods in 65% nitric acid. When the tests were completed, it was discovered that the samples representative of a certain lot of fittings had clearly failed. The poor performance of these samples appeared to be due to the presence of chromium carbides in the austenitic microstructure. The configuration of a typical fitting is shown in Figure 4.

Two problems presented themselves. First, some of the fittings had already been welded into the piping system. Second, during preparation and clean-up for welding all fittings had lost their lot identification. The problems, therefore, were to determine which fittings were usable and whether any defective fittings had already been welded into the system. It was decided that the most efficient and rapid test available was nondestructive metallography and each of approximately fifty fittings was examined for evidence of carbide precipitation.

An initial screening was carried out on the outside diameters of the fittings, as these were immediately accessible to a portable microscope (point A in Figure 4). The portable electropolisher proved to be very efficient and enabled the large number of fittings to be prepared rapidly. Screening revealed that while most fittings showed fully annealed austenite grains, free of carbide precipitation, one size consistently revealed a significant amount of grain boundary chromium carbide. The next step was to verify that the structures seen on the external diameter also appeared consistently throughout the fitting.

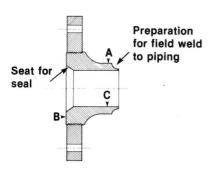 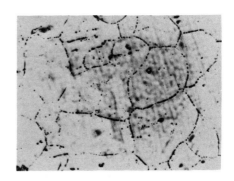

Figure 4. (Left) Configuration of AISI 316L fitting showing location of examination points.

Figure 5. (Right) Replica of sensitized structure in 316L stainless steel. 10% oxalic acid etch (140X).

Fittings of each size were, therefore, polished and etched on their internal diameters and adjacent to the seal seat. The flexibility of the polishing probe served excellently in reaching these locations and, after polishing and etching, a replica was taken for examination (points B and C in Figure 4).

The work confirmed that the structures were consistent throughout each fitting and that the suspect fittings had sensitized structures on the crucial interior and joining surfaces. Figure 5 shows a typical sensitized structure found in the bore of fitting. This has been photographed directly from a replica.

Fortunately, the fittings which had already been welded into the system were of good quality so no rework was necessary. In addition, because the acceptable unwelded fittings had been separated out, it was possible to begin welding acceptable fittings into position while the replacements for the rejected items were being manufactured. In this way, a major loss of time was avoided during an important stage of the project.

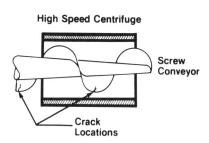 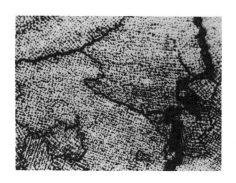

Figure 6. (Left) Sketch of centrifuge and screw conveyor showing location of suspected stress corrosion cracks.

Figure 7. (Right) Replica of cracked surface showing typical shrinkage cracks in the case structure of hard faced deposit. 10% oxalic acid etch (140X).

Case II: Cracks on a Screw Conveyor in a Plastics Plant

Routine inspection of a type 316L stainless steel screw conveyor in a plastics plant revealed cracks in the surfaces transporting the product. Because of the presence of stress, a temperature above 80 C, and the presence of both chlorides and moisture, stress corrosion cracking was a serious possibility. The location of the conveyor inside a high speed centrifuge meant that consequential damage to the system might be severe (Figure 6). However, nondestructive examination using the same technique as described above revealed that the cracks were restricted to corrosion and wear-resisting weld metal that had been deposited on the stainless steel. The photomicrograph in Figure 7 shows that the cracks are typical of hot tearing, a process that frequently takes place during the solidification and contraction of austenitic "hard faced" deposits. There is no evidence of the transgranular, branching cracks which are typical of chloride stress corrosion in austenitic steels. As the cracks had formed during the fabrication of the screw conveyor and there was no evidence that they had propagated in service, it was decided to allow the equipment to remain in operation. This decision avoided expensive plant downtime and was later justified by the fact that subsequent failure did not occur.

ACKNOWLEDGEMENT

The author would like to thank AECI Ltd., Northrand, Transvaal, South Africa, for permission to publish this paper and in particular, Dr. A. J. P. Tucker and Mr. R. E. Leyman for their guidance when this work was being done.

REFERENCES

1. M. E. Blum, "The Use of Non-destructive Metallography to Inspect Weldments in a New High Purity Stainless Steel," Practical Metallography, Vol. 13, pp. 23-30 (1976).

2. P. Lobert, "Application Possibilities of In-Situ Metallography for the Non-Destructive Evaluation of Surface Defects and Microstructural States," Maschinenschaden, Vol. 55, No. 2, pp. 136-144 (1982).

3. M. E. Blum, "Non-destructive Metallography and Corrosion," Proceedings, International Symposium, Metallography and Corrosion, 1983 (to be published).

4. W. Strohfeld, "Further Developments in Non-Destructive Mechanical and Electrolytic Grinding and Polishing Techniques and In The Production of Replicas of Metal Surfaces," Practical Metallography, Vol. 13, pp. 534-548 (1976).

SULFIDATION CHARACTERISTICS OF INCOLOY 800

G. R. Smolik*
and
D. V. Miley*

ABSTRACT

The corrosion behavior of Incoloy 800 has been investigated at temperatures and oxygen and sulfur potentials representative of coal gasification processes. The corrosion coupons had carbon contents and grain sizes representing the range for the different grades of Incoloy 800. The coupons were examined by optical metallography and scanning electron microscopy.

INTRODUCTION

Coal gasifier environments are aggressive due to low oxygen activities at the reaction temperatures and significant sulfur activities from the sulfur in the coal. Activities of these two species have been calculated based on the thermodynamic equilibria for gas compositions produced by various gasification processes [1]. Corrosion behavior in both process and experimental environments has been examined and compared with thermostability diagrams for various alloys [1-4]. These investigations show time-dependent oxide/sulfide boundaries representing protective chromium oxide or external sulfide formation which are displaced from the equilibrium boundaries for Fe-Cr-Ni alloys, as illustrated in Figure 1. Displacements to approximately 10^5 and 10^3 times higher oxygen partial pressures were observed at 923 and 1143 K, respectively [2].

---

* E.G.& G. Idaho, Inc., P.O. Box 1625,
  Idaho Falls, ID 83415 USA.

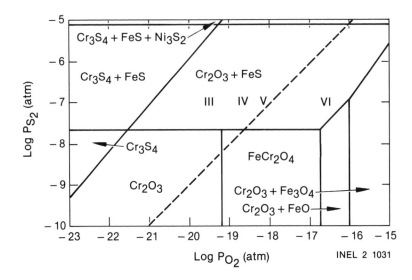

Figure 1. Thermochemical diagram [2] for an Fe-Ni-Cr alloy at 1143 K illustrating a kinetic oxide/sulfide boundary (dashed line) associated with the formation of Cr_2O_3 and showing environments III-VI of this investigation.

The above analyses considered only the chemical differences between alloys. However, some studies have suggested various microstructural influences. For example, small grains have been reported to promote the migration of chromium to the surface, with the resultant establishment of Cr_2O_3 under oxidizing conditions [5,6]. Carbides have been reported to promote attack in sulfidizing salt baths [7-10]. Internal sulfidation was also reported to occur preferentially at grain boundary carbides in Incoloy 800H when exposed in a simulated coal gasification environment [11]. Therefore, the current program examined the influences of grain size and carbides on corrosion mechanisms, particularly internal penetration, in environments which span the oxidizing/sulfidizing regime for Incoloy 800.

EXPERIMENTAL PROCEDURE

Experimental heats of Incoloy 800 were homogenized by alternate cycles of arc remelting and mechanical reduction. Three heats were produced with carbon contents of 0.016,

0.069, and 0.119 wt. %, which span the specification limits
for the different grades of this alloy. The alloys were
worked into 3.2-mm thick sheets which contained 30 % residual
cold work from the final rolling cycle. The sheets were given
1-h anneals in argon at selected temperatures to provided
grain sizes of ASTM 7-8 (0.022 - 0.032 mm diameter) or ASTM
0-2 (0.180 - 0.359 mm diameter) for the low-carbon heat No. 1,
and ASTM 1 (0.254 mm diameter) for Heat Nos. 2 and 3. Coupons
measuring 29.9 x 12.7 x 3.1 mm were sectioned from the sheets,
surface ground to 600 grit, and cleaned in acetone.

The coupons were exposed for 100 h in a test system which
used H_2, a H_2/H_2S mixture, and water vapor to provide targeted
activities for oxygen and sulfur. The H_2 and H_2/H_2S gases
were blended and passed through a temperature-controlled water
bath at a total flow rate of 200 cm^3/min. These gases were
heated in a quartz test chamber, providing the environments
listed in Table 1. Figure 1 shows the locations of the 1143 K
environments relative to the observed oxide/sulfide boundary
[2] upon the thermostability diagram. Voltage drop through a
yttria-stabilized zirconia tube was used to verify the oxygen
activities of these test environments.

Surfaces of the coupons were examined after exposure
by optical microscopy and scanning electron microscopy (SEM)
with energy dispersive x-ray spectrometry (EDS). The surfaces
were examined for appearance, uniformity, spalling, and the
existence and composition of reaction products. Metallographic
sections were prepared and examined by optical metallography
and SEM/EDS. These sections revealed external corrosion
products, the thickness, uniformity, and nature of surface
layers, and penetration depths of internal oxidation and
sulfidation. Constituents of these products were determined
by EDS.

OBSERVATIONS AND DISCUSSION

Coupons representative of all material conditions
exposed to environments I and II, respectively, are shown in
Figure 2. The coupons exposed in the more sulfidizing environment I developed a loosely adherent product over most of the
surfaces. However, some regions possessed a black, tenacious
layer typical of chromium oxide. Exposures in environment II,
with lower sulfur activity, caused less disruption of the
oxide layers and only a few localized regions of external
corrosion products developed. The nearly continuous, ∼90
micrometre thick scale produced in environment I and a nuclei

Table 1. Test Environments

Environment	Temperature C	Temperature K	PO_2, atm	PS_2, atm
I	650	923	3×10^{-24}	1×10^{-8}
II	650	923	3×10^{-24}	1×10^{-9}
III	870	1143	3×10^{-20}	1×10^{-7}
IV	870	1143	3×10^{-19}	1×10^{-7}
V	870	1143	1×10^{-18}	1×10^{-7}
VI	870	1143	3×10^{-17}	1×10^{-7}

of corrosion product which formed in environment II are shown in SEM micrographs in Figures 2c and 2d. EDS analyses determined that the outer regions of the scale consisted of (Fe,Ni) sulfide. Chromium was detected at locations within the scale, and was the major metallic constituent near the scale/metal interface. No evidence of internal sulfidation could be detected along grain boundaries beneath the sulfide scale. These observations agree with the results of similar tests conducted by Tiearney and Natesan [2] on Incoloy 800 and AISI 310 stainless steel, i.e., a shift in oxygen potentials, defining an apparent oxide/sulfide boundary, which are $\sim 10^5$ greater than the equilibrium value. In addition, these exposures did not reveal any differences in sulfidation attack due to grain size or carbon content.

All coupons exposed at 1143 K in environment III, with the lowest oxygen activity, displayed similar surface features. Typical features, shown in Figure 3a, include: (1) an irregular scale over the majority of the surface, (2) globular formations indicating prior melting of external corrosion products, (3) minor regions of oxide, and (4) some spalling. EDS analysis of the metallographic sections confirmed the oxide layer to be a chromium oxide. Intergranular precipitation and voids were observed under the oxide layer, extending to a depth of 0.08 mm. Some of the internal precipitates were found to be either chromium or manganese sulfides. Sections through the highly reacted regions showed external formations,

Sulfidation of Incoloy 800 / 155

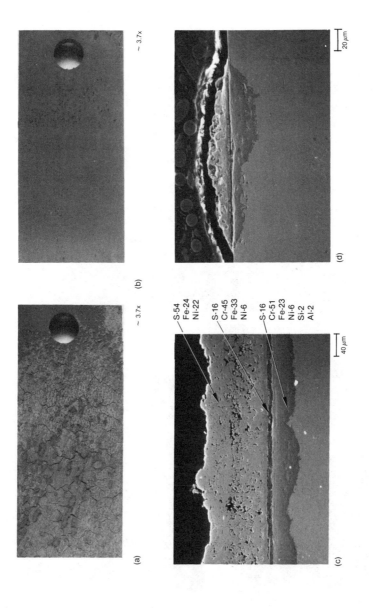

Figure 2. Coupons of heat No. 2, 0.254 mm diameter grain size exposed in environments I (a and c) and II (b and d). SEM micrograph of sulfide scale (c) and an external sulfide formation (d).

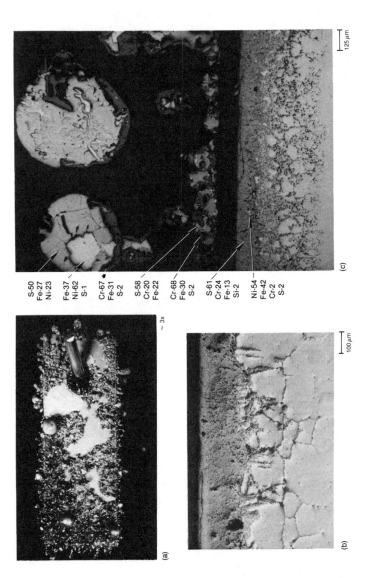

Figure 3. Sulfidation attack of coupons exposed in environment III. Macrograph (a) and optical micrograph (b) of heat No. 3, (0.254 mm diameter grains). Micrograph (c) showing external sulfides, sulfide scale, and internal intergranular sulfidation of heat No. 1 coupon (0.022-0.032 mm diameter grains).

a fairly continuous scale, and intergranular penetration, as illustrated in Figures 3b and 3c. EDS showed that the external formations consisted of (Fe,Ni) sulfides and (Ni,Fe) and (Cr,Fe) oxide or alloy phases, while the scale contained (Cr,Fe) sulfides and (Ni,Fe) alloy or oxide inclusions. Internal sulfides which formed along grain boundaries provided spectra containing primarily chromium with some titanium and iron.

Environment III proved to be severely sulfidizing, with surface scales ranging from 0.18 to 0.25 mm and intergranular penetrations of 0.29 to 0.32 mm. The total attack measured for all coupons was in the range of 0.47 to 0.55 mm, and no significant differences could be attributed to grain size or carbon content.

Exposures in environment IV showed that this environment was primarily oxidizing, with some formation of internal and external sulfides. Coupons of all conditions exhibited an adherent oxide with localized external corrosion products, such as shown in Figure 4a. The extent of breakdown of the oxide could not be related to either grain size or carbon content. Cross sections of the thin oxide shown in Figure 4b show the layer to be approximately 0.002 mm thick. Voids and some internal chromium and manganese sulfides were noted at depths of up to 0.05 mm beneath this oxide layer. A localized corrosion area is shown in the SEM micrograph of Figure 4c. The internal sulfidation illustrated here extended to a depth of 0.25 mm. No influence of carbon content could be related to the magnitude of internal penetration.

All coupons exposed in environments V and VI (partial pressures of oxygen of 1×10^{-18} and 3×10^{-17} atm, respectively) experienced similar oxidizing behavior. The surfaces basically possessed adherent oxide layers with external, crystalline or porous reaction products at localized regions. The external reaction products were less prevalent than those developed in environment IV. SEM examinations of transverse sections demonstrated the existence of the chromium-rich oxide layer and internal oxidation to depths of 0.04 mm. EDS spectra of the grain boundary precipitates did not reveal sulfur, but indicated the presence of titanium, aluminum, and chromium, in addition to the iron and nickel which were probably contributed by the surrounding matrix. Environment V lies near the sulfide/oxide boundary observed by Tiearney and Natesan [2], and the present observations confirmed the shift of this boundary by a factor of $\sim 10^3$ at these conditions.

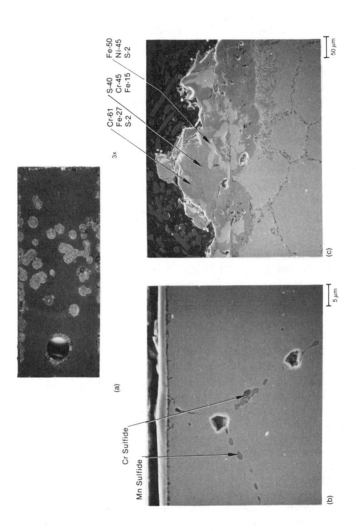

Figure 4. Macrograph (a) of heat No. 1 (low carbon) coupon with 0.254 mm diameter grains exposed in environment IV and SEM micrographs (b and c) illustrating sections through Cr-oxide layer and disrupted oxide region having external sulfides.

Constituents present at the localized surface defects of coupons exposed in environments IV, V and VI were determined by EDS. Various sulfur contents were detected within the spalled regions, Figures 5a and 5b, with some areas showing up to 50 at. % sulfur. It is believed that these sulfur concentrations resulted from external sulfides which formed at the test temperature and spalled off upon cooling, since no evidence of inherent sulfide inclusions could be found in metallographic sections. Corrosion products near the spalled region on samples exposed in environment IV contained sulfide crystals with varying amounts of chromium, iron, and nickel. No significant indications of sulfur were observed in the reaction products formed upon coupons exposed in environments V and VI. Rather, these apparent oxides, which were sometimes porous appearing as shown in Figures 5c, contained primarily iron with lesser amounts of chromium, nickel, manganese, and silicon. It appears, however, that these formations are not impervious to sulfur; rather sulfur may penetrate, form sulfides between the oxide formations, and eventually lead to the breakdown of the protective oxide layer.

Enhanced sulfidation may follow the breakdown of the oxide layers described above. The attack may include the formation of external sulfides, sulfide scale, and internal sulfidation. Eutectic reactions, such as $Ni_3S_2:Ni$, may also contribute to the enhanced attack as illustrated for environment III in Figure 3. However, depths of reaction appeared similar, 0.48 to 0.60 mm, for all material conditions subjected to this mode of attack, i.e., no dependence upon grain size or carbon content was observed within the scope of this investigation.

CONCLUSIONS

This investigation showed the following behavior of Incoloy 800 coupons, with varied microstructure and carbon content, in low-oxygen sulfur-bearing environments.

1. The oxidizing/sulfidizing behavior resulting from exposure in the environments of this study was not significantly influenced by grain size or carbon content.

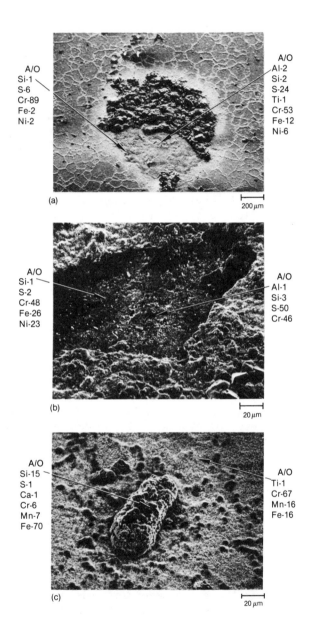

Figure 5. SEM examination of defect areas formed on oxide layers of coupons exposed in environments IV (a), V (b), and VI (c) at 870 C.

2. The development of protective Cr_2O_3 layers demonstrated the influence of kinetic processes on the formation of surface scales, and the resulting phases differ from those predicted by thermodynamic equilibria. Higher oxygen potentials were required for a protective Cr_2O_3 layer to develop than those indicated by thermostability diagrams. The magnitude of shifts to higher oxygen potentials agrees with those of other investigators.

3. Deterioration of oxide layers occurred at localized regions. These disruptions in the oxide appeared to be associated with the formation of porous-appearing high-iron oxides and ensuing sulfide formation.

ACKNOWLEDGEMENT

This work was supported by the U.S. Department of Energy, Office of Energy Research, Office of Basic Energy Sciences, under DOE Contract No. DE-AC07-76ID01570.

REFERENCES

1. K. Natesan and O. K. Chopra, "Corrosion Behavior of Materials for Coal-Gasification Applications," Properties High Temperature Alloys, Z. A. Foroulis and F. S. Pettit (eds.), The Electrochemical Society, pp. 493-510 (1976).

2. T. C. Tiearney, Jr. and K. Natesan, "Metallic Corrosion in Simulated Low-Btu Coal-Gasification Atmospheres," J. Mat. for Energy Systems, Vol. 1, pp. 13-29 (1980).

3. R. A. Perkins, "High Temperature Gaseous Corrosion of Incoloy 800 in Sulfidizing Environments," Alloy 800, W. Betheridge, et al. (ed.), pp. 213-229 (1978).

4. F. H. Stott and S. Smith, "High-Temperature Corrosion of an Iron-Nickel-Chromium Alloy in Gases," International Conference on the Behavior of High Temperature Alloys in Aggressive Environments, Petten (NH), The Netherlands (October 15-18, 1979).

5. C. S. Giggins and F. S. Pettit, "The Effect of Alloy Grain Size and Surface Deformation on the Selective Oxidation of Chromium in Nickel-Chromium Alloys at Temperatures of 900 and 1100 C," Trans. Met. Soc. AIME, Vol. 245, pp. 2509-2514 (1969).

6. M. D. Merz, "The Oxidation Resistance of Fine-Grained Sputter-Deposited 304 Stainless Steel," Met. Trans., Vol. 10A, pp. 71-77 (1979).

7. H. vonE. Doering, "Electrode Potential Measurements of Nickel-Base Alloys in Molten Salts," J. Materials, Vol. 4, No. 2, pp. 457-472 (1969).

8. T. T. Huang and G. H. Meier, An Investigation of the Initiation Stage of Hot Corrosion in Ni-Base Alloys, Progress Report NASA CR-159616 (March 31, 1979).

9. G. H. Meier and E. A. Gulbransen, Hot Corrosivity of Coal Gasification Products on Gas Turbine Alloys, Summary Report for Period April 15, 1978 through November 15, 1979, DOE/ET/13547-T1.

10. J. A. Goebel, F. S. Pettit, and G. W. Goward, "Mechanisms for the Hot Corrosion of Nickel-Base Alloys," Met. Trans., Vol. 4, pp. 261-278 (1973).

11. R. A. Page, "Microstructural Study of Failure Mechanisms During Mechanical Testing in Coal Gasification Environments," A Program to Discover Materials Suitable for Service Under Hostile Conditions Obtained in Equipment for the Gasification of Coal and Other Fuels, The Metals Properties Council, Inc., Quarterly Report, pp. 3-15 to 3-21 (Jan.-Mar. 1982).

NONDESTRUCTIVE ROOM TEMPERATURE TESTS FOR METALLIC CORROSION ON SPACE HARDWARE

P. Mahadevan[*]
and
P. Breisacher[*]

ABSTRACT

Multimetallic junctions (weld joints) inside space hardware are susceptible to corrosion, due, inter alia, to imperfect cleaning prior to storage, residual cleaning compounds and solvents left in the tank and galvanic activity at intermetallic junctions. The weld joints are inaccessible to visual examination or probing.

A nondestructive test procedure was developed to detect very small traces of rust on such a system with the use of hydrazine as a probe gas. The technique is based on the rapid catalytic decomposition of N_2H_4 in contact with iron oxides. The decomposition proceeds to completion in the presence of a catalyst as per the reaction

$$2N_2H_4 \xrightarrow{Fe_xO_y} 2NH_3 + N_2 + H_2$$

Initial tests of the method with pure iron oxides and rusted specimen samples from a fuel tank diaphragm showed that this reaction is selective enough to detect the presence of trace amounts of rust. The probe gas does not interact with the surface either. The decomposition products of hydrazine were analyzed with a time of flight mass spectrometer.

---

[*] Aerophysics Laboratory, Laboratory Operations, The Aerospace Corporation, El Segundo, CA 90245 USA.

Details of a laboratory test facility operating
with hydrazine vapor (low pressure) at room temperature, test constraints and analytical procedures
will be discussed. The technique has potential
for broader commercial applications such as pressure
vessels, highly stressed tankage, and nuclear
reactor vessels.

INTRODUCTION

A method for using hydrazine vapor as a nondestructive
probe for detection of corrosion on stainless steel surfaces
is outlined in this paper. A feasibility study was undertaken
prior to the design and assembly of a laboratory test facility
to be used for examining a number of Reaction Control Systems
(RCSs) fuel tank assemblies for corrosion. The RCS flight
hardware under inspection is a propellant tank assembly with
a metallic diaphragm expulsion device for the controlled
release of hydrazine to a reaction control thruster. The
spherical tank and the diaphragm are made of 304-L stainless
steel. A set of metallic reinforcement rings, decreasing in
diameter from the outer periphery of the diaphragm, are brazed
to the side that is pressurized. Thus, multimetal junctions
are formed at the ring contours on the diaphragm surface (the
brazing compound itself, gold-nickel wire, provides a bimetallic
interface). These contact points on the diaphragm surface
are more susceptible to corrosion because of galvanic and
other effects than the normal inner wall of the tank assembly.
Corrosion effects at these multimetal junctions can be
accentuated by the presence of trace quantities of water or
other rinsing solvents and possible active contaminants in
the Freon pressurizing gas itself.

VISUAL AND X-RAY RADIOGRAPHIC EXAMINATION
OF THE FUEL TANK

The fuel tank was examined for dents and other impact
damage during shipment. The Freon and hydrazine ports on
the tank were unplugged and inspected. A reddish brown
discoloration/stain was observed on the internal threads of
the Freon ports. The diaphragm was observed to be in its
normal position with one of the reinforcement ribs close to
the bottom of the porthole. Thus, visual inspection of the

diaphragm was limited to the cross-sectional area of the porthole only. No attempt was made to clean the stain on the threads. The port on the fuel side of the tank appeared to be clean. An x-ray radiograph of the inner tank configuration was taken by an outside service laboratory. No structural damage to the diaphragm, rib assembly, or weld joints was observed.

PRELIMINARY ACCEPTANCE TESTS

Pump-Down and Leak Check

During pump-out of the tank, it is required that the pressure differential between the large fuel section and the small Freon section at no time exceeds 50-Torr. This requirement is to prevent undue stressing of the expulsion diaphragm. In the normal unextended position of the diaphragm, the ratio of the volumes of the two sides is about 20. A pumping system was designed and built using a common pump and sets of fixed and variable valves on each side to regulate the pump rate.

The procedure can be explained with the detailed schematic of the test setup shown in Figure 1. The crescent shaped section of the spherical tank is the pressurized side P_5. The fuel segment is P_6. The metering valve V_4 on the pumping manifold is adjusted manually while all the other valves V_{17}, V_{14}, V_{15}, V_3, V_5, V_{13}, V_8, V_9, V_{11}, V_{10}, V_1, and V_2 are fully open. A pressure differential of <50-Torr is maintained between the two chambers during pump-down. Both sides are pumped down to <5-Torr pressure.

The tanks are then pressurized with helium through the same pumping lines, maintaining the Δp requirement as just discussed. The settings on V_4 remained the same. The pump and fill sequence was then repeated several times to ensure good control of the metering valve settings.

Leak Checking

Each chamber is checked for leaks by monitoring the pressure change as a function of time after the pump is shut off for about 2 h. Subsequently, the fuel tank is pressurized

Schematic of Test Setup

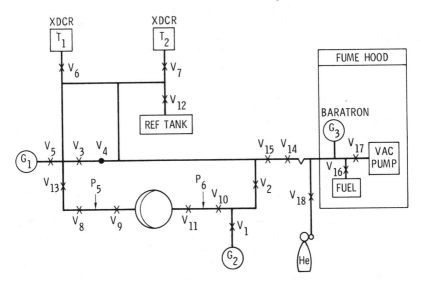

Figure 1. Schematic of test setup.

to about 20-Torr with helium. Valves V_2, V_3, and V_6 are closed (see Figure 1). The pressure rise in the Freon tank as a function of time is observed. If the pressure rise exceeds the normal rate recorded earlier, we should suspect a leak(s) across the diaphragm. If a leak occurs, the test is to be repeated at 30- and 40-Torr pressure differentials, respectively.

HYDRAZINE PROPERTIES RELEVANT TO THESE TESTS

The decomposition of N_2H_4 to NH_3 and N_2 on an iron oxide catalyst proceeds accordng to

$$3N_2H_4 \xrightarrow{Fe_xO_y} 4NH_3 + N_2 \qquad (1)$$

Therefore, if N_2H_4 is sealed within a closed volume at room temperature and is permitted to decompose completely on the catalyst, a pressure rise to 1.67 times the initial N_2H_4 pressure should occur. If the reaction proceeds further by partially decomposing the NH_3 to H_2 and N_2, then the reaction can be written

$$3N_2H_4 \rightarrow 4(1-X)NH_3 + (1+2X)N_2 + 6X\ H_2 \qquad (2)$$

where X is the fraction of NH_3 decomposed. For example, if 50% of the NH_3 is decomposed, then

$$3N_2H_4 \rightarrow 2NH_3 + 2N_2 + 3H_2 \qquad (3)$$

and a pressure change of 2.33 times the initial value should result. If all of the NH_3 is decomposed, then

$$3N_2H_4 \rightarrow 3N_2 + 6H_2 \qquad (4)$$

and a threefold increase in pressure should result.

Since the vapor pressure of hydrazine is only 10-13 Torr near room temperature, a maximum final pressure of 30-39 Torr can be expected if N_2H_4 decomposes entirely to N_2 and H_2. The maximum allowable differential pressure across the diaphragm is about 50-Torr. Therefore, the unlikely but possible complete decomposition of all of the N_2H_4 that contributes to pressurization of the tank would yield an acceptable pressure difference across the diaphragm.

Hydrazine readily absorbs on clean metallic surfaces. In order to obtain meaningful indications of pressure change, it is necessary to thoroughly passivate the surfaces with N_2H_4 prior to the decomposition test. This condition is assumed to be achieved when a fresh N_2H_4 charge placed into the expulsion gas volume does not reveal a demonstrated decrease in pressure over 5 min.

APPARATUS

As indicated previously, a schematic of the test assembly is shown in Figure 1. A photograph of the apparatus is shown in Figure 2. The reference tank is a new one litre 304-L S.S. bottle. The Heise gauges G_1 and G_2 are isolated from the hydrazine environment by valves V_5 and V_1, respectively. During tests, an absolute pressure gauge G_3 (MKS Baratron, Type 170M-34B 0-1000 Torr) designed for operation in a toxic gas environment was used. A pair of matched transducers T_1 and T_2 were mounted as shown to monitor continuously the pressure differential between the two chambers. The voltage outputs of T_1 and T_2, through a low-gain amplifier, were arranged to cancel each other. A strip chart trace of the null signal with the two sections of the fuel tank at common

168 / Microstructural Science, Volume 12

Figure 2. Photograph of test set-up.

pressure would then be on center zero. After establishing long-term signal stability, the transducer assembly was calibrated with known overpressure of helium on either side. The transducer locations were interchanged to check for any cumulative signal drift.

The hydrazine source was kept under a fume hood in the laboratory, which also handled the exhaust from the vacuum pump.

RESULTS

The pressurant tank was conditioned with hydrazine, as discussed previously. After checking stability of the transducer system output, a test run with approximately 12-Torr of fuel vapor in the Freon chamber was started. The fuel side of the tank was pumped down and sealed off. A strip chart trace of a typical test run is shown in Figure 3. The run lasted 23 h. The differential Δp is estimated to be about 21-Torr. At the end of this run, the expulsion gas side of the tank was sealed off, and the whole system moved to the analytical laboratory. The residual gas in this side of the tank was analyzed with a Bendix time-of-flight mass spectrometer. Residual fuel was still found in the tank along with NH_3, N_2, and H_2. Accurate quantitative estimates of the extent of decomposition of N_2H_4 are difficult to obtain with the analytical procedure adopted here. A clean reference tank with a few rusted test pieces was substituted for the fuel tank in a subsequent test run over approximately the same amount of time. The pressure increase recorded was almost one and a half times that for the tank. However, this confirmatory test cannot be used to estimate the extent of corrosion in the fuel tank assembly.

Several test runs were made subsequently confirming the presence of rust on the tank. The discoloration on the inner threads of the Freon port was later cleaned with isopropyl alcohol, and the tests were repeated. The tank still exhibited a signal indicating rust spots inside.

CONCLUSIONS

A sensitive, nondestructive test (summarized in the Appendix) for the detection of rust in flight hardware has been developed and tested. Quantitative estimates of corrosion are very difficult to obtain even under normal circumstances.

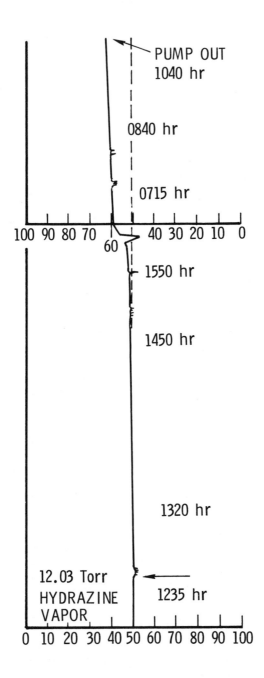

Figure 3. Strip chart trace of differential pressure.

The problem is even greater in this situation because of configurational inaccessibility of the tank inner surface for visual examination.

Commercial applications of this approach for component testing in systems such as nuclear reactor vessels, highly stressed tankage, and pressure vessels should be explored.

APPENDIX

PROCEDURE FOR CORROSION TEST WITH HYDRAZINE

A step-by-step test procedure for corrosion detection in the fuel tank assembly, based on the schematic in Figure 1 is given below.

1. Open V_1, V_2, V_3, V_5, V_6, V_7, V_8, V_9, V_{10}, V_{11}, V_{12}, V_{13}, and V_{14}.

2. Close V_6. Open V_{15}. Control pressure difference between G_1 and G_2, to be less than 50-Torr, by controlling pumping with V_4.

3. Pump down entire system to less than 5-Torr, as read on G_1, G_2, T_1, and T_2.

4. Close V_1 and V_5 to prevent damage to gauges G_1 and G_2 from contact with hydrazine.

5. Close V_2 and V_{17}. Open V16 (under fume hood). Read G_3. This should read in the range 10 to 15 Torr at room temperature.

6. Passivate system for 10 min. Close V_{16}. Read steady reading of G_3. If it reads less than 10-Torr, open V_{16} again. Repeat until hydrazine pressure in system is steady and greater than 10-Torr.

7. Close V_7, V_3, V_5 and V_{16}. Open V_{17}. Pump out.

8. Record start time and hydrazine initial pressure on strip chart recorder for differential pressure readings.

9. Record differential pressures for at least 4 h.

10. Record elapsed time and T_1 and T_2 readings.

11. Close V_8, V_9, V_{10} and V_{11}.

12. Disconnect fittings at ports P_5 and P_6.

13. Connect mass spectrometer (Bendix time-of-flight or equivalent) to P_5.

14. Prepare mass spectrometer for analyzing sample before opening V_9.

15. Record sample gas data.

16. Close V_9.

17. Disconnect mass spectrometer at P_5.

18. Reconnect tank to test system at P_5 and P_6.

19. Go to step 1 and repeat entire test one additional time.

20. After completion of test, go to step 3.

21. Close V_{17}. Open V_{18} through regulator and backfill entire system with helium to a pressure of 760-Torr. Ensure that at no time during the backfilling process are G_1 and G_2 reading 50-Torr or more apart.

CORROSION STUDIES IN COAL LIQUEFACTION PLANTS

James R. Keiser* and Arnold R. Olsen*

ABSTRACT

During the past few years, four direct coal liquefaction pilot plants have been operated in the United States to evaluate several liquefaction processes. Oak Ridge National Laboratory has assisted pilot plant operators by assessing materials' performance through supply and examination of corrosion samples, on-site examination of equipment, and analysis of failed pilot plant components in our laboratory. This paper describes these examinations, which have revealed chloride and polythionic acid stress corrosion cracking, water-side pitting, sulfidation, and a chloride-related acid attack. These analyses have helped identify corrosion problems and make proper materials selections or design changes, and have provided designers with information useful in selecting materials for proposed plants.

INTRODUCTION

To provide an alternative to scarce and increasingly expensive petroleum and its products, the U.S. Department of Energy provided partial funding for the construction and operation of several direct coal liquefaction pilot plants.

---

* Metals and Ceramics Division, Oak Ridge National Laboratory, Oak Ridge, Tennessee 37830 USA.

Information about four of the plants is given in Table 1. These plants were intended to permit evaluation of the processes, but they also provided an opportunity to assess the performance of materials in coal liquefaction environments.

Table 1. Major Coal Liquefaction Pilot Plants in the United States

Process	Location	Coal Capacity (tons/day)	Operating Period
SRC[a]-I	Wilsonville, Alabama	6	1974-present
SRC[a]-II	Fort Lewis, Washington	50	1974-1981
EDS[b]	Baytown, Texas	250	1980-1982
H-Coal	Catlettsburg, Kentucky	600	1980-1982

[a] Solvent Refined Coal

[b] Exxon Donor Solvent

Oak Ridge National Laboratory (ORNL) has been involved in the evaluation of materials performance in all four of the pilot plants. This involvement included trips to the pilot plants to perform on-site component examinations and examination of failed pilot plant components at ORNL. In addition, thousands of corrosion samples were supplied and these provided additional, extensive information on materials behavior in coal liquefaction environments.

These assessments of materials performance identified several degradation mechanisms that are common to the coal liquefaction processes. Corrosion mechanisms include general sulfidation, chloride and polythionic acid stress corrosion cracking, pitting by aqueous and hydrocarbon liquids, and severe chloride-related corrosion in the 215 to 260 C (419-500 F) temperature range of fractionation columns. The most serious materials degradation was slurry erosion, but this will not be addressed here. The remainder of this paper will describe our studies of several of the corrosion problems encountered at the coal liquefaction pilot plants.

CORROSION MECHANISMS

Chloride Stress Corrosion Cracking

Instances of chloride stress corrosion cracking have been encountered at all four pilot plants; an example of cracking that initiated on the cooling water side of an alloy 800 heat exchanger was described previously [1]. Failures originating on the process side have also been observed; on several occasions these occurred in "dead legs", where no material was flowing through the pipes and temperatures dropped low enough that aqueous condensate formed. A piece of a failed dead leg, which came from the Wilsonville preheater-to-dissolver transfer line, was analyzed. Metallographic examination of a cross-section of the pipe showed extensive cracking, which apparently originated at the inside wall. Examination of the crack tip on an etched section revealed transgranular cracking with the branching pattern characteristic of chloride stress corrosion cracking (Fig. 1). Wavelength dispersive x-ray analysis showed sulfur in the wider cracks and both sulfur and chlorine in the leading narrow cracks. This supported the hypothesis that the cracking was related to chlorine rather than the ubiquitous sulfur.

A sulfur corrosion scale is found on most components exposed at elevated temperature to coal liquefaction process streams. When these sulfide scales are exposed to moist air during plant shutdowns, polythionic acid is formed. When this acid contacts a sensitized stressed component, stress corrosion cracking can result. An example of this type of attack was encountered during an on-site examination of the Wilsonville pilot plant. The original Wilsonville fractionation column had a carbon steel shell with a 2-mm (5/64-in.) thick AISI 304 stainless steel cladding in the lower half of the column. Because of serious corrosion problems near the middle of the column, ORNL metallographers performed an on-site examination. A small area of the cladding was ground, polished, and etched in situ, and a room-temperature vulcanizing (RTV) rubber replica was made of the prepared surface. Microscopic

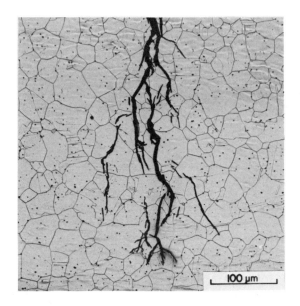

Figure 1. Etched transverse section showing transgranular cracks in AISI 316 stainless steel vent line attached to the slurry transfer line from the Wilsonville preheater to the reactor.

examination of the replica showed that the AISI 304 stainless steel liner was cracked intergranularly (Fig. 2). Taking into consideration the materials and conditions, this cracking was likely the result of polythionic acid stress corrosion cracking. Similar cracking was found in the 304 stainless steel cladding of three Fort Lewis pilot plant flash vessels that had been sensitized in fabrication [2].

Pitting

Pitting corrosion has been seen in both cooling water and process stream applications. Inadequately treated cooling water caused pitting of carbon steel tubing at two of the pilot plants. Of greater frequency is pitting of steels in sour water environments. Examples were seen in a carbon steel vessel shell and U-bend specimens exposed in the Catlettsburg pilot plant and also in duplex tubing examined at ORNL after exposure at the Baytown pilot plant.

Figure 2. Room-temperature vulcanizing rubber replica of the exposed surface of the AISI 304 stainless steel liner in the lower half of the Wilsonville fractionation column.

Following termination of the Catlettsburg pilot plant operation in November 1982, the ammonia stripping tower (N-402) was one of several vessels examined by ORNL personnel during an on-site inspection. As shown in Figure 3, the carbon steel shell of N-402 was seriously pitted. The U-bend specimens, although exposed in the ammonia sour water stripper system only 1870 h, showed similar pitting of carbon steel and AISI 409 stainless steel samples. Figure 4 shows the inside bend of six of the ten alloys exposed as U-bends at the middle manway of N-402. Alloys shown are (left to right): carbon steel, AISI 409 stainless steel, 18 Cr-2 Mo, SAF 2205 (a duplex alloy), AISI 304 stainless steel, and the austenitic alloy 904L. No pitting or stress corrosion cracking was found on the last four alloys.

Figure 3. Carbon steel shell and the bottom of an AISI 410 stainless steel tray in the ammonia stripper column in the sour water treatment area of the Catlettsburg pilot plant. Holes in the tray are 35-mm in diameter.

Another example of sour water pitting was observed on duplex alloy 3RE60 tubing from the solvent hydrogenation hot separator condenser of the Baytown pilot plant. Pitting was observed during the inspection that followed the end of plant operations. Pieces of several tubes that had been exposed about 10,500 h were examined at ORNL. Figure 5 shows a pit in a transverse cross-section of a tube. In this figure, the gray continuous phase is austenite and the white islands are ferrite. Examination of a longitudinal section (Fig. 6) shows that the ferrite was preferentially attacked so that bands of austenite were left behind. Clearly, additional studies of duplex alloy performance in coal liquefaction plant sour water environments are needed.

Sulfidation

The presence of eastern bituminous coals of 2 to 5 wt % sulfur, some of it in reactive forms, results in sulfidation of most internal surfaces of a coal liquefaction plant. The extent of sulfidation is affected by many factors including

Figure 4. Inner surface of U-bend samples after 1870 h at about 146 C (295 F) in the Catlettsburg ammonia stripper column. Left to right: carbon steel, AISI 409 stainless steel, 18Cr-2Mo, SAF 2205, AISI 304 stainless steel, 904L.

the temperature, sulfur activity, alloy composition, and stream velocity. Sulfidation was especially severe in the reactor vessels where sulfur activity was relatively high and the temperature reached 460 C (860 F). Carbon steel, low-chromium steel, and molybdenum-free, high-nickel alloys like alloy 800, were often totally consumed by corrosion. Figure 7 shows a cross-section of an Incoloy 800H U-bend specimen that was welded in the middle with Inconel 82 weld wire and exposed 2600 h in the Fort Lewis reactor vessel. The center section of the U-bend containing the Inconel 82 weld is very badly sulfidized, while the other sections underwent significant but lesser amounts of corrosion. Even more severe sulfidation occurred with Inconel 600, but molybdenum-containing nickel-base alloys like Inconel 625 and Hastelloy C-276 were not badly attacked.

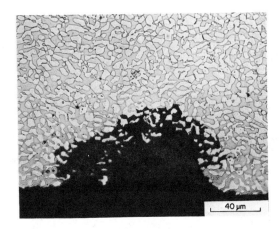

Figure 5. Transverse section through a pit in duplex alloy 3RE60 exposed 10,500 h to a sour gas and sour water environment in the Baytown pilot plant solvent hydrogenation hot separation condenser.

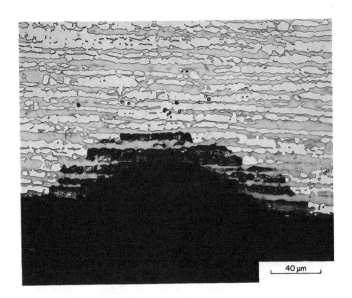

Figure 6. Longitudinal section through a pit in 3RE60 alloy from the Baytown condenser tubing showing preferential corrosion of the ferrite phase.

Figure 7. Incoloy 800H U-bend sample welded with Inconel 82 showed severe sulfidation attack after 2600 h in the Fort Lewis vessel at approximately 450 C (840 F).

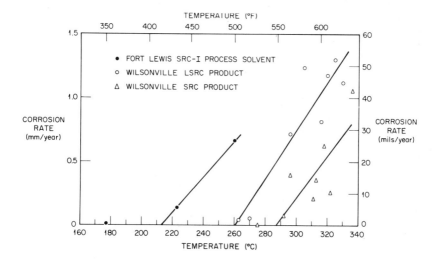

Figure 8. Static laboratory tests of sulfidation corrosion rates of carbon steel show a significant effect of temperature and the oil fraction.

To assess the relative sulfidation potential of several pilot plant process streams on carbon steel, corrosion tests were performed with: (1) SRC-I process solvent from the Fort Lewis pilot plant, (2) SRC product from the Wilsonville pilot plant, and (3) LSRC (light SRC) also from the Wilsonville plant. The results of these tests, shown in Figure 8, show that the sulfidation rate was linear at the test temperature and was strongly influenced by the oil source. For the three oils studied, the lower the boiling point, the more corrosive the oil.

The effect of stream velocity on sulfidation rate was demonstrated in our analysis of a failed pipe from the Kerr-McGee Critical Solvent Deashing area of the Wilsonville pilot plant. The pipe section examined was part of the line transferring a mixture of deashing solvent and LSRC from a settler vessel to a flash tank. Flow in this jacketed transfer line was sporadic, so velocities are not known. However, in the area of the failure, considerable turbulence was probably caused by two unrelated factors: poor matchup between the pipe and a flange and localized heating of the pipe caused by the heat transfer fluid, which entered the jacket very near the point where the pipe failed. The inside of the pipe containing the area of the failure is shown in Figure 9. A transverse section of the pipe through the failure is shown in Figure 10. Note the ledges, which are indicative of erosive wear caused by turbulent flow. The cause of this failure was attributed to erosion-corrosion (i.e., sulfidation accelerated by turbulent flow and consequent scale removal).

Chloride-Related Fractionation Column Corrosion

Perhaps the most serious, and certainly the most unexpected corrosion problem encountered in the pilot plants, occurred in the fractionation columns used to separate the medium- and high-boiling point liquid products. The mechanism [3] and the role of chlorine [3-5] in this corrosion problem have been described. This corrosion, which was most severe in the 215 to 260 C (420-500 F) range, resulted in corrosion rates as high as 25 mm/year (1.0 in/year) on ferritic and martensitic steels and up to 6.3 mm/year (0.25 in/year) on austenitic stainless steels. The localized nature of the corrosion is clearly shown in Figure 11, which pictures five corrosion coupons that were

Coal Liquefaction Plants / 183

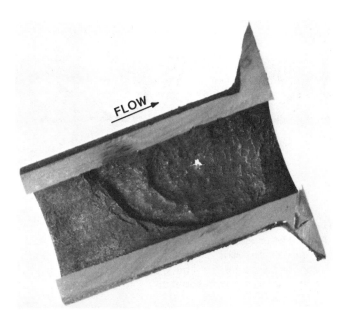

Figure 9. Failed 3/4-in. schedule-80 carbon steel pipe that carried a mixture of LSRC and deashing solvent in the Wilsonville pilot plant.

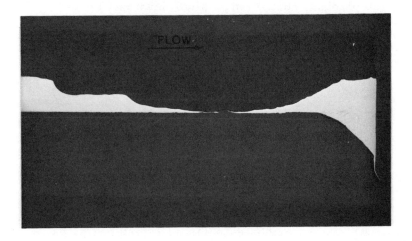

Figure 10. Longitudinal section through the pit and pin hole in the pipe shown in Figure 9.

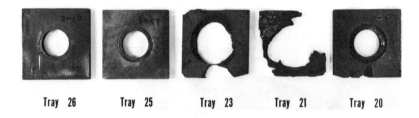

Figure 11. Type 410 stainless steel corrosion coupons exposed 4920 h in the most severely corroded area of the Baytown fractionation column. Similar localized patterns were seen in coupons exposed in the other three plants. Coupons are approximately 5-cm squares.

exposed on the indicated trays in the Baytown fractionation column. Corrosion was obviously more severe on trays 21 through 23 [reportedly operated around 230 C (450 F)] than on trays above or below. In general, this corrosion was most severe on trays and tray valves with appreciable liquid flow. This necessitated frequent tray replacement at all pilot plants. Furthermore, significant attack of the column walls was noted, with penetration of the carbon steel walls at Wilsonville and Baytown. The entire column or portions were replaced at Wilsonville and Fort Lewis, and austenitic stainless steel linings were added in the critical temperature region at Baytown and Catlettsburg.

SUMMARY AND CONCLUSIONS

Although this paper has presented examples of corrosion-related failures from the coal liquefaction pilot plants, it should be recognized that similar problems with sulfidation, pitting, and stress corrosion cracking caused by chlorides and polythionic acid also occur in the petroleum refining industry. We feel the pilot plant experience, coupled with the petroleum experience, has established a firm basis for selection of appropriate materials and process conditions to alleviate these problems in larger plants.

Although not discussed in this paper, an extremely serious materials degradation problem in the pilot plants was erosion where slurries are handled. For this problem, as for the unique chlorine-related corrosion in the fractionation towers, an understanding of causes, process condition limitations, and the relative resistance of a number of materials have been identified. However, additional testing of proposed process modifications to remove chlorine and reduce the corrosivity of the oils as well as improved component design and testing of improved materials to control erosion are still desirable.

ACKNOWLEDGEMENTS

The microprobe analysis of the vent line was performed by R. S. Crouse, and the examination of the 3RE60 tubing was conducted by R. J. Gray. Metallographers and photographers who provided the photographs for this report are M. D. Allen, C. W. Houck, B. C. Leslie, G. C. Marsh, J. R. Mayotte, and J. W. Nave. The corrosion coupon tests and laboratory experiments were performed by M. Howell.

This work was sponsored by the U.S. Department of Energy, Oak Ridge Operations Solvent Refined Coal Projects and AR&TD Fossil Energy Materials Program (DOE/FE AA 15 10 10 0, WBS Element ORNL-3.4) under contract W-7405-eng-26 with the Union Carbide Corporation.

REFERENCES

1. R.J. Gray, et al., "Metallurgical Study of a Failed Heat Exchanger in an H-Coal$^{(R)}$ Plant," Microstructural Science, Vol. 10, Elsevier Science Publishing Co., Inc., N.Y., pp. 183-193 (1982).

2. "Solvent Refined Coal (SRC) Process, Final Report," Pittsburg and Midway Coal Mining Company, DOE/ET/10104-46, pp. 158-159 (May 1982).

3. J.R. Keiser, et al., "Control of Corrosion in Coal Liquefaction Plant Fractionation Columns," Journal of Materials for Energy Systems, Vol. 3, No. 3, pp. 48-55 (1981).

4. T. Johnson, et al., Distillation Tower Corrosion: Synergistic Effects of Chlorine and Phenolic Compounds in Coal Liquids, Institute for Mining and Minerals Research, University of Kentucky, Lexington, Kentucky, IMMR48-PD23-80 (March 1980).

5. F. F. Lyle, Jr., "Corrosion of Metals in Coal Liquefaction Process Liquids," paper 85 presented at Corrosion/83, Anaheim, California, National Association of Corrosion Engineers, Houston, Texas (April 1983).

THE IDENTIFICATION OF THE ORIGINS OF
SOME DEPOSITS IN A STEAM TURBINE

J. C. Thornley*
and
J. K. Sutherland**

ABSTRACT

The intermediate pressure (IP) turbine of a thermal generating station is driven by steam from the boiler's reheater. On one particular IP turbine, a thick deposit was found on the insides of the rotor blade shrouds first in 1977 and then again in late 1979.

The source of the deposits was not known; bulk chemical analysis had simply shown that iron was a major component. Optical microscopy and electron microprobe analysis were used to identify the deposits. That in 1977 was found to be of debris left in the reheater tubes during boiler modification and swept to the turbine by the steam.

There were still some of these debris particles in 1979 but generally the 1979 deposit was found to be of two layer oxide particles which were shown to have spalled from 2-1/4% chromium reheater tube surfaces.

INTRODUCTION

During the course of a substantial routine inspection of a large turbine from an oil-fired generating station in Maritime Canada in 1977, and then again in 1979, a substantial

---

* New Brunswick Research and Productivity Council,
P. O. Box 6000, Fredericton, New Brunswick,
Canada E3B 5H1.
** New Brunswick Electrical Power Commission.

(a) (b)

Figure 1. (a) Some blades belonging to the first two rows of blades of the IP turbine. (b) Mounds of unidentified deposits on the inside of the shroud of the first row of blades on the IP turbine in 1979.

deposit was found adhering to the inside of the rotor blade shrouds. The deposit was only on the first two or three rows of blades. Some distortion of the shroud was, at the time, thought to be associated with this deposition, although subsequent observations suggest that this association is probably incorrect. The metallographic work enabled the source of the 1977 and of the 1979 deposits to be determined.

TURBINE DEPOSITS

General Description

A photograph of the outside of the first two rows of rotor blades is shown in Figure 1. Deposition occurred underneath the shroud and was up to about 10-mm (0.4-inches) thick. Concurrent distortion of the shroud can be seen in Figure 1, but the shroud distortion may not be a product of the deposit. The turbine blades were not severely damaged. There was no evidence of erosion on the blades although the blades' leading edges did have some indentations in them. Each of these indentations was like a small ball-peen hammer mark in that the pit was accompanied by spreading of material around the pit. It appears that some relatively large particles have hit the blades to cause these indentations.

Steam Turbine Deposits / 189

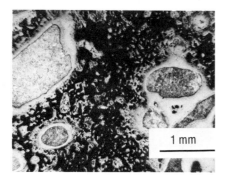 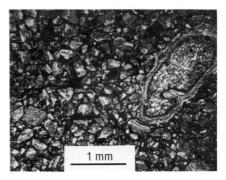

Figure 2. (Left) Metal particles set in an oxide matrix; this is the 1977 turbine deposit.

Figure 3. (Right) The 1979 deposit contained metal particles as did the 1977 deposit. One particle is shown at the top right here. There were, however, far fewer metal particles in the 1979 deposit than there were in the 1977 deposit.

 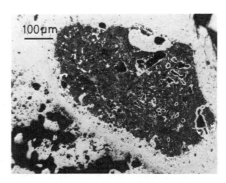

Figure 4. (Left) A spherical nonmetallic inclusion in apparently a mild steel particle in the 1979 deposit. Spherical inclusions like this are found in a particle after solidification from liquid.

Figure 5. (Right) A metal particle in the 1977 deposit. The steel appears to be an alloy steel and the shape of the porosity demonstrates that this particle formed directly from liquid steel, e.g., from weld spatter or in thermal cutting.

Some of the deposit was removed and is referred to subsequently as the 1979 deposit. When the turbine was inspected in 1977, a deposit was also seen and some of this was available for examination; it is referred to as the 1977 deposit.

Deposit Examination

The two deposits were examined by optical metallography and by electron microprobe analysis. Since both deposits were porous, they were vacuum impregnated with an epoxy compound before standard encapsulation in Bakelite and diamond polishing for optical metallography. They had been analyzed chemically earlier (elsewhere) but the only result available from that analysis was that the predominant element in each deposit was iron.

There were considerable numbers of metal particles in the 1977 deposit and a few metal particles in the 1979 deposit. Optical micrographs of the deposits are shown in Figures 2 and 3. The metal particles were not all of the same material and most showed evidence of having formed into their present shape by solidification from the molten state. Evidence of this is shown in Figures 4 and 5. Some of the metal particles had the typical ferrite/pearlite structure of mild steel (Figure 4), others contained acicular transformation products and were considered to be alloy steel (Figure 5), and still others appeared to be stainless. It is thought that these particles produced the indentations on the leading edges of the turbine blades which were referred to earlier.

The material between the metal particles was principally iron oxide but the oxide in the two deposits was different. That in the 1979 deposit was of two layers and an example is shown in Figure 6, and on careful examination, the two layer nature of the oxide can also be seen in Figure 3. The oxide in the 1977 deposit was not arranged in such a regular manner although it too consisted of two types of oxide. Some of the oxide in the 1977 deposit had clearly been formed by the oxidation of the metal particles. Figure 7 shows a site where a metal particle appears to have been completely converted to oxide.

 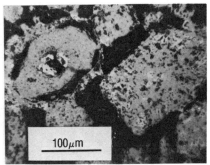

Figure 6. (Left) Oxide particles in the 1979 deposit. Note the layered nature of the deposit. These particles constituted the bulk of the 1979 turbine deposit.

Figure 7. (Right) Oxide in the 1977 deposit. The particle at center left appears to have formed by the complete oxidation of a small metal particle. No layered oxide was seen except that surrounding metallic particles.

The electron microprobe results extended the optical observations. Thus, three metal particles in the 1977 deposit were analyzed for chromium and nickel; one contained 1.2% Cr, 0% Ni; another, 2.2% Cr, 0% Ni; and a third, 15% Cr, 10% Ni. Presumably these are from 1-1/4 Cr 1/2 Mo and 2-1/4 Cr 1/2 Mo steels and an austenitic stainless steel, respectively. Similarly, in addition to iron oxides between the metal particles, particles of alumina (Al_2O_3) were identified by x-ray diffraction and the presence of silicon was seen.

The most valuable microprobe data was, however, that of the distribution of chromium in the oxide in the 1977 and 1979 deposits. Distribution in the 1979 deposit was the most straightforward. A typical two layer oxide particle in the 1979 deposit is seen in Figure 8. All of the particles, including the large rectangular one in the center, contain iron in a fairly uniform distribution. Chromium, however, is restricted to the pale, essentially dense part of the oxide and there is a slight suggestion of a higher chromium layer being present just at the interface between the chromium containing and the chromium free parts of the oxide.

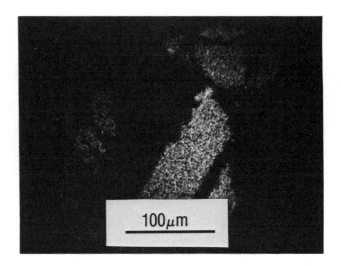

Figure 8. Electron microprobe scan for chromium. This is the same area as in Figure 6 and is from the 1979 deposit. The chromium is concentrated in the nonporous part of the oxide and there is a suggestion of a higher concentration of chromium at the interface between the porous and the nonporous oxides.

In the 1977 deposit, iron was again liberally and fairly uniformly distributed throughout the whole oxide. The chromium distribution, however, was not regular as in the 1979 deposit. Some particles were chromium free, some were very rich in chromium and yet others contained intermediate amounts. This, of course, corresponds with the distribution of chromium within the metallic parts of the deposit. The layering of the oxide within the 1977 deposit was only seen in oxide surrounding metal particles.

SOURCES OF THE DEPOSITS

The 1977 Deposit

The 1977 deposit consisted principally of metallic particles of varying compositions and degrees of oxidation and those particles showed clear evidence of having been formed directly by the solidification of a liquid particle. An intermediate pressure turbine is fed by superheated steam from

 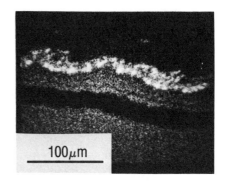

Figure 9. Optical (left) and electron microprobe scanning photographs (right) of the inner scale from the lower reheater inlet on a tube sample removed in 1979 for metallographic examination. Note that chromium is limited to the inner scale and there is a chromium rich band in the center of the scale.

the reheater. The reheater in turn heats steam which exhausts from the high pressure turbine so that any particle entering the intermediate pressure turbine must have come from the reheater tubes or have come right through the high pressure turbine and the reheater from some part of the boiler such as the superheater.

In 1974, a major modification had been performed on this boiler when a substantial amount of reheater tubing was removed. This necessarily involved oxyacetylene flame cutting and arc welding operations, both of which give rise to molten metal droplets. It is fairly clear then that the 1977 deposit is oxidized debris from the 1974 reheater modification. Particles rich in silicon and aluminum would correspond to welding slag. Some of this debris is also still to be seen turning up in the 1979 deposit. Most of the 1979 deposit is, however, very different in character to that of 1977 and we must look at the detailed structure of oxide layers on alloy steels to identify the source of that deposit correctly.

The 1979 Deposit

Boiler tubes are made from a variety of steels and those operating in the creep range almost invariably contain some chromium. Reaction with the steam inside the tube results in the formation of a layer of magnetite on the tube bore. The oxide grows to become two-layered; the steam-side layer is porous and the metal-side layer is generally dense. There is, of course, some chromium in the oxide as there is in the alloy steel, but the chromium is not distributed uniformly within the oxide. The porous steam-side layer is almost chromium free while the denser metal-side layer contains a greater concentration of chromium than does the alloy steel substrate. These features are illustrated in Figure 9 for a 2-1/4% Cr 1% Mo steel tube. It will also be seen that there is a band between the porous and the dense layers with an increased concentration of chromium in it.

It appears that the 1979 deposit is predominantly of oxide from boiler tube surfaces. The reheater has 1-1/4% Cr 1/2% Mo, and 2-1/4% Cr 1% Mo, and 5% Cr 1% Mo steel tubes in it. The concentration of chromium in the metal-side (dense) layer of the oxide growing on each of these steels was measured and is shown in Table 1. Twenty oxide particles in the 1979 deposit were selected and, again using the electron microprobe for quantitative analysis, the chromium concentration in the dense layer of each particle was measured. The values were between 3.1% Cr and 4.4% Cr. The metal-side layer of the oxide growing on a 2-1/4% Cr 1% Mo tube contained 3.0% to 3.4% Cr.

Table 1. Chromium Content of Chromium-Rich Layers in Magnetite Scales

Source	Chromium Content of the Chromium-Rich Layer
Oxide grown on 1-1/4% Cr 1/2% Mo steel	2 to 2.1%
Oxide grown on 2-1/4% Cr 1% Mo steel	3 to 3.4%
Oxide grown on 5% Cr 1% Mo steel	5.5 to 8.5%
1979 turbine deposit	3.2 to 4.4%

Some 2-1/4% Cr 1% Mo tubing had been removed from the lower reheater inlet in connection with some other metallographic work performed on this boiler at the same time as the work described here. Internal examination of this tubing revealed quite clearly that the surface oxide was spalling from the bore surface.

The source of the 1979 deposit then is the lower reheater inlet tubing of 2-1/4% Cr 1% Mo composition. Some oxide was also spalling from the bore of the 1-1/4% Cr 1/2% Mo upper reheater inlet tubing. None of this material was found in the deposit.

CONCLUSIONS

The 1977 deposit in the turbine was due to metal particle debris from a major reheater modification performed in 1974. Some debris from 1974 was present in the 1979 deposit but most of the 1979 deposit was oxide which had spalled off the bore of 2-1/4% Cr 1% Mo reheater tubing in the lower reheater inlet area of the boiler.

ACKNOWLEDGEMENTS

The authors would like to thank the boiler's owners for permission to publish these results. Comments on the manuscript by Mr. H. L. J. Drummong and the technical assistance of Mr. J. A. Grimm are also acknowledged.

IN-SITU OBSERVATION OF WHISKERS, PYRAMIDS AND PITS
DURING THE HIGH TEMPERATURE OXIDATION OF METALS

G. M. Raynaud* and R. A. Rapp**

ABSTRACT

A hot-stage environmental scanning electron microscope has been used to observe the in-situ development of oxide whiskers, pyramids and pits in the oxidation of copper and nickel at elevated temperatures. The effects of oxidation temperature, metal deformation and the presence of water vapor on these irregular oxidation features were studied. In each case, the feature results from the presence of a central screw dislocation which provides ledges for the extension of the oxide lattice, but the specific geometries are decided by factors such as surface diffusion along the dislocation core, the rate of the molecular dissociation step, and the balance of surface energy and dislocation line tension forces.

INTRODUCTION

At high temperatures and large scale thicknesses, the rate limiting step for pure metal oxidation is the lattice diffusion of point defects through the scale [1]. The crystallographic defects present in the scale (i.e., dislocations or dislocation arrays) should not greatly influence diffusion through thick scales at high temperatures. Consequently, an oxide scale grown by cation diffusion at high temperature would be expected to

---

  * IREQ, Varennes, Quebec, Canada   J0L 2P0
 ** Department of Metallurgical Engineering,
    The Ohio State University, Columbus, Ohio   43210   USA.

exhibit a flat scale/gas interface and be composed of columnar grains. However, grain boundary diffusion becomes important at lower temperatures and for smaller grain sizes [2]. The oxide grains may then be shallow-dish shaped, with slight ridges at their boundaries [3].

The influence of dislocations during scale growth is a subject of continuing interest. For example, in the intermediate temperature regime the rate of nickel oxidation is believed to be controlled by transport along defects in the oxide (dislocations or grain boundaries) [4,5], whereas lattice diffusion is the important transport mechanism above 1200 C [6]. Dislocations also provide transport paths for blades or whiskers formed on oxide scales. Voss et al. [7] concluded that the primary mechanism for Fe_2O_3 (hematite) blade growth is surface diffusion along a tunnel centered on one dislocation or a bundle of parallel dislocations [8].

The observations of whiskers have been particularly numerous for the oxidation of copper; it is commonly agreed that whiskers consist of CuO, grow along the <110> direction, and contain an axial screw dislocation. Hardy [9] suggested that dislocations ending in the metal surface serve as oxide nucleation points and that the dislocation would be continued into the oxide layer. Appleby and Tylecotte [10] showed more generally that the formation of CuO whiskers is a consequence of stresses arising during the oxidation process. Pfefferkorn [11] showed that CuO whiskers or platelets grow at their tips, and Jaenicke and Albert [12] concluded that whisker growth occurs by surface diffusion obeying a parabolic rate law.

Irregular scale growth has been observed during NiO growth at the scale/gas interface. Armanet et al. [13] observed a very flat oxide scale at 750 - 1250 C when the oxygen was carefully dried. On the contrary, in wet oxygen facetting occurred. Furthermore, the scale/gas interface was covered with whiskers when cold worked nickel was oxidized in air.

The present investigation of the growth of Cu_2O on OFHC copper and of NiO on pure Ni was intended to clarify these aspects of irregular scale growth.

EXPERIMENTAL EQUIPMENT AND PROCEDURES

The present in-situ study of metal oxidation was carried out in a hot-stage environmental scanning electron microscope (HSESEM) for which a detailed description was given by Verma et al. [14]. Briefly, a Cambridge S-4 SEM was adapted with two major modifications: the capacity to heat a specimen by radiation from a W filament to temperatures approaching 1300 C, and to subject that specimen to the impingement of a molecular beam of any desired gaseous atmosphere at an equivalent pressure which is several orders of magnitude higher than that for the surrounding vacuum chamber.

Well-annealed OFHC copper cylinders (1-cm dia. by 3-cm tall), and 99.999% polycrystalline nickel disks were prepared using metallographic and electropolishing methods to obtain flat and clean surfaces. Oxygen was passed at a known flow rate to the specimen through a 0.6-mm ID platinum pipe or a 0.35-mm ID stainless steel pipe to produce an equivalent oxygen pressure at the specimen surface less than or equal to 3×10^{-4} atm. The specimens were heated to the oxidation temperature (or higher) in the chamber vacuum or in a dry hydrogen atmosphere, and then exposed to oxygen.

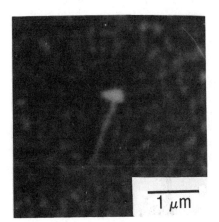 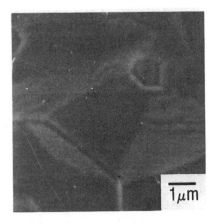

Figure 1. (Left) Oxide film formed on electropolished OFHC copper at 300 C in $P_{O_2}=3\times10^{-4}$ atm after 3 days for 2.5 h.

Figure 2. (Right) Oxide formed on electropolished OFHC copper at 500 C in $P_{O_2}=3\times10^{-4}$ atm after 20 h, 50 min.

RESULTS

Oxidation of OFHC Copper

The details of Cu_2O nucleation on OFHC copper are presented elsewhere [15], and the results shown here are restricted to irregular growth at the oxide/gas interface for thick scales. Figure 1 shows the morphology of electropolished OFHC copper oxidized at 300 C in $P_{O_2} = 3 \times 10^{-4}$ atm for 3 days and 2.5 h. The center of oxide grains exhibited a dark depression and the scale was covered with CuO whiskers already after 30 h of oxidation. These whiskers apparently nucleated at random sites in the underlying oxide film. The whiskers were found to grow with time at their tips, and to develop lateral projections at their tips.

At 500 C after 20 h of oxidation in $P_{O_2} = 3 \times 10^{-4}$ atm, the oxide scale on OFHC copper developed pyramids of relatively large size (~3 micrometers) composed of flat faces, except for the existence of macroscopic spiral ledges (Figure 2). After 44 h of oxidation at 500 C, the scale became relatively flat and uniform. After longer times, oxide cracking and whisker and platelet formation were observed. To study the effect of stresses on the oxidation behavior at 520 C, OFHC copper specimens were subjected to a shot-peening surface treatment to produce a highly cold worked, 20-micrometer thick, surface layer. Upon oxidation, tiny crystals of Cu_2O nucleated randomly on the surface without any obvious growth habit relation to the substrate. After 3 h, the cuprous oxide crystallites had an average diameter of 1-micrometer (Figure 3a). Subsequently, the grains grew in height with small steps on their faces (Figures 3b and c). After 1 day and 5 h oxidation, deep holes were clearly seen at the top of the crystallites (Figure 3d). After 1 day, the oxide grains grew by the advance of steps on pyramid faces, very similar to the Cu_2O growth mechanism observed on copper at 500 C; after 2 days the oxide scale was rather flat.

In the case of hydrogen-heated, undeformed samples, secondary oxide grains formed on an initial porous scale after 11 h of oxidation at 550 C in $P_{O_2} = 3 \times 10^{-4}$ atm. The new grains grew very slowly and after 1 day, 8 h of oxidation,

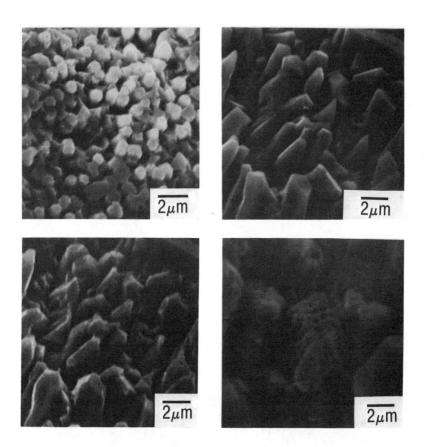

Figure 3. Oxidation of cold worked OFHC copper at 520 C in $P_{O_2}=3\times10^{-4}$ atm after (a) 3 h, 14 min; (b) 23 h and 55 min; (c) 1 day, 1 h, 10 min; (d) 1 day, 5 h, 40 min.

they appeared as uniform dodecahedra with a 1-micrometer edge (Figure 4a). Upon coalescence of these secondary oxide grains after four days of oxidation, a large grain size (40-micrometers) was observed (Figure 4b). Then, oxide growth occurred by the translation of macroscopic surface ledges similar to the ones shown previously.

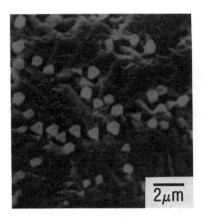 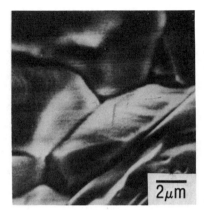

Figure 4. Copper oxide formed on OFHC copper at 550 C in $P_{O_2}=3\times10^{-4}$ atm; (a) after 1 day and 8 h, (b) after 4 days.

 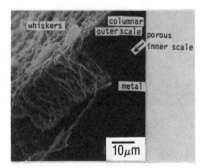

Figure 5. Oxidation morphology of mechanically polished hydrogen heated polycrystalline 99.999% nickel after 6 h at 1110 C in $P_{O_2}=3\times10^{-4}$ atm; (a) surface, (b) cross section.

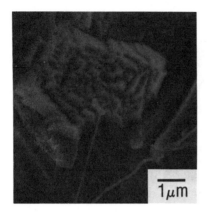 

Figure 6. (Left) Oxidation morphology observed on 99.99% nickel after 5 h of oxidation at 1110 C in wet oxygen ($P_{O_2}=3\times10^{-4}$ atm).

Figure 7. (Right) Oxidation morphologies obtained when a 99.99% nickel grid is oxidized in super dry oxygen at 1000 C in $P_{O_2}=3\times10^{-4}$ atm after 2 h and then 5 min. in moist O_2.

Oxidation of Nickel

Nickel samples (99.99%) were heated and held for 30 min. at the reaction temperature in dry hydrogen prior to oxidation. After 6 h of oxidation at 1110 C in moist oxygen of an equivalent pressure of 3×10^{-4} atm, the scale/gas interface exhibited a rough morphology (top right of Figure 5a) and was covered with whiskers and platelets. The scale consisted of a 20-micrometer thick outer layer and a porous inner layer, 4-micrometers thick (Figure 5b). Initially (after 10 min. of oxidation), oxide whiskers nucleated at the surface intersections of metal grain boundaries, but after 30 min., whiskers covered the scale almost uniformly. When oxidized in dry oxygen, the oxide/gas interface showed a rough morphology composed of pyramids and pits, but no whisker formation was detected even after 5 h of oxidation. However, when the vapor pressure of water was increased, not only were randomly distributed whiskers formed on the surface, but the oxide grains became facetted along preferred crystallographic directions (Figure 6). Armanet et al. [13] observed similar facetting when nickel was oxidized in moist oxygen.

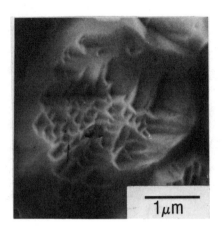

Figure 8. Oxidation of 99.999% nickel in $P_{O_2}=3 \times 10^{-4}$ atm at 1350 C after 15 min.

The growth of whiskers was also observed on a cold worked 99.99% nickel grid oxidized in dry oxygen (Figure 7). These short, stubby whiskers were quite numerous with a rectangular or square cross-section (Figure 7a). A typical length was 30-micrometers and each side was about 2-micrometers wide. The whisker growth rate was approximately 2-micrometers min^{-1}. When water vapor was added to oxygen during the experiment, a different type of whisker was formed (seen also in Figure 7a). These whiskers were very long (200-micrometers) and extremely small in cross-section. Their growth rate was as high as 50-micrometers min^{-1}.

During the growth of NiO at 1100 C, $P_{O_2}=3 \times 10^{-4}$ atm, apparent new nucleation, similar to Figure 4a on copper, was observed. Even at temperatures as high as 1350 C, the NiO scale was not flat but each pyramidal grain (3-micrometers average diameter) was indented with an array of aligned pits of 0.15-micrometer average diameter (Figure 8).

DISCUSSION

From observations of copper and nickel oxidation inside the HSESEM, ideal morphologies consisting of large columnar grains were only seen after long oxidation times and at film thicknesses well beyond those for which the thick scale growth mechanisms should prevail.

Morphologies and Growth Mechanisms

Four different irregular growth morphologies have been observed: concave oxide surfaces, oxide pyramids, whiskers and platelets, and pitted oxide grains. Concave surfaces of grains with grain boundaries delineated by shallow grooves (Figure 1) were observed. Fast grain boundary diffusion in the intermediate range of temperature ($\sim$300 C for Cu and 700 C for Ni) has already been suggested by several authors [3,16] to rationalize this morphology.

Oxide pyramids were observed on copper at 500 C, on polycrystalline Ni at 1300 C, and on a Ni grid at 1100 C. More generally, protuberances were observed at the scale/gas interface on cold worked OFHC copper oxidized at 520 C (Figure 3) and on nickel oxidized at 1110 C (Figure 5). In general, the growth of oxide pyramids could be supported by either lattice or grain boundary diffusion through the overgrowth followed by surface diffusion to the steps created by the screw dislocation.

At sufficiently high temperatures, lattice diffusion is the rate determining step for the oxidation of pure metals. However, screw dislocations probably still provide the surface steps necessary for growth. For example, the growth of flat pyramids (Figure 2) cannot be supplied solely by surface diffusion along the central dislocation tunnel. When the dislocation glides or climbs out of the grain, no further steps are created, so further growth must occur by a lateral advance of the existing steps, and the top of the pyramid flattens (Figure 2). The appearance of "new" oxide grains at the scale/gas interface must occur when dislocations within existing oxide grains no longer create steps so that new grains arise from fresh dislocations.

Oxide grain boundaries may act as continual sources of dislocations, and nucleation events should be favored at these sites (Figure 4a). Grain boundary diffusion is predominant during the oxidation of nickel below 1000 C [3] and for the oxidation of OFHC copper at 500 C. The value of the nickel tracer lattice diffusion coefficient in NiO at 1100 C (for $P_{O_2} = 3 \times 10^{-4}$ atm) is [17] $D_{Ni} = 2 \times 10^{-12}$ cm^2 s^{-1}. According to a calculation of the parabolic rate constant from Wagner's theory, this corresponds to a scale thickness of about 5-micrometers after 6 h, a factor of 4 smaller than the observed thickness (Figure 5b). However, the grain boundary diffusion coefficient for Ni in NiO extrapolated to 1110 C from the values given by Atkinson and Taylor [3] provides a calculated rate constant which agrees rather well with the experimental value. Then, grain boundary cation diffusion followed by rapid surface diffusion to steps maintained by screw dislocations is the likely mechanism for NiO growth on pure Ni for the temperatures of 1100 C and lower. On the contrary, lattice diffusion prevails for the growth of NiO exhibiting pyramids and pits at 1350 C (Figure 8).

Whiskers and platelets were grown on OFHC copper oxidized at 300 C (Figure 1), and at 500 C, and on polycrystalline nickel oxidized at 1110 C in moist oxygen (Figure 5), and finally on a nickel grid oxidized at 1000 C. One mechanism proposed for whisker growth from the vapor is the so-called Frank mechanism, extended by Burton et al. [19]. A screw dislocation emerging on the face of a crystal provides it with a reproducible surface step needed for growth. For elemental crystals grown from the vapor, the adsorbed species are directly available for further step growth. On the contrary, for oxidation of most metals, cations must be supplied via diffusion through the oxide scale. The diffusion of metal species to support the growth of a whisker (or a pyramid) may follow distinct paths shown schematically in Figure 9. Surface diffusion by cations to the tip of a whisker can occur along the sides of the central tunnel at the dislocation core, followed by surface diffusion down the outside of the whisker until it is incorporated with oxide ions into surface steps. This mechanism was first proposed by Tallman and Gulbransen [20] for the growth of alpha-Fe_2O_3 whiskers on iron at 400 C in dry oxygen. TEM studies have confirmed the existence of a central hollow tunnel within the alpha-Fe_2O_3 blades or whiskers grown on iron from 600 to 800 C in 20 torr of oxygen [7].

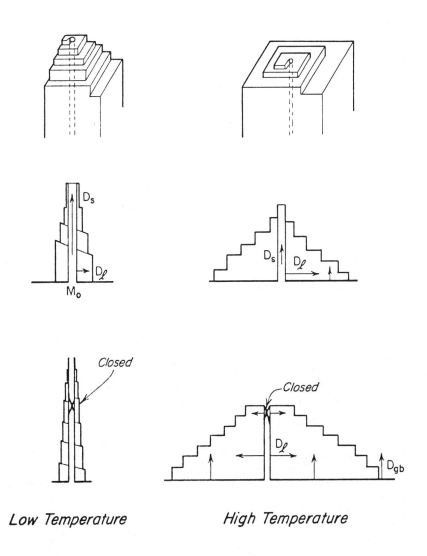

Figure 9. Schematic model for whisker growth (low temperature) and pyramid growth (high temperature).

Pits were observed in the growth of NiO on nickel oxidized at 1110 C (Figure 5), and at 1350 C (Figure 8). According to Frank's model [8], pits are formed in crystal growth to minimize the sum of dislocation line tension and surface energy. Thus, the pits observed on NiO at really high temperatures (Figure 8) are believed to be located at dislocations intersecting the surface and aligned along sub-boundaries. Jungling and Rapp [21] have observed large, regular pits for the growth of wustite on iron at elevated temperatures.

Monatomic steps can obviously not be resolved in the HSESEM, but several effects can explain the presence of macroscopic steps on pyramid faces. For example, the pile-up of monatomic steps to form macroscopic ledges through the presence of surface impurities is well known. Macrosteps could also arise from dislocation bundles or multiple screw dislocations [22]. These dislocations may be introduced during scale growth by a misarrangement upon the impingement of adjacent grains [23].

Influence of Temperature

As just explained, lattice, grain boundary and surface diffusion can all contribute to the growth of oxide pyramids and whiskers in metal oxidation. As indicated by Le Claire [24], if the rates of these three types of diffusion in polycrystals are characterized by diffusion coefficients, D_1, D_{gb}, and D_s and activation energies Q_1, Q_{gb}, and Q_s, then $D_1 < D_{gb} < D_s$ and $Q_1 > Q_{gb} > Q_s$. No data are available for grain boundary or surface diffusion for Cu in Cu_2O. However, Dhallenne et al. [25] measured the lattice and surface diffusion coefficients for Ni on NiO using a grain boundary grooving technique at 1400 and 1520 C. A comparison between their values and the values for lattice and grain boundary diffusion from Volpe and Reddy [17] and Atkinson and Taylor [3] give $D_s \simeq 10^4 D_1$, and $D_s \simeq 3 D_{gb}$. Although the diffusion distance is longer for whiskers than for bulk oxide, the local flux associated with surface diffusion along with a central whisker core can be several orders of magnitude larger than the flux associated with volume diffusion. Because the activation energy for surface diffusion is smaller than that for lattice diffusion, the growth associated with steps created by a screw dislocation will lead to whisker growth at low temperatures. On the contrary, as the temperature is increased, lateral lattice diffusion becomes relatively more favorable and oxide growth leads to flatter morphologies such as pyramids (Figure 9). Thus, for otherwise similar oxidation conditions, whisker growth was observed for OFHC copper oxidation at low temperatures (T ≃ 300 C, Figure 1) whereas pyramids were observed at higher temperatures (T > 400 C, Figure 2). Analogous phenomena occurred for nickel oxidation (at higher temperatures) although the morphologies were influenced greatly by the experimental conditions. For example, pyramid and whisker growth at dislocations was more prevalent when the metal was initially cold-worked. Whiskers were observed on a nickel grid (cold drawn wire) oxidized at temperatures as high as 1000 C. But only pyramids were observed for Ni oxidation at 1350 C.

Influence of Water Vapor

Whisker formation was enhanced when water vapor was added to oxygen. A flat oxide grown in super-dry oxygen would become covered with whiskers as soon as water vapor was added during the course of oxidation. This suggests that tunnels are always available in the scale. Similarly, although stubby whiskers (similar to pyramids) were grown at high temperature in super-dry oxygen on the deformed nickel grid, NiO whiskers were nucleated and grown on an annealed Ni crystal only after water vapor was mixed with oxygen (Figure 7). Whiskers grown in oxygen containing water vapor were extremely long and thin, so water vapor promotes the growth along the axis. Armanet et al. [13] observed similar phenomena on a Ni-20Cr alloy oxidized at 1000 and 1200 C in air containing water vapor.

Water vapor does not accelerate the diffusion of vacancies within the oxide [26]. Similarly, water vapor cannot influence the surface diffusion coefficient along the central tunnel. Consequently, the transport of metal species via surface diffusion is not the rate limiting step for whisker growth, but rather the breakup of the oxidant molecules near the whisker tip must limit the kinetics. Wagner [27] showed that dissociation of molecular O_2 (as a step in series with lattice diffusion) was the rate determining step for the growth of compact NiO at 250 C. The transition temperature between diffusion as the rate limiting step and the breakup of the molecular oxidant as the rate limiting step would be much higher for the more rapid surface diffusion (whisker growth). Further, the breakup of a water molecule is generally found to be faster than that for most other oxidant molecules. For example, the interfacial reaction in the oxidation of iron to wustite and for wustite to magnetite in the temperature range 850-1150 C, is faster for water vapor-hydrogen mixtures than for equivalent carbon dioxide-carbon monoxide gas mixtures [28,29].

In conclusion, because surface diffusion is so much faster than lattice diffusion, the rate limiting step for whisker growth is the breakup of the oxidant gas molecules. Then, the easier breakup of water vapor compared to oxygen accelerates the kinetics of whisker growth. Lattice diffusion is still the rate limiting step at high temperatures for thick scales where the scale does not grow by a tunnel mechanism.

CONCLUSIONS

Oxide whiskers grow predominantly by surface diffusion of cations along a tunnel centered on the core of a screw dislocation or a bundle of dislocations. In pure O_2, especially at reduced pressure, the rate limiting step for whisker growth is the breakup of the adsorbed oxygen. Thus, the growth of whiskers is enhanced when water vapor is mixed with oxygen. Growth of a whisker near its tip with little thickening maximizes the extension rate; this occurs when the oxidant is rapidly dissociated so that the axial cation flux is constrained to lattice growth at ledges near the tip. When the temperature is increased, this surface diffusion mechanism leads to the formation of pyramids since the competitive lateral lattice diffusion is enhanced (Figure 9). Pyramidal (or stubby whisker) growth is obtained from lattice or grain boundary diffusion, but a screw dislocation or a bundle of dislocations still provides the steps required for growth. When the screw dislocations become inactive (gliding or climbing out of the crystal, or healing of the central tunnel), scale growth proceeds by the consumption of existing steps. Then, the oxide scale becomes flat and diffusion through the lattice leads to columnar grains.

ACKNOWLEDGEMENTS

This research was supported by Division of Material Research, Department of Energy, under contract No. AC02-79ER10404.

REFERENCES

1. C. Wagner, "Diffusion and High Temperature Oxidation of Metals," Atom Movements, Amer. Soc. Metals, pp. 153-173 (1951).

2. G. Martin and B. Perraillon, J. Phys., Paris, Vol. 36, pp. C4 165-C4 190 (1975).

3. A. Atkinson and R. I. Taylor, Phil. Mag. A, Vol. 43, pp. 979-998 (1981).

4. J. V. Cathcart, G. F. Petersen, and C. J. Sparks, Jr., J. Electrochem. Soc., pp. 664-668 (1969).

5. J. M. Perrow, W. W. Smeltzer, and J. D. Embury, Acta Met., Vol. 16, pp. 1290-1218 (1968).

6. M. J. Graham, D. Caplan, and M. Cohen, J. Electrochem. Soc., Vol. 119, pp. 1265-1267 (1972).

7. D. A. Voss, E. P. Butler, and T. E. Mitchell, Met. Trans., Vol. 13, pp. 929-935 (1982).

8. F. C. Frank, Acta Cryst., Vol. 4, pp. 497-501 (1951).

9. H. K. Hardy, J. Inst. Metals, Vol. 79, p. 497 (1957).

10. W. K. Appleby and R. F. Tylecotte, Corr. Sci., Vol. 10, p. 325 (1970).

11. G. Pfefferkorn, Proc. of the 3rd Int. Conf. on Electron Microscopy, London, p. 491 (1954).

12. W. Jaenicke and L. Albert, Naturwiss, Vol. 46, p. 491 (1959).

13. F. Armanet, G. Beranger, and D. David, Proc. 8th Int. Cong. on Metallic Corr., DECHEMA, Frankfurt, Germany, pp. 735-738 (1981).

14. S. K. Verma, G. M. Raynaud, and R. A. Rapp, Oxid. Met., Vol. 15, p. 471 (1981).

15. G. M. Raynaud and R. A. Rapp, submitted for publication to Metall. Trans.

16. J. Van Landuyt, R. Gevers, and S. Amelinckx, Phys. Stat. Soc., Vol. 10, p. 319 (1965).

17. M. L. Volpe and J. Reddy, Journal of Chem. Phys., Vol. 53, No. 3, pp. 1117-1125 (1970).

18. A Atkinson, R. I. Taylor, and A. E. Hughes, Phil. Mag. A, Vol. 45, pp. 823-833 (1982).

19. W. K. Burton, N. Cabrera, and F. C. Frank, Phil. Trans. Roy. Soc., Vol. 243, pp. 299-358 (1951).

20. R. L. Tallman and E. A. Gulbransen, J. Electrochem. Soc., Vol. 114, pp. 1227-1230 (1967).

21. T. L. Jungling and R. A. Rapp, Proc. of Symp. on High Temperature Materials Chemistry, Z. A. Munir and D. Cubiciotti, eds., Electrochem. Soc., Pennington, N.J. (1983).

22. S. Mardix, A. R. Lang, and I. Blech. Phil. Mag., Vol. 22, pp. 683-693 (1971).

23. W. Gotch and H. Komatsu, J. Cryst. Growth, Vol. 54, p. 163 (1981).

24. A. D. LeClaire, Prog. Met. Phys., Vol. 4, p. 1263 (1953).

25. G. Dhallenne, A. Revcolevschi, and G. Monty, Phys. Stat. Sol. (A), Vol. 56, p. 623 (1979).

26. C. W. Tuck, M. Odgers, and K. Sachs, Corr. Sci., Vol. 9, pp. 271-285 (1969).

27. C. Wagner, Corr. Sci., Vol. 10, pp. 641-647 (1970).

28. E. T. Turkdogan, W. M. McKewan, and L. Zwell, J. Phys.Chem., Vol. 69, pp. 327-334 (1965).

29. P. J. Meschter and H. J. Grabke, Metall. Trans., Vol. 10B, pp. 323-329 (1979).

EFFECTS OF A FLOWING LITHIUM ENVIRONMENT ON THE SURFACE MORPHOLOGY AND COMPOSITION OF AUSTENITIC STAINLESS STEEL

P. F. Tortorelli* and J. H. DeVan*

ABSTRACT

Optical and electron microscopical techniques were used to study the surfaces of AISI 316 stainless steel following exposure to molten lithium. In nonisothermal flowing systems, the major corrosive attack by lithium occurred through the transport of nickel and chromium from hotter to cooler parts of the systems. At the higher temperatures, surfaces were depleted of nickel and chromium and porosity formed in the iron-enriched near-surface layer. Small nodules enriched in molybdenum were also observed. At lower temperatures, net deposition of chromium, nickel, and iron from the lithium occurred such that the composition of the deposits varied with time. Qualitative correlations between porosity variations and surface concentrations as a function of temperature indicated that porosity development may be more related to a nickel or chromium reaction with impurities than with the actual nickel depletion process.

INTRODUCTION

Like other liquid metals, molten lithium is an excellent coolant for high power density systems because of its very favorable heat transfer properties. Additionally,

---

* Metals and Ceramics Division, Oak Ridge National Laboratory, Oak Ridge, Tennessee 37830 USA.

lithium is required in deuterium-tritium-fueled fusion reactors for breeding of tritium by the reaction of fusion-produced neutrons with lithium atoms. Liquid lithium is therefore attractive because it has the potential to serve several purposes in future fusion reactors. Consequently, the compatibility of structural materials with molten lithium is a critical concern in fusion system design and alloy selection. Corrosion by molten lithium may take a variety of forms [1]; the type and extent of such corrosion must be known and characterized for each candidate alloy in order to define the operating limits based on compatibility criteria.

The present study is concerned with the corrosion of AISI 316 stainless steel that occurs under thermal gradient conditions in a closed loop. The resulting mass transfer (dissolution, deposition) reactions can have significant effects on reactor operation [1,2]. Prior work has dealt with the dissolution and deposition behavior of lithium-exposed AISI 316 stainless steel with time, the general effects of such exposure on surface morphology and phase stability, and the effect of nickel concentration of the structural alloy on its corrosion response in lithium [1-3]. This paper reports the comprehensive characterization of the effects of a thermally convective lithium environment on the surface morphology and composition of AISI 316 stainless steel and compares these results with findings from earlier, less detailed studies.

EXPERIMENTAL PROCEDURES

The analyzed surfaces were of AISI 316 stainless steel specimens that were exposed to lithium in a thermal convection loop (TCL) of the type shown schematically in Figure 1. The lithium density gradient caused by the imposition of a temperature difference of 150 C (600-450 C) across the loop resulted in a lithium velocity of approximately 25 mm/sec. The loop is designed to allow the coupons to be inserted and removed from the hot and cold legs without stopping the lithium flow. In this way, specimen weight and microstructural changes can be measured as a function of exposure time, although, in the present case, all results are for one exposure time (7488 h). The placement

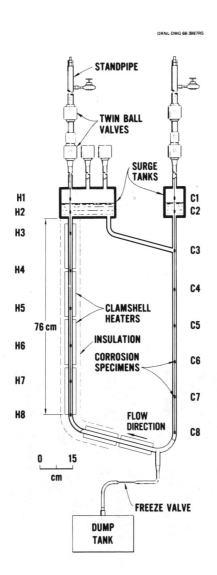

Figure 1. Thermal convection loop with accessible corrosion coupons.

of specimens around most of the loop (see Fig. 1) permits the measurement of any mass transfer tendencies. In order to avoid any corrosion effects due to the presence of dissimilar metals in the same lithium system, the loop and specimen rods were constructed of the same alloy as the specimens.

The lithium used in the loop was purified by cold trapping and subsequent heating at 815 C (1499 F) for 100 h in a titanium-lined pot containing zirconium foil. Typical impurity concentrations of the purified lithium were 30 to 80 wt ppm of nitrogen and 30 to 130 wt ppm of oxygen. The nitrogen and oxygen concentrations of the loop lithium were subsequently measured several times during the course of this study and did not significantly vary from the above ranges.

The fully annealed AISI 316 stainless steel loop specimens were rectangular coupons (measuring ∿ 25 x 8 x 1 mm) composed of Fe-17 %Cr-11 %Ni-2 %Mo-0.08 %C (wt %). The coupons were electropolished and then placed at the indicated positions

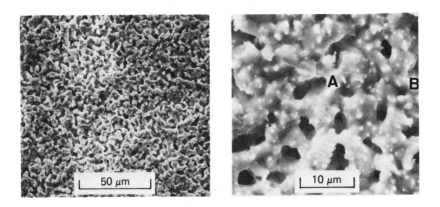

Figure 2. AISI 316 stainless steel exposed to thermally convective lithium for 7488 h at 600 C (maximum loop temperature).

in the hot (H) and cold (C) legs of the TCL (see Figure 1). After 7488 h of actual exposure to the lithium, the specimens were removed from the loop for the final time and destructively examined. A cross-section of one part of each specimen was polished and etched while the exposed surface of the other part of each coupon was analyzed using scanning electron microscopy and associated energy dispersive x-ray analysis. The results of such observations for the specimens in the H3 (maximum temperature) through H8 loop positions are presented below. The evaluation of specimens in the cold leg (C3 through C8) is not complete and will be published at a later date. However, the temperature range of the hot leg specimens covers parts of both the dissolution and deposition zones and thus allows the study of the effects of dissolution and deposition.

RESULTS

The AISI 316 stainless steel specimens examined were exposed between 600 and 500 C (1112-932 F) for 7488 h. Scanning electron microscopy of the surface of the coupon at the maximum temperature position revealed considerable porosity (Figure 2) while energy-dispersive x-ray analysis (EDS) showed that the surface was depleted in both nickel and chromium relative to the composition of the starting material (Figure 3). Additionally, small, more highly emitting areas (such as "A" and "B" in Figure 2) were observed on the corroded surface. These "nodules" were found to be somewhat enriched in molybdenum relative to the underlying matrix. A cross-sectional view of the exposed surface of the 600 C (1112 F) specimen is shown in Figure 4 and reveals that the observed porosity is interconnected and extends rather deeply ($\sim$50 micrometer) into the alloy.

Results from the examination of the other specimens (H4 through H8) showed an interesting change in surface morphology with decreasing exposure temperature in the loop. This is illustrated in Figure 5, which includes typical scanning electron micrographs of the respective specimens. Note how the size and distribution of the porosity decreased with decreasing temperature until, at and below 520 C (968 F), porosity is minimal and deposits are the predominant surface feature. The change from net specimen weight loss to net weight gain after 7488 h also occurred at this point (520 C). The change in surface topography shown in Figure 5 was accompanied by significant changes in the surface compositions

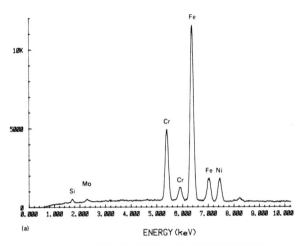

UNEXPOSED TYPE 316 STAINLESS STEEL

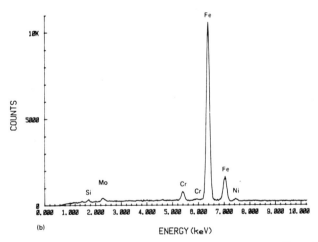

TYPE 316 STAINLESS STEEL EXPOSED TO Li, 600°C, 7488h

Figure 3. Comparison of typical energy-dispersive x-ray spectra for (a) unexposed AISI 316 stainless steel and (b) AISI 316 stainless steel exposed to thermally convective lithium at 600 C for 7488 h. Note depletion of nickel and chromium.

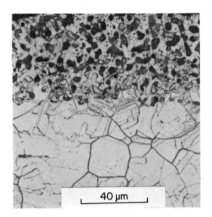

Figure 4. Polished and etched cross-section of AISI 316 stainless steel exposed to thermally convective lithium for 7488 h at 600 C (maximum loop temperature).

of the various specimens. As seen from the qualitative energy-dispersive x-ray analysis data in Table 1, the hotter specimen surfaces were depleted in nickel and chromium while the cooler ones were enriched.

Table 1. K_α Peak Intensity Ratios from Energy Dispersive X-ray Analysis Spectra of Type 316 Stainless Steel Exposed to Flowing Lithium for 7488 h

Exposure Temperature (C)	K_α Peak Intensities	
	Cr/Fe	Ni/Fe
Unexposed	0.40	0.13
600	0.05	0.01
575	0.08	0.01
560	0.10	0.01
540	0.11	0.02
520	0.58	0.14
500	4.16	0.31

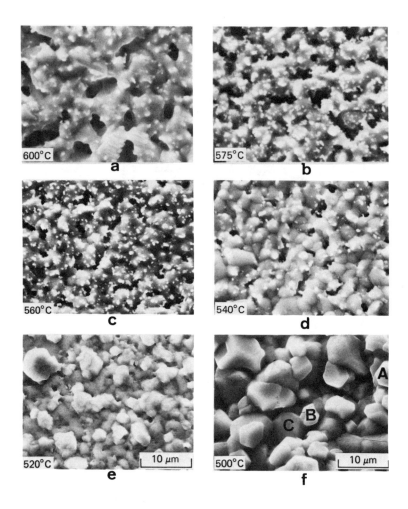

Figure 5. Comparison of topography of surfaces of AISI 316 stainless steel exposed to lithium for 7488 h in a thermal convection loop with a maximum temperature of 600 C and a ΔT of 150 C (a) 600 C, (b) 575 C, (c) 560 C, (d) 540 C, (e) 520 C, (f) 500 C.

The observed enrichment in nickel and chromium at the lower temperatures was not uniform. Spectra taken from different deposition zones on a surface varied significantly, particularly in their peak intensities for chromium. Detailed analyses of these surfaces indicated that some underlying deposits tended to be more highly enriched in chromium than

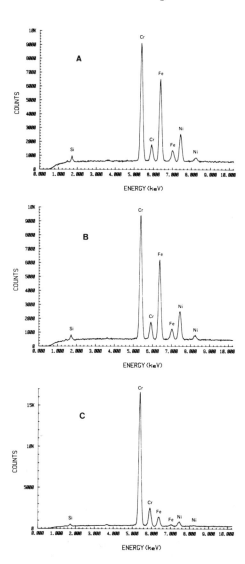

Figure 6. Energy-dispersive x-ray spectra for AISI 316 stainless steel exposed at 500 C to thermally convective lithium in a loop with a maximum temperature of 600 C. Letters A, B, and C denote areas shown in Figure 5(f).

those resting on top (for example, see Figure 6). Such observations of variations in deposit composition with depth from the original specimen surface indicate a time-dependent deposition process.

DISCUSSION

Thermal gradient mass transfer is the process whereby a net movement of material occurs from the hotter part of a fluid circuit to the cooler area. It is governed by the mechanisms and kinetics of the dissolution and deposition reactions, which in the present case, result in significant changes in the surface morphology and composition.

Table 2. Approximate Compositions from Standardless Analysis of Energy-Dispersive X-ray Data Using ZAF Corrections

Exposure Temperature (C)	Concentration (wt %)				
	Fe	Cr	Ni	Si	Mo
600	92.9	2.8	1.6	1.2	2.0
575	89.0	5.0	1.8	1.1	2.9
560	88.0	5.8	2.0	1.1	2.6
540	87.1	6.8	2.5	1.0	2.3
520	58.3	25.2	12.0	0.9	1.3
500	23.0	65.9	8.7	0.8	0.0

As reported above, preferential leaching of nickel and chromium occurred in the hot zone of the loop. Approximate compositions for these corroded surfaces can be obtained from the energy-dispersive x-ray data using normalization and correction factors as listed in Table 2. The quantitative accuracy of such concentration determinations, however, suffers from the roughness of the specimen surfaces and the corresponding uncertainties in the extent of electron and x-ray absorption paths. Nevertheless, the relative depletion of nickel and chromium noted from these values is accurate. While the chromum concentration of the surface increased steadily with decreasing temperature between 600 and 540 C (1112-1004 F), the surface nickel concentration remained low (<2.5 wt %) throughout most of this temperature interval. An extended temperature range over which nickel remains depleted to a low level has also been observed using x-ray fluorescence on austenitic stainless steel specimens exposed to lithium in a different set of shorter-term TCL experiments [4]. In addition, significant depletion

of nickel, but not chromium, has been observed in lithium
loop systems operating at a maximum temperature of 500 C
(932 F)[3,4]. Such results, therefore, indicate the very
serious susceptibility of Fe-Ni-Cr alloys to preferential
leaching of nickel during exposure to molten lithium, and
that the activation energy for this reaction is low
due to the relative insensitivity of the nickel depletion
process to temperature.

Consideration of Figure 5 shows that the pore size and
density change continually between 600 and 520 C (1112-968 F).
If a correlation is made between these porosity variations
with temperature and the changes in surface concentration
between 600 and 520 C, it would appear that there is no direct
relationship between the amount of nickel and chromium depletion
and the porosity since changes in these surface concentrations
did not correlate correctly with changes in the pore density and
size. In addition to vacancy agglomeration at preferential
lattice sites in response to the leaching of nickel or
chromium out of the alloy, other possible mechanisms for
the porosity formation include the development of localized,
soluble corrosion products (which may be what is seen
in some of the pores in Figure 4) or the redistribution of
surface atoms by continual dissolution-deposition to create
such cavities. Our present data are not sufficient to determine
a definitive mechanism for porosity formation (work in this
regard is still in progress), but the second mechanism could
involve reaction products of nickel or chromium (such as
Li_9CrN_5 [5]), that initially formed at localized sites
such as grain boundaries.

One definite consequence of the preferential leaching of
nickel from AISI 316 stainless steel is that the surface nickel
concentration is reduced to a level below that necessary for
the stability of austenite; a phase transformation to ferrite
therefore ensues. This ferrite phase can be seen in Figure 4,
which shows this layer as containing the porous zone situated
above the austenite matrix.

The relative enrichment in molybdenum of the nodules
on the surfaces suffering dissolution has not been noted
previously in lithium although such an observation has been
made for one case of a sodium-exposed austenitic alloy [6].
Because of the very low solubility of molybdenum in
lithium [7], it would appear that these nodules are areas
of molybdenum-containing precipitates that recede only

slowly relative to the remainder of the matrix. (If the precipitate was Mo_2C, the carbide would decompose in lithium [8] but the molybdenum would remain on the surface). The lack of any evidence of molybdenum on the surfaces that experienced deposition (as noted from energy-dispersive spectra) is consistent with the fact that these areas are dissolution-resistant. Such behavior, however, would eliminate redeposition in the hot zone as a possible mechanism for the porosity formation (see above).

As discussed in the Results section, the surface enrichment at lower temperature, which is caused by deposition of dissolved species from the lithium, was not uniform. As shown by the example in Figure 6, spectra taken from different surface deposits varied significantly. There was a gradient in deposit composition between the original specimen surface and the final outer layer such that the chromium concentration of the deposits decreased with distance above the original surface. Such observations are consistent with findings from earlier work with lithium-AISI 316 stainless steel thermal convection systems [9]: deposits of pure chromium formed during the initial stages of loop operation, while at later exposure times (greater than 5000 h) the deposits contained significant quantities of nickel and iron in addition to chromium. The reason for the initial formation of pure chromium deposits is not known, although it may be related to the temperature dependence of impurity reactions in the lithium (particularly reactions with nitrogen [5,10]).

The present data, from a more comprehensive surface analysis of AISI 316 stainless steel exposed to thermally-convective lithium, has confirmed several findings from earlier, less-detailed work with lithium. This includes the observation of porosity formation, nickel and chromium depletion as a function of loop position, and the time-dependence of the deposition process. In addition, the above results yield new information about topological differences among hot-leg specimens and about the deposition of molybdenum on the corroded surfaces.

SUMMARY

Analyses of AISI 316 stainless steel specimens exposed to thermally convective lithium confirmed earlier observations regarding preferential leaching of nickel and chromium and deposition of pure chromium during the initial part of the exposures. In addition, the present results showed that exposure temperature in the thermal convection loop strongly affects porosity and that molybdenum enrichment occurred on surfaces undergoing dissolution. Qualitative correlations between porosity variations and surface concentrations as a function of temperature indicate that porosity development may be more related to a chromium or nickel reaction with impurities than with the actual nickel depletion process. Further work is in progress to rigorously determine the mechanism of porosity formation.

ACKNOWLEDGEMENT

This research was sponsored by the Office of Fusion Energy, U.S. Department of Energy, under contract W-7405-eng-26 with the Union Carbide Corporation.

REFERENCES

1. P. F. Tortorelli and O. K. Chopra, "Corrosion and Compatibility Considerations of Liquid Metals for Fusion Reactor Applications," Journal of Nuclear Materials, Vols. 103 and 104, pp. 621-632 (1981).

2. J. H. DeVan, "Compatibility of Structural Materials with Fusion Reactor Coolant and Breeder Fluids," Journal of Nuclear Materials, Vols. 85 and 86, pp. 249-256 (1979).

3. P. F. Tortorelli and J. H. DeVan, "Effect of Nickel Concentration on the Mass Transfer of Fe-Ni-Cr Alloys in Lithium," Journal of Nuclear Materials, Vol. 103 and 104, pp. 633-638 (1981).

4. P. F. Tortorelli, J. H. DeVan, and J. E. Selle, "Corrosion in Lithium-Stainless Steel Thermal-Convection Systems," Proceedings of the Second International Conference on Liquid Metal Technology in Energy Production, U.S. Department of Energy, CONF-800401-P2, 13-44-54 (1980).

5. M. G. Barket, et al., "The Interaction of Chromium with Nitrogen Dissolved in Liquid Lithium," Journal of Nuclear Materials, Vol. 114, pp. 143-149 (1983).

6. J. H. DeVan and C. Bagnall, to be published.

7. V. A. Maroni, E. J. Cairns, and F. A. Cafasso, A Review of the Chemical, Physical, and Thermal Properties of Lithium That are Related to Its Use in Fusion Reactors, Argonne National Laboratory, ANL-8001 (March 1973).

8. T. L. Anderson and G. R. Edwards, "The Corrosion Susceptibility of 2 1/4 Cr-1 Mo Steel in a Lithium-17.6 wt pct Lead Liquid," Journal of Material Energy Systems, Vol. 2, pp. 16-25 (1981).

9. P. F. Tortorelli and J. H. DeVan, "Mass Transfer Deposits in Lithium-Type 316 Stainless Steel Thermal-Convection Loops," Proceedings of the Second International Conference on Liquid Metal Technology in Energy Production, U.S. Department of Energy, CONF-800401-P2, 13-55-63 (1980).

10. W. F. Calaway, "The Reaction of Chromium with Lithium Nitride in Liquid Lithium," Proceedings of the Second International Conference on Liquid Metal Technology in Energy Production, U.S. Department of Energy, CONF-800401-P2, 18-18-26 (1980).

METALLOGRAPHY OF CORROSION AND HYDRIDING OF ZIRCONIUM ALLOY NUCLEAR REACTOR PRESSURE TUBING

Kim J. Chittim* and Derek O. Northwood*

ABSTRACT

The CANDU-PHW nuclear reactor uses zirconium alloy pressure tubes as the pressure vessels, natural uranium fuel, and heavy water (deuterium oxide) as both moderator and coolant. During reactor operations there is a corrosion reaction between the zirconium alloy pressure tube and the hot ($\sim$300 C) pressurized ($\sim$9 MPa) heavy water moderator/coolant. As a result of this corrosion reaction, there is oxidation of the zirconium alloy pressure tubing and release of hydrogen (deuterium) which can be "taken-up" (hydriding) by the pressure tube material. This corrosion/hydriding reaction has been studied in Zr-2.5wt%Nb and ExCel alloy (Zr-3.3wt%Sn-0.8wt% Nb-0.8wt%Mo) pressure tubing using accelerated corrosion tests in a pressurized LiOH solution. The variation in degree of corrosion/hydriding is related to the thermo-mechanical processing schedules. The hydrides produced by the take-up of hydrogen have been characterized by optical and electron metallography, and x-ray diffraction.

INTRODUCTION

In the CANDU-PHW (CANadian Deuterium Uranium-Pressurized Heavy Water) nuclear reactor, the calandira, or reactor vessel, is a cylindrical tank laid on its side with its end faces forming vertical planes. This tank is filled with the heavy water moderator. Several hundred tubes (pressure tubes)

---

* Department of Engineering Materials, University of Windsor Windsor, Ontario, Canada N9B 3P4

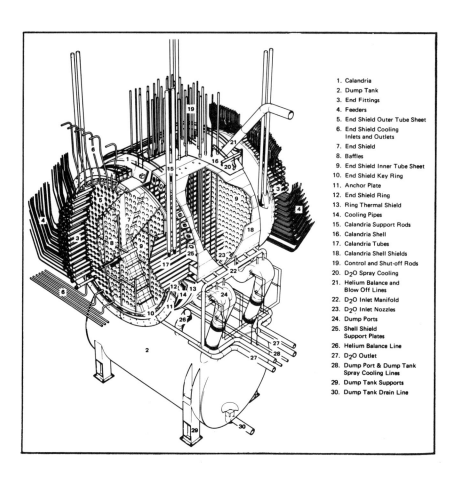

Figure 1. Cut-away view of a CANDU-PHW reactor assembly showing the reactor core.

penetrate the tank and contain the natural uranium fuel. A coolant (heavy water) is pumped past uranium fuel within the pressure tubes, and the heat of fission is transferred to the coolant. The coolant flows onto the boilers or steam generators where it gives its heat to the ordinary, or light water, to produce steam. Figure 1 is a cut-away view of a CANDU-PHW reactor assembly showing the reactor core.

Since the CANDU reactors produce economic electrical power by utilizing natural uranium fuel, structural materials in the core of the reactor must have a low cross-section for thermal neutrons. Zirconium alloys have a much lower neutron absorption per unit strength than other commercially available structural materials, with the exception of beryllium which is unsuitable for nuclear applications due to poor ductility, particularly after irradiation [1]. Thus, with the exception of the calandria vessel which is made from stainless steel, most of the in-core components are fabricated from zirconium alloys.

Pressure tubes are the pressure vessels of the CANDU reactors. They are about 10-cm. in dia., 0.3- to 0.5-cm thick by 6-m long and contain the fuel and coolant operating at about 9.6 MPa and 300 C. A pressure tube alloy must: be able to be fabricated into tubes; have adequate neutron economy; have adequate in-reactor creep strength; resist corrosion by the reactor coolant; have adequate tensile strength; and, retain adequate ductility in the reactor. Pressure tubes in the following alloys and metallurgical conditions are now in service in CANDU power reactors: Zircaloy-2, cold-worked; Zr-2.5wt%Nb, quenched and aged; and, Zr-2.5wt%Nb, cold-worked. Cold-worked Zr-2.5Wt%Nb is the current reference tube material. Tubes from a new alloy, ExCel (Zr-3.3wt%Sn-0.8Wt%Nb-0.8wt%Mo), are now being developed [2].

During reactor operations, there is a corrosion reaction between the zirconium alloy pressure tube and the heavy water moderator/coolant:

$$Zr + 2H_2O \rightarrow ZrO_2 + 2H_2 \text{ (in CANDU-PHW, H is replaced by D)}$$

As a result of this corrosion reaction, there is a release of hydrogen (deuterium) and this can be "taken up" by the pressure tube material. Hydrogen pick-up above a certain level will give rise to solid hydrides at reactor operating temperatures. These hydrides are brittle in nature and could lead to failure of the pressure tube by brittle fracture, particularly when the reactor is shut-down, i.e., cooled down [3,4].

Earlier work by one of the authors [5] has shown that a suitable method of simulating the long term corrosion/hydriding behavior of the pressure tubes in reactor is by autoclave tests in pressurized lithium hydroxide solution at 300 C. Such a test has already been used to investigate the effects of various thermomechanical processing schedules on the corrosion/hydriding behavior of Zr-2.5wt%Nb alloy pressure tubing [6,7], the current design material. In this study, further results are presented for Zr-2.5wt%Nb and these are compared to those for pressure tubing fabricated from the ExCel alloy. The general aim of this work is to develop thermomechanical processing schedules for the pressure tubing that will give rise to low rates of both corrosion and hydriding. The particular emphasis of the present paper is the metallographic aspects of the hydrides produced during the corrosion reaction.

EXPERIMENTAL DETAILS

Materials and Thermomechanical Processing Schedules

The starting materials were commercial Zr-2.5wt%Nb alloy pressure tubing supplied by Ontario Hydro and ExCel pressure tubing supplied by Atomic Energy of Canada Limited. The chemical analyses are given in Tables 1 and 2. The pressure tubing was then subject to a series of thermomechanical processing schedules as detailed in Table 3. Heat treatments were performed in evacuated quartz capsules in order to avoid oxidation during the heat treatment cycle. Cold-working was done on a small laboratory rolling machine at room temperature. The 560 C heat treatment was chosen on the basis of mechanical property/microstructure considerations [8] and the fact that this type of heat treatment has given the lowest overall corrosion/hydriding rates in the preliminary tests on Zr-2.5wt%Nb [6,7].

Table 1. Chemical Analysis of Zr-2.5wt%Nb Alloy

Weight Percent		
Nb	Fe	O
2.50	0.055	0.110

Impurity Analysis (ppm by weight)								
Al	B	C	Cd	Cr	Co	Cu	H	Hf
35	<0.2	120	<0.2	74	<10	30	10	44

Table 1. (Continued)

Impurity Analysis (ppm by weight)

Mg	Mn	Mo	N	Ni	Pb	Si	Sn	Ta
<10	<25	<25	50	<35	<50	59	15	<200

Impurity Analysis (ppm by weight)

Ti	U	V	W
<50	<0.5	<25	25

Table 2. Chemical Analysis of ExCel Alloy

Weight Percent

Sn	Mo	Nb
3.70	0.80	0.71

Impurity Analysis (ppm by weight)

H	O
14	1120

Table 3. Summary of Thermomechanical Processing Schedules

Primary Heat Treatment	Cold Work (pct)	Secondary Heat Treatment
As Received	0	1 hr/560 C, air cooled
7 hr/850 C, air cooled	0	2 hr/560 C, air cooled
7 hr/850 C, air cooled	0	10 hr/560 C, air cooled
7 hr/850 C, air cooled	0	24 hr/560 C, air cooled
1/2 hr/750 C, air cooled	0	24 hr/400 C, air cooled
1/2 hr/750 C, air cooled	20	24 hr/400 C, air cooled
1/2 hr/750 C, air cooled	30	24 hr/400 C, air cooled
1/2 hr/750 C, air cooled	40	24 hr/400 C, air cooled

Corrosion/Hydriding Procedures

Hydriding was carried out in stainless steel (AISI 304) autoclaves at an internal pressure of approximately 9 MPa at 300 C. Lithium hydroxide solution containing 100 gm of LiOH per litre was used. The volume of the autoclave was approximately 40 cc and approximately 25 cc of LiOH solution was used. Specimens were small being approximately 1.5-cm x 1.5-cm x thickness of the pressure tubing. At least three specimens from each thermomechanical processing route were made. Two specimens were run in the autoclave tests and one specimen was retained for examination of the microstructure prior to hydriding/corrosion.

Hydrogen Analysis and Metallography of Hydrides

One specimen from each hydriding/corrosion run was used for hydrogen analysis by the vacuum hot-extraction method. The hydrogen analyses were performed by Luvac Inc., Boylston, MA, 01505, U.S.A. The duplicate specimen was used for a metallographic study of the hydride form and distribution. The samples for metallographic examination were wet polished with silicon carbide papers down to 600 grit, and swab etched using a solution of 45 parts HNO_3, 45 parts H_2O, and 5 to 10 parts HF. The samples were examined by optical metallography using bright field illumination. The same samples were used for x-ray diffraction analysis. X-ray diffraction analysis was performed using graphite monochromated CuK_α radiation on a Philips Powder Diffractometer operating at a scanning speed of 2 degrees (2θ) per minute. Phase identification was performed using data from reference 7.

A few thin foils for transmission electron microscopy were also prepared from selected Zr-2.5wt%Nb samples. Details of the specimen preparation techniques are described elsewhere [9]. The thin foils were examined in a JEOL 100-cx electron microscope operated at 100 kV.

RESULTS

Hydrogen Pick-Up Data and General Appearance of Specimens

This section of the results will be divided into two sub-sections, one dealing with the effects of secondary heat treatment at 560 C, and the second dealing with the effect of an intermediate cold-work operation.

Table 4. Effect of Time at Secondary Heat Treatment Temperature of 560 C on the Hydrogen Pick-Up and Corrosion of Zr-2.5Wt%Nb Exposed to LiOH Solution of 100g/l at 300 C and 9 MPa Pressure for 500 Hours

Metallurgical Condition	Hydrogen Content (ppm by weight)	Hydrogen Flux at 300 C (g/m^2s)	Observation (Physical Appearance)
As Received (Present Fabrication Route)	96	4.6×10^{-7}	1.6-mm laminated white oxide
850 C/7hr AC + 560 C/1hr AC	37	1.9×10^{-7}	Thin white oxide layer (0.5-mm)
850 C/7hr AC + 560 C/2hr AC	41	2.1×10^{-7}	Thin white oxide layer (0.4-mm)
850 C/7hr AC + 560 C/5hr AC	43	2.2×10^{-7}	Thin white oxide layer (0.3-mm)
850 C/7hr AC + 560 C/10hr AC	50	2.6×10^{-7}	Thin white oxide layer (0.3-mm)
850 C/7hr AC + 560 C/24hr AC	55	2.8×10^7	Top layer is thin grey oxide and underneath is brown oxide layer (0.2-mm)

+ AC - Air Cooled

The effect of the 560 C treatment is shown in Table 4 for Zr-2.5wt%Nb and in Table 5 for the ExCel alloy. For Zr-2.5wt%Nb there was a minimum rate of hydrogen pick-up for the material heated for 2 h at 560 C but the general level of corrosion was the least for the material heated the longest time (24 h) at 560 C. The ExCel alloy showed a quite different behavior in that increasing secondary heat treatment times at 560 C gave rise to an increase both in the general corrosion rate (thicker, laminated oxide formed) and in the rate of hydrogen pick-up, Figure 2. The present fabrication route for ExCel tubing appears to give both a low level of hydrogen pick-up and a low level of corrosion.

The effects of an intermediate cold-working step have, so far, only been determined for Zr-2.5wt%Nb. In general, a high level of intermediate cold-work gave rise to both a high level of hydrogen pick-up and a high level of general corrosion, Table 6.

Table 5. Effect of Time at Secondary Heat Treatment Temperature of 560 C on the Hydrogen Pick-Up and Corrosion of ExCel Exposed to LiOH Solution of 100g/l at 300 C and 9 MPa Pressure for 500 Hours

Metallurgical Condition	Hydrogen Content (ppm by weight)	Hydrogen Flux at 300 C (g/m^2s)	Observation (Physical Appearance)
As Received (Present Fabrication Route)	19	0.92×10^{-7}	Light brown thin oxide layer
850 C/7hr AC + 560 C/1hr AC	71	3.32×10^{-7}	Slight lamination of oxide
850 C/7hr AC + 560 C/2hr AC	124	5.69×10^{-7}	Laminated
850 C/7hr AC + 560 C/5hr AC	363	18.53×10^{-7}	Very laminated
850 C/7hr AC + 560 C/10hr AC	411	20.75×10^{-7}	Very laminated
850 C/7hr AC + 560 C/24hr AC	463	23.42×10^{-7}	Very laminated

+ AC - Air Cooled

Metallography of Hydrides

No evidence was found for a hydride layer at the corroding surface but rather there was a relatively even distribution of hydrides through the thickness of the pressure tubing. The size of the hydrides and their distribution varied both with the level of hydrogen pick-up and with the thermomechanical processing route (microstructure of sample).

As has been noted previously for the Zr-2.5wt%Nb alloy [6,7], all samples having the 560 C secondary heat treatment had small evenly distributed hydrides. Although all hydrides were small, there was a tendency to an increasing amount of radially oriented hydride with increasing time at 560 C.

Table 6. Effect of Intermediate Cold-Work on the Hydrogen Content and Corrosion Behaviour of Zr-2.5wt%Nb Pressure Tubing Exposed to LiOH Solution of 100 g/l at 300 C and 9 MPa Pressure for 500 Hours

Metallurgical Condition+	Hydrogen Content (ppm by weight)	Hydrogen Flux at 300 C g/m^2s	Physical Appearance of Oxide	Original* Thickness (mm)	Final* Thickness (mm)	% Reduction in Thickness
750 C/ 1/2hr AC + 400 C/24hr AC	63	3.2×10^{-7}	Black brittle oxide layer; underneath is white oxide layer (total oxide thickness is 0.65 mm)	4.42	4.19	5.2
750 C/ 1/2hr AC + 400 C/24hr AC	81	4.1×10^{-7}	1 mm thick of white and grey laminated oxide layers	3.48	3.10	10.9
750 C/ 1/2hr AC + 30% CW + 400 C/24 hr AC	121	6.1×10^{-7}	1.5 mm thick of white and grey laminated oxide layers	3.05	2.54	16.6
750 C/ 1/2hr AC + 40% CW + 400 C/24hr AC	650	32.7×10^{-7}	1.8 mm thick of white, grey and tarnish laminated oxide	2.67	0.84	68.5

+ AC - Air Cooled; CW - Cold-Work
* Applies to Thickness of Metal Before and After Corrosion Reaction

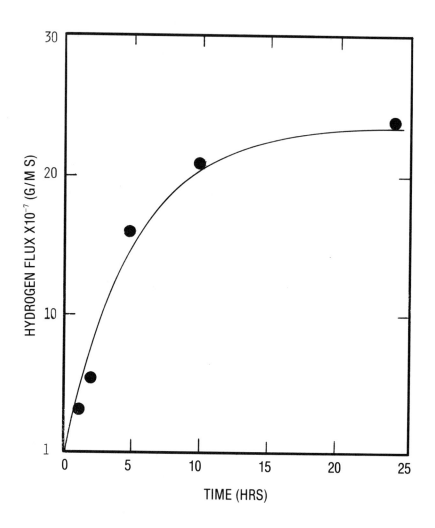

Figure 2. Effect of secondary heat treatment time (hours) at 560 C on the hydrogen pick-up of ExCel alloy.

Obviously, radial (through wall) hydride should be avoided if possible. The new results for the ExCel alloy show a similar trend with increasing amount of radial hydride with increase in time at 560 C. The difference between Zr-2.5wt%Nb and ExCel was that for ExCel, not only was the proportion of radial hydride increasing with time at 560 C but also the total amount of hydride was also increasing.

Rather than just use the qualitative observation that the amount of radial hydride was increasing, an attempt was made to quantify the amount (fraction) of radial hydride in the ExCel alloy. In order to do this, both bright field and polarized light images were used to determine f_{45}, the fraction of hydrides whose traces in the plane of observation make an angle with a reference direction (longitudinal-circumferential) in the range 45 to 90 degrees. A high value of f_{45} would indicate a large amount of hydride with a tendency to orient in the radial direction. A series of bright field micrographs illustrating the change in hydride morphology with secondary heat treatment time at 560 C for ExCel alloy are given in Figure 3. A representative polarized light micrograph is given in Figure 4. f_{45} was determined using a Ladd L40000 Microcomputer Image Analyzer. The results are illustrated in Figure 5, where it can readily be seen that starting with a totally circumferential hydride (f_{45} = 0) in the as-fabricated tubing, the amount of radial hydride increases with time at 560 C and for times greater than about 10 h, approximately half of the hydrides are oriented in a radial direction.

Large interconnected hydrides were formed in the Zr-2.5wt%Nb alloys having an intermediate cold-work step. However, all these hydrides formed in a circumferential orientation [7].

The x-ray diffraction analysis and limited selected area diffraction analysis using the TEM indicated that the predominant hydride in all specimens was the γ-hydride (ZrH) with δ-hydride ($ZrH_{1.66}$) occurring in some specimens. This is in agreement with previous data for zirconium alloys (including Zr-2.5wt%Nb) where a lower hydrogen content and air cooling favor the γ-hydride over the δ-hydride [4,10]. The high hydrogen content, ε-hydride (ZrH_2) was only noted in one specimen containing extremely high (650 ppm) levels of hydrogen.

DISCUSSION OF RESULTS

The results for Zr-2.5wt%Nb show the disadvantage from both corrosion and hydrogen pick-up standpoints of an intermediate cold-work step. Tests are underway to determine whether this is also true for ExCel alloy. Even if the intermediate cold-work cannot be eliminated, the degree of cold-work should be reduced to a minimum.

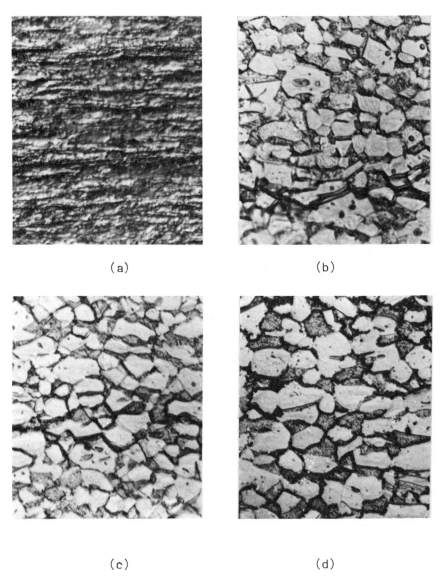

Figure 3. Series of bright field micrographs for ExCel showing change in hydride (black phase) morphology with time at 560 C secondary heat treatment temperature. (a) As received, (b) 1 h/560 C, (c) 2 h/560 C, (d) 5 h/560 C, (e) 10 h/560 C, (f) 24 h/560 C.

Corrosion and Hydriding of Zirconium Alloy / 239

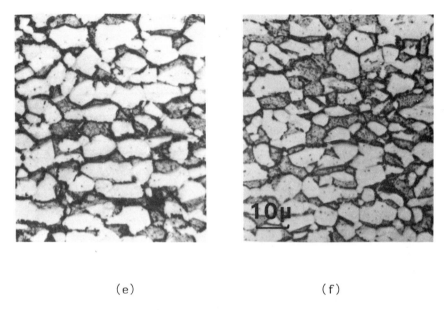

(e) (f)

Figure 3. (Continued)

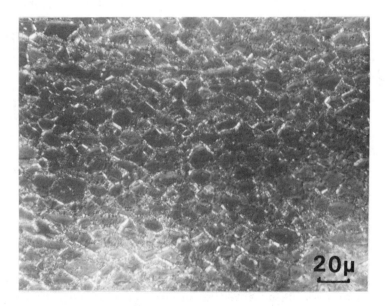

Figure 4. Typical polarized light micrograph showing distribution of hydride phase (white) produced during corrosion testing. ExCel specimen given secondary heat treatment of 24 h/560 C prior to corrosion testing.

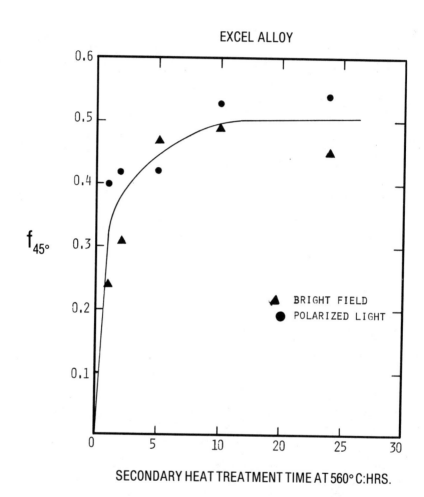

Figure 5. Effect of secondary heat treatment time (hours) at 560 C on the orientation of the hydride expressed as f_{45}, the fraction of hydride oriented at greater than 45 degrees to the circumferential direction.

Although the secondary heat-treatment at 560 C gives rise to a generally low level of hydrogen pick-up and general corrosion for Zr-2.5wt%Nb, it is generally detrimental to the ExCel alloy. This may be a general reflection of the fact that ExCel is generally less resistant to hydrogen pick-up and general corrosion than Zr-2.5wt%Nb. This is somewhat unfortunate since ExCel has other advantages, such as higher strength [2] and lower level of irradiation growth [11], which would make it a very promising material for future generations of CANDU reactors. However, these are only preliminary corrosion/hydriding results and it may be possible to develop a thermomechanical processing schedule for ExCel which will give good resistance to corrosion/hydriding whilst retaining high strength levels and resistance to irradiation growth.

The tendency towards a higher fraction of radial hydride for both Zr-2.5wt%Nb and ExCel alloy as the secondary heat treatment time at 560 C is increased is cause for concern. However, at least for Zr-2.5wt%Nb, the hydrogen pick-up levels are low enough that the hydrides are small and isolated and as such cannot provide a continuous crack path through the thickness of the pressure tubing.

ACKNOWLEDGEMENTS

The authors are pleased to recognize the financial support of the Natural Science and Engineering Research Council of Canada through a research grant (A4391). They are also indebted to Dr. B. A. Cheadle of the Chalk River Nuclear Laboratories of Atomic Energy of Canada Limited for the supply of the ExCel pressure tube material, and to Dr. R. G. Fleck of Ontario Hydro for the supply of the Zr-2.5wt%Nb pressure tube material.

REFERENCES

1. D. O. Northwood, "Zirconium Alloy Structural Components in the Core of a CANDU-PHW Nuclear Reactor," The Production and Applications of Less Common Metals Conference, Hangzhou, Peoples Republic of China, 8-11 November 1982. Proceeding published by Metals Society (UK) and The Chinese Society of Metals, Book 7, pp. 21/1-21/12 (1982).

2. C. E. Ells and W. Evans, "The Pressure Tubes in the CANDU Power Reactors," CIM Bulletin 74, pp. 105-110 (1981).

3. P. A. Ross-Ross, J. T. Dunn, A. B. Mitchell, G. R. Towgood and T. A. Hunter, "Some Engineering Aspects of the Investigation into the Cracking of Pressure Tubes in the Pickering Reactors," Atomic Energy of Canada Ltd. Report AECL-5261 (1976).

4. D. O. Northwood and U. Kosasih, "Hydrides and Delayed Hydrogen Cracking in Zirconium and its Alloys," International Metals Reviews, Vol. 28, No. 2, pp. 92-121 (1983).

5. D. O. Northwood and M. Thit, "Hydriding of Zirconium Alloys in Lithium Hydroxide Solutions," TMS Paper Selection A80-31, The Metallurgical Society of AIME, Warrendale, PA 15086, U.S.A. (1980).

6. D. O. Northwood and U. Kosasih, "Microstructural Control of Corrosion and Hydriding Behavior of Zr-2.5wt%Nb Nuclear Reactor Pressure Tubing," 2nd Asian-Pacific Corrosion Control Conference, Kuala Lumpur, Malaysia (June 28-July 3, 1981).

7. D. O. Northwood and U. Kosasih, "Corrosion and Hydriding Behavior of Zr-2.5wt%Nb Alloy Nuclear Reactor Pressure Tubing," J. Materials for Energy Systems, Vol. 4, No. 1, pp. 3-15 (1982).

8. D. O. Northwood and W. L. Fong, "Modification of the Structure of Cold-Worked Zr-2.5wt%Nb Nuclear Reactor Pressure Tube Material," Metallography, Vol. 13, pp. 97-115 (1980).

9. D. O. Northwood and R. W. Gilbert, "Comments on the Preparation of Thin Foils of Zirconium and Zirconium Alloys for Electron Microscopy," J. Australian Institute Metals, Vol. 18, pp. 158-161 (1973).

10. D. O. Northwood and D. T. H. Lim, "A TEM Metallographic Study of Hydrides in a Zr-2.5wt%Nb Alloy," Metallography, Vol. 14, pp. 21-35 (1981).

11. R. A. Herring and D. O. Northwood, "Neon Ion Simulation of Neutron Induced Irradiation Growth in Zirconium Alloys," TMS Paper Selection A83-8, The Metallurgical Society of AIME, Warrendale, PA 15086, U.S.A. (1983).

SURFACE CORROSION COMPARISONS OF SOME ALUMINUM ALLOYS IN 3.5% NaCl SOLUTION

W. J. D. Shaw*

ABSTRACT

The short term corrosive effect of an aerated 3.5% NaCl solution on the surface of a number of different aluminum alloys (7075-T6, 2024-T4, 6061-T6511, and IN-9021-T452) is documented as a function of time. Attack as related to microstructure is shown to occur in an ordered progressive manner. The IN-9021 material is very susceptible to pitting and exhibits a high general corrosion rate, being three times the corrosion rates of 7075 and 2024 alloys.

INTRODUCTION

The corrosion behavior of aluminum alloys has been studied for many years. The effect of various heat treatments on the susceptibility to intergranular attack is well documented [1]. Additionally, information on the compatibility of aluminum alloys with each other and different materials in 3.5% NaCl solutions is available [2]. However, little progress has been made in understanding the corrosion attack at the microstructural level. Most studies to date have been concerned with the morphological appearance of corrosion products after a considerable period of surface exposure time which results in an obscuring of the microstructural features of the material [3-6]. Other studies are mainly concerned with the examination of pitting attack [7,8].

---

* Department of Mechanical Engineering, University of Calgary, Calgary, Alberta, Canada T2N 1N4.

The purpose of this study is twofold. First, it is intended to follow the surface corrosion progression of a number of common aluminum alloys as related to microstructural features. Second, it is intended to compare the corrosion behavior of a new alloy material, IN-9021, to other common aluminum alloys.

IN-9021 material is a P/M mechanically alloyed analog of 2024 aluminum alloy. However, because of carbide and oxide dispersoids and an extremely fine grain structure it achieves a much higher strength than the 2024 alloy while still maintaining a high degree of ductility.

EXPERIMENTAL

Two sets of specimens for each alloy were prepared for optical examination; one set was unetched, the other etched with Keller's reagent. Both longitudinal and transverse sections were mounted and polished to a 0.05 micrometer finish. Two micro-indentation marks were placed on each surface for a reference so that the same area could be followed as corrosion progressed. Photographs of the surface morphology corrosion progression were taken at 1000X on an optical microscope (Zeiss ICM-405 metallograph). Pictures were taken using polarized light and Nomarski interference in order to sharpen the base metal surface morphology as contained beneath a thin transparent film. Pitting and final corrosion morphology was also studied on a Cambridge S100 scanning electron microscope. The surfaces were placed in an aerated 3.5% NaCl solution at 20 C for periods of 20 min, 1 h, 3 h, 9 h, 27 h, 81 h and 168 h. The surfaces were removed, rinsed, dried and examined at the end of each time period. The chemical compositions of the alloys studied are listed in Table 1.

The mechanical properties were measured using four half-size standard specimens of each alloy and testing was conducted in a 20 KIP MTS machine according to ASTM E-8. Hardness was measured using a standard Rockwell tester.

A corrosion coupon test was conducted according to ASTM G-31. Six coupons of each alloy were positioned in separate retorts with aerated 3.5% NaCl solution. The coupons were weighed to the nearest 1-mg. A short two-week test was conducted on these coupons, after which weighing and progressive cleaning was conducted using concentrated nitric acid at room temperature.

Table 1. Nominal Chemical Content, wt. %

Element	Material			
	7075	2024	6061	IN-9021
Si	0.4	0.5	0.6	0.05
Fe	0.5	0.5	0.7	0.02
Cu	1.6	4.3	0.3	4.0
Mn	0.3	0.6	0.15	0.01
Mg	2.5	1.5	1.0	1.5
Cr	0.3	0.1	0.2	0.01
Zn	5.6	0.25	0.25	0.01
Ti	0.2	0.15	0.15	-
O	-	-	-	0.8
C	-	-	-	1.1

RESULTS AND DISCUSSION

Interest in IN-9021 alloy stems primarily from its superior strength and good ductility as seen in Table 2. All alloys tested show good agreement with commonly accepted values and can be considered to be typical. The IN-9021 shows slightly less ductility than what has been reported previously [9] and has a slightly higher strength in the transverse direction.

Typical microstructures, as shown in Figure 1, are compared for these materials. The standard aluminum alloys show typical grain size and complex particles. The IN-9021 microstructure shows an exceedingly small grain size with dispersoids of carbide and oxide.

The corrosion progressions for the aluminum alloys are shown in Figures 2-6. The progressions start with the area as-polished (or etched) prior to corroding and progresses from 20 min, 3 h, 27 h and 81 h. The other times between 0 to 81 h exhibit a normal progression of what is seen in these figures. The 168 h time period morphology appeared identical in all cases to the 81 h time period.

Table 2. Mechanical Properties

	Measured	Standard Value+
7075-T6		
yield, MPa	490.9 ± 4.4	503
ultimate, MPa	537.2 ± 5.6	572
elongation, %	9.2 ± 0.8	11.0
hardness, BHN	164.4 ± 1.8	150
2024-T4		
yield, MPa	334.6 ± 6.7	324
ultimate, MPa	442.0 ± 3.0	469
elongation, %	13.4 ± 1.7	20
hardness, BHN	130.3 ± 1.9	120
6061-T6511		
yield, MPa	272.1 ± 2.9	276
ultimate, MPa	300.0 ± 3.7	311
elongation, %	11.2 ± 0.7	12
hardness, BHN	91.5 ± 0.7	95
IN-9021 Longitudinal		
yield, MPa	528.6 ± 7.8	591*
ultimate, MPa	590.1 ± 5.9	612*
elongation, %	9.8 ± 0.9	14*
hardness, BHN	151.6 ± 2.8	145*
IN-9021 Transverse		
yield, MPa	547.1 ± 5.0	584*
ultimate, MPa	609.0 ± 4.1	597*
elongation, %	8.1 ± 0.4	11*
hardness, BHN	150.6 ± 1.6	145*

+ ASTM accepted standard unless noted otherwise
* reference [9].

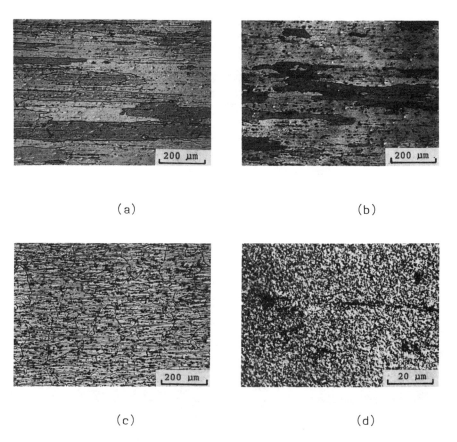

Figure 1. Longitudinal sectioned microstructures etched in Keller's reagent. (a) 7075-T6, (b) 2024-T4, (c) 6061-T6511, (d) IN-9021-T452.

Figures 2 and 3 show the 7075 alloy in the etched and unetched condition respectively. There is no difference seen between these two conditions. A general matrix corrosion attack occurs being fairly even and showing a gradually increasing amount of damage. Generally neither the grain boundaries, the precipitates, nor the areas immediately adjacent to the precipitates were attacked. It was found that a temporary slight attack at the grain boundaries occurred within 9 h on the longitudinal sections and within 27 h in the transverse sections, but this effect was quite small and quickly disappeared as corrosion progressed. This is in difference to

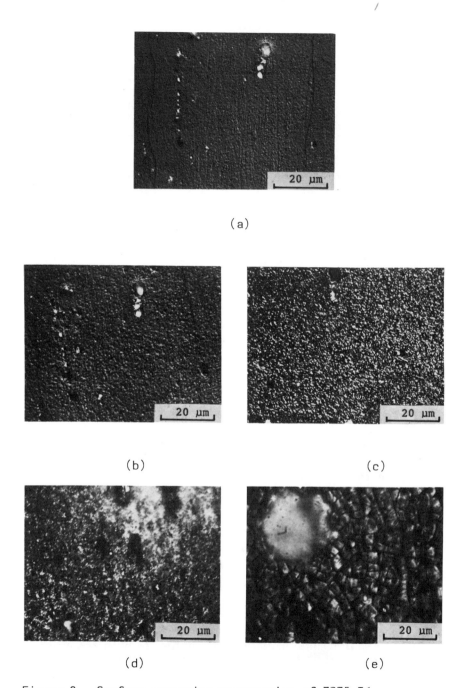

Figure 2. Surface corrosion progression of 7075-T6 alloy, longitudinal section, etched in Keller's reagent. (a) 0 minutes, (b) 20 minutes, (c) 3 hours, (d) 27 hours, (e) 81 hours.

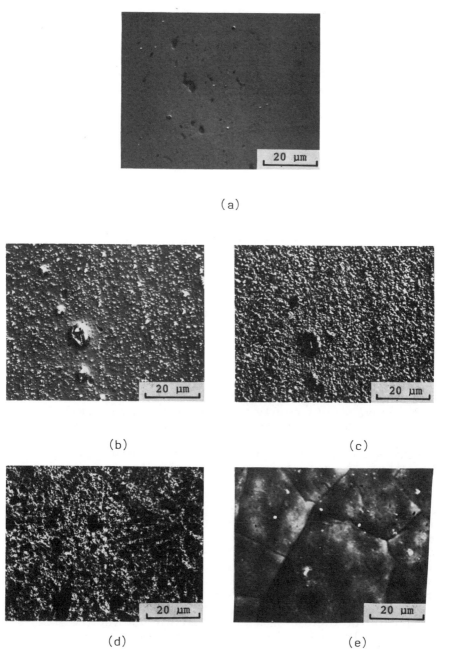

Figure 3. Surface corrosion progression of 7075-T6 alloy, longitudinal section, unetched. (a) 0 minutes, (b) 20 minutes, (c) 3 hours, (d) 27 hours, (e) 81 hours.

findings by others who have reported preferential grain boundary attack [4]. A buildup of corrosion products at select sites occurred within 9 h along with the initiation of some pits although generally pitting did not normally occur in a noticeable way until 27 h. A checkered network of a corrosion film occurred after 81 h. The size of the checkered network was quite variable.

The 2024 alloy, as seen in Figure 4 in the unetched condition, shows an initial attack of the areas immediately adjacent to the large particles. This attack is in preference to general matrix attack which takes over a period of 1 h before the matrix begins to experience noticeable attack. There is no specific attack at the grain boundaries, rather the areas surrounding the large particles are further attacked until they become quite large or the particle is dislodged. Checkered networks develop after 81 h and localized pitting occurs within 27 h.

The 6061 alloy follows a somewhat similar pattern to the 2024 alloy as can be seen in Figure 5. The areas around the large particles experience attack. A very small amount of localized checking is present after 81 h. This material appears to stabilize after 27 h and shows remarkably little matrix attack.

The IN-9021 material, being chemically similar to 2024 alloy, would normally be expected to follow a similar corrosion behavior. However, as seen in Figure 6, the corrosion behavior is similar to the 7075 alloy with a checkered network occurring after 81 h. A high activity of pitting occurs in this material with pits forming within the first few hours.

Some general observations from these corrosion progression results are that in IN-9021 and 7075 alloys no differences occur between etched and unetched surfaces. However, slight differences do occur in the 2024 and 6061 alloys, with the matrix in the unetched material showing a delay in attack while preference of corrosion attack occurs in the areas surrounding the particles. This difference evened out after 9 h. Attack on the transverse and longitudinal sections were identical for the 6061 and IN-9021 alloys but for the 2024 and 7075 alloys the attack on the transverse section was slightly less initially, balancing out after 9 h. All specimens showed large variations in corrosion activity in some local regions. Contrasts between fast and slow corrosion activity could be found in comparison to the norm

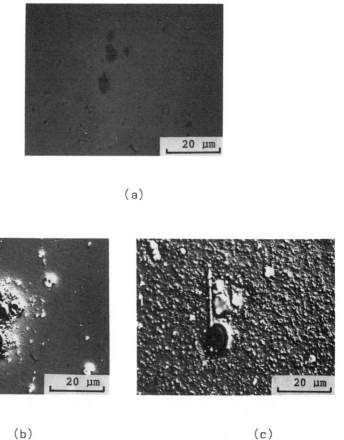

Figure 4. Surface corrosion progression of 2024-T4 alloy, longitudinal section, unetched. (a) 0 minutes, (b) 20 minutes, (c) 3 hours, (d) 27 hours, (e) 81 hours.

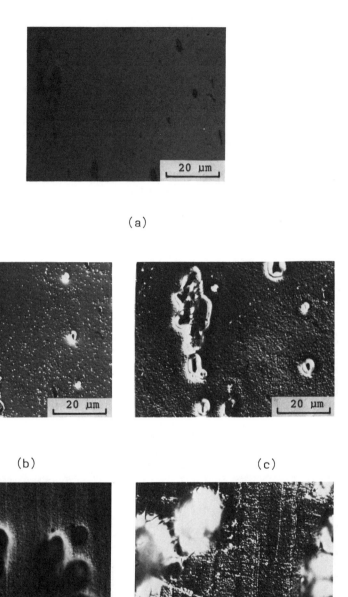

Figure 5. Surface corrosion progression of 6061-T6511 alloy, longitudinal section, unetched. (a) 0 minutes, (b) 20 minutes, (c) 3 hours, (d) 27 hours, (e) 81 hours.

Corrosion of Aluminum Alloys / 253

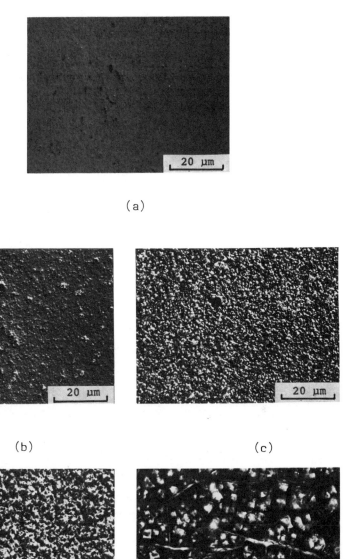

Figure 6. Surface corrosion progression of IN-9021-T452 alloy, longitudinal section, unetched. (a) 0 minutes, (b) 20 minutes, (c) 3 hours, (d) 27 hours, (e) 81 hours.

for the alloy. An example of pitting variability for the 7075 alloy is shown in Figure 7. The area in Figure 7(a) shows a corrosion product buildup surrounding the pit, which was the usual case, while it was also found that a denuded area with no signs of surface corrosion could also occur, as seen in Figure 7(b).

Corrosion product buildup occurred within 9 h on all alloys and developed to a marked extent after a period of one week, as seen in Figure 8. The buildup occurred both around pits and similarly on flat surface areas without pits.

Pitting occurred quickly and was quite extensive in the IN-9021 alloy. Figure 9 shows a pit after 3 h as well as the development of a pit with surface checking after 81 h. The pits in the IN-9021 material were considerably more plentiful, developed quicker, and were much deeper than those that occurred in the 7075 alloy.

The attack of the matrix surrounding the particles in the 2024 alloy can be seen in Figure 10. This checkered network was observed to be a film that was transparent at first but became translucent as it grew in thickness. This corrosion product film is seen to have cracked due to the drying of the specimen and may well remain intact for longer periods of time under conditions of continuous immersion. This type of cracking has been seen previously [5,7,8] under different conditions and has been referred to as "mud cracking" [7,8]. Because the Pilling-Bedworth ratio is greater than unity for aluminum oxide, it is not directly known why this cracking occurs. Figure 10(b) also shows that this product can come loose and fall away exposing fresh surface directly to the corrosive media. The surface below the film is fairly rough. This is confirmed by removal of the corrosion deposits on the coupon specimens.

Even though the 6061 alloy did not exhibit a general tendency to checking, given a sufficient period of time it is felt that checking would develop. Figure 11 shows the initiation of checking or cracking of an oxide film with the origin occurring in the corroded valleys surrounding the large particles. Thus, it can be stated that all of the aluminum alloys tested exhibit the same overall corrosion behavior. Corrosion progresses gradually with surface or localized attack leading to the buildup of corrosion products on select sites with the eventual development of a thick uniform corrosion film which eventually cracks into a checkered network.

Corrosion of Aluminum Alloys / 255

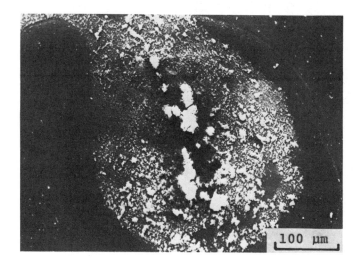

(a)

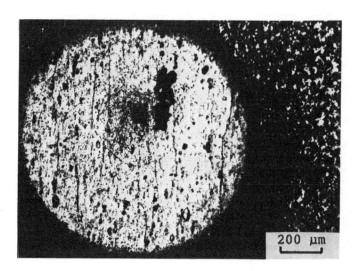

(b)

Figure 7. Pitting variability in 7075-T6, after 27 hours (a) corrosion product buildup, (b) denuded area.

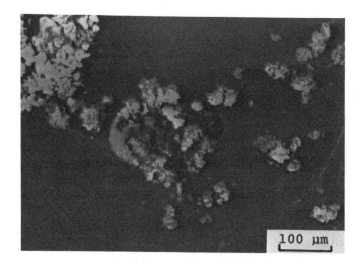

(a)

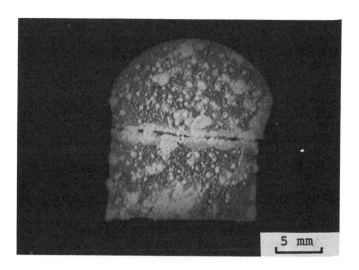

(b)

Figure 8. Corrosion product buildup. (a) 7075-T6 after 9 hours, (b) 2024-T4 after one week.

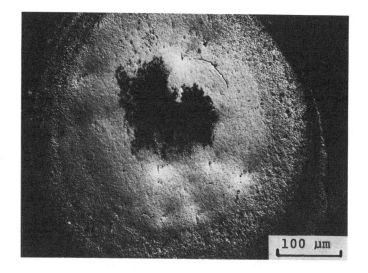

(a)

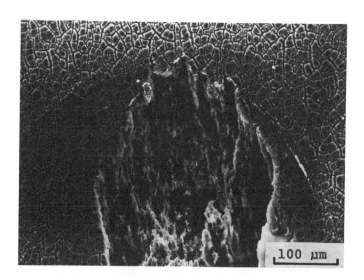

(b)

Figure 9. Pitting in IN-9021 alloy. (a) 3 hours, (b) 81 hours.

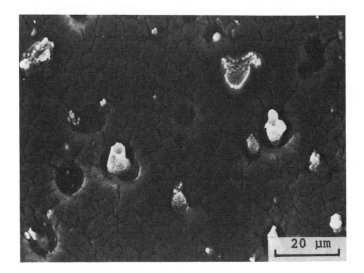

(a)

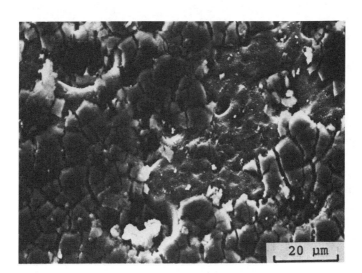

(b)

Figure 10. Particles and checkered network in 2024 alloy. (a) particle and checkered network, (b) break away of checkered network.

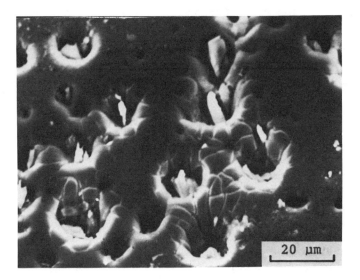

Figure 11. Particle and start of checkered network in 6061 alloy after 81 hours.

Measurements of the comparative corrosion rates of these alloys, as determined from the coupon tests, are given in Table 3. These rates indicate that the IN-9021 material corrodes three times faster than its counterparts of 2024 or 7075. The 6061 alloy, known to be fairly corrosive resistant, shows a very low corrosion rate as expected. The measured corrosion rate of 7075 coupons, even though being obtained from a short term test, agree favorably with those given elsewhere [9]. However, the IN-9021 values are much higher than has been reported previously [9]. It was normally thought that the corrosion resistance of IN-9021 was slightly better than that of 7075. This study shows that this is not the case. In fact, the high corrosion rate measured, combined with the active pitting of this material, makes it very undesirable when used in salt environments.

Table 3. Corrosion Rate Comparison
(Two Week Test)

Material	Mean Value, mpy	Standard Deviation
7075-T6	6.49	± 0.47
2024-T4	5.49	± 0.65
6061-T6511	2.11	± 0.19
IN9021-T452	17.47	± 6.43

Based upon the overall chemical compositions of the alloys, we would expect the IN-9021 to corrode more readily than the other elements due to the high carbon and copper contents; following this, the 2024 with its high copper content, then the 7075, and finally 6061. The results do not bear this out entirely.

The area surrounding the particles in the 2024 material would be expected to be anodic compared to the $CuMgAl_2$ particles and would thus corrode preferentially which, in fact, occurs. However, the 6061 material with its Mg_2Si particles are not expected to be cathodic compared to the immediately surrounding matrix; yet the results indicate this to be the case. Thus, it would appear that further work needs to be done on a microscopic level, possibly by using micro-electroprobes, to determine the electro-kinetics of the particles and surrounding matrix in order to better predict and understand the corrosion behavior of aluminum alloys.

CONCLUSIONS

The conclusions pertain to the material-environmental combination of 3.5% NaCl to which these alloys were exposed:

1. Aluminum alloys exhibit the same corrosion development, i.e.,

    a. progressive general surface attack
    b. buildup of corrosion products at select sites
    c. development of a corrosion film eventually cracking into a checkered network.

2. Particles affect the corrosion attack in 2024 and 6061 materials but do not affect the attack in 7075 and IN-9021 materials.

3. Pitting is very active in IN-9021 alloy, followed by 7075 and 2024.

4. The corrosion rate of IN-9021 is three times higher than 2024 or 7075.

ACKNOWLEDGEMENT

The author is grateful to the Natural Sciences and Engineering Research Council of Canada for support of this work.

REFERENCES

1. B. W. Lifka and D. O. Sprowls, "Significance of Intergranular Corrosion in High Strength Aluminum Alloy Products," ASTM STP 516, pp. 120-144 (1972).

2. F. Mansfeld and J. V. Kenkel, "Laboratory Studies of Galvanic Corrosion of Aluminum Alloys," ASTM STP 576, pp. 20-47 (1976).

3. J. Perkins and R. A. Bornholdt, "The Corrosion Product Morphology Found on Sacrificial Zinc Anodes," Corrosion Science, Vol. 17, pp. 377-384 (1977).

4. H. P. Van Leeuwen, J. A. M. Boogers, and C. J. Stentler, "The Contribution of Corrosion to the Stress Corrosion Cracking of Al-Zn-Mg Alloys," Corrosion, Vol. 31, pp. 23-29 (1975).

5. P. Neufeld and A. K. Chakrabarty, "The Corrosion of Aluminum and its Alloys in Anhydrous Phenol," Corrosion Science, Vol. 12, pp. 517-525 (1972).

6. R. C. Furneaux et al., "The Application of Ultramicrotomy to the Electronoptical Examination of Surface Films on Aluminum," Corrosion Science, Vol. 18, pp. 853-881 (1978).

7. S. M. DeMicheli, "The Electrochemical Study of Pitting Corrosion of Aluminum in Chloride Solutions," <u>Corrosion Science</u>, Vol. 18, pp. 605-616 (1978).

8. J. A. Richardson and G. C. Wood, "A Study of the Pitting Corrosion of Al by Scanning Electron Microscopy," <u>Corrosion Science</u>, Vol. 10, pp. 313-323 (1970).

9. D. L. Erich and S. J. Donachie, "Benefits of Mechanically Alloyed Aluminum," <u>Metal Progress</u>, pp. 22-25 (Feb. 1982).

AMELIORATION OF ALLOY OXIDATION BEHAVIOR
BY MINOR ADDITIONS OF OXYGEN-ACTIVE ELEMENTS

Haroun Hindam[*]
and
David P. Whittle[**]

ABSTRACT

The influence of minor additions of oxygen-active elements (Al, Hf and Ta) on the high temperature oxidation behavior of several Fe- and Ni-based alloys was investigated. Structural modification of the otherwise relatively planar alloy/scale interface to a tortuous morphology, as revealed by a devised "alloy deep-etching" technique, leads to a remarkable improvement in scale adhesion and an ensuing better athermal oxidation resistance. The development of such growth configuration is interpreted by either the classical internal oxidation theory or a model based on "short-circuit diffusion" of oxygen in the scale. Its influence on scale adhesion is tersely discussed.

INTRODUCTION

Alloy protection against aggressive environments relies on the formation of a continuous, slow-growing, adherent oxide film. During service, operating conditions are seldom isothermal; hence, considerable athermal stresses, in addition to growth stress, may develop. Exfoliation of the protective film leads eventually to accelerated or "break-away attack"

---

[*] Canada Centre for Mineral and Energy Technology, Ottawa, Ontario K1A 0G1. (On leave to National Research Council, Montreal, Quebec, Canada H4C 2K3).
[**] Lawrence Berkeley Laboratory, University of California, Berkeley, California 94720 USA.

which can be catastrophic. Thence, material wastage and
process interruption, induced by corrosion, are major concerns
in materials and energy processing systems functioning at
elevated temperatures.

Several decades ago, it was discovered that small additions
of metallic elements, having high affinity towards oxygen (e.g.,
Ca, Y, Ce, etc.), remarkably improve scale adhesion to alloys
exposed to isothermal as well as athermal conditions [1].
The models [1-6], which were advanced to interpret this effect,
are listed in Table 1. For a detailed discussion, the reader
is referred to a recent comprehensive review [7].

Table 1. Adhesion Models

Mechanical pegging	Lustman	(1950)
Graded seal	Pfeiffer	(1957)
Chemical bonding	McDonald	(1965)
Vacancy sink	Stringer	(1966)
Enhanced scale plasticity	Antill	(1967)
Alteration of scale growth	Stringer	(1972)

In this paper, typical metallographic observations are
presented to substantiate the "mechanical pegging" mechanisms,
originally proposed by Lustman [2]. The amelioration of scale
adhesion is attributed to the development of inward-growing,
continuous oxide particles, referred to as "pegs", which
interlock the scale and the alloy. Two distinct types of pegs
are identified based on morphological and growth criteria. The
mechanisms and conditions of peg formation are defined. The
influence of this morphology on scale adhesion is discussed
based on the fracture behavior of adherent scales which were
subjected to microindentation.

EXPERIMENTAL PROCEDURES

The compositions investigated consisted of Ni with
one of the following nominal additions: 2 or 4 wt. % Al,
1 wt. % Hf or Ta in addition to 2 wt. % Al, and an Fe-10 wt. %
Al alloy containing 1 wt. % of either Cu or Hf. The ingots %
were cast by either electrical resistance (Ni alloys) or ts
induction (Fe alloys) melting of 99.99% pure metal pellets

under an ultrapure He atmosphere and homogenized overnight at 1300 C. Following the standard preparation procedure, the at samples were suspended in a tube furnace and reacted at 1200 C in static, dry air.

Specimen characterization included x-ray diffraction, optical and scanning electron microscopy, and electronprobe microanalysis. A devised "alloy deep-etching" technique [8,9] was extremely useful in revealing microstructural details[8,9] which otherwise would not be discernible. Subsequent to the conventional sectioning and polishing procedures, the demounted specimens were immersed carefully in a warm ($\sim$30 C) bromine (10 V.%) - methanol solution for few minutes. They were transferred to cleaning solutions consisting of alternate trays of methanol and distilled water, and finally kept in a vacuum dessicator for few hours before further examination.

RESULTS AND DISCUSSION

Type I Microstructure

This distinct peg morphology was observed in the dilute NiHf and NiAl [8,9] alloys which satisfied the criteria for internal oxidation [10,11]. Typical microstructures, revealed by the deep-etch technique, are included in Figures 1 and 2. The internal oxide particles extended in the alloys as filaments (NiHf), cylindrical rods (NiAl) or rectangular platelets (NiAlTa and NiAlHf). This illustrates three important features for the particles to act as effective pegs. First, they should extend as a continuous "phase" throughout the internal oxidation zone; second, they should be incorporated into the surface scale as the alloy/surface scale interface recedes; and third, a better bond is achieved if there is an interaction between the internal oxide and the surface scale or the alloy through the formation of ternary compounds as in the present cases.

Peg growth in all alloys conformed to the parabolic law. In the case of the binary Ni-Al alloys, the rate constants decreased slightly with increasing aluminum content. Nonetheless, the presence of the minority active elements (Ta or Hf) had no effect despite the marked change in particle shape. The rate constants, along with other experimental and calculated data, are listed in Table 2.

Figure 1. Peg morphology in the Ni-2Hf alloy. The duplex external scale formed on this alloy consisted of outer and inner layers of NiO and NiO/Hf_3NiO aggregate, respectively. Note how firmly the filaments (Hf_3NiO) are attached to the external scale.

Alloy Oxidation Behavior / 267

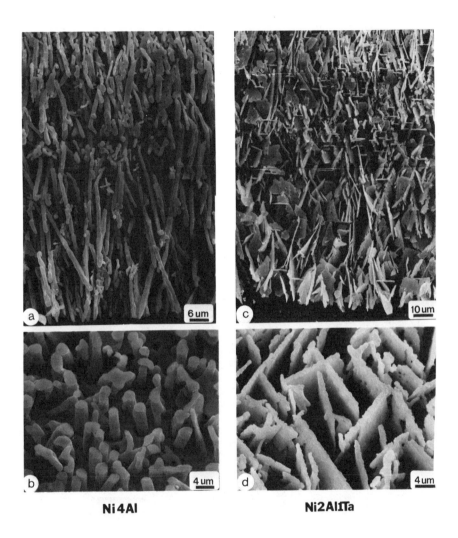

Figure 2. Longitudinal (a,c) and transverse (b,d) sections illustrating rod and plate-like pegs in the Ni-4Al (a,b) and Ni-2Al-1Ta (c,d) alloys, respectively. The rods and platelets exhibited an identical duplex ($NiAl_2O_4/Al_2O_3$) composition; the latter, innermost phase accounting for ~35% of the total thickness of the precipitation zone [9].

Table 2. Experimental and Calculated Data for Peg Formation in the Ni-Hf and Ni-Al Alloys

Alloy	Composition N_B	$k \times 10^4$ cm/sec$^{1/2}$	$\bar{d}$ micrometer	$\bar{s}$ micrometer	V_f calc.	V_f meas.	$N_o^S D_o \times 10^{10}$ cm^2/sec
Ni-2Al	0.043	0.83	1.25	3.3	0.12	0.27	1.3
Ni-4Al	0.083	0.61	4.15	6.6	0.24	0.35	1.8
Ni-2Al-1Ta	0.043	1.04	$\bar{l}$ = 5.5 $\bar{w}$ = 0.5	-	~0.12	0.36	1.8
Ni-2Al-1Hf	0.043	1.04	-	-	~0.12	0.40	1.8
Ni-2Hf	0.007	2.05	<1	-	-	-	-

N_B: active solute atom fraction, k: parabolic growth rate constant,
$\bar{d}$: average particle diameter, $\bar{l}$: average length, $\bar{w}$: average width,
$\bar{s}$: average particle spacing, V_f: precipitate volume fraction (as $NiAl_2O_4$)

Aluminum enrichment in the internal oxidation zone inferred that growth of the $NiAl_2O_4/Al_2O_3$ particles was controlled by the simultaneous diffusion of aluminum outwardly and oxygen inwardly [9]. This enrichment was ascertained by comparing the measured precipitate volume fraction and calculated values assuming no aluminum diffusion in the alloy (Table 2) since aluminum depletion ahead of the precipitation front occurred over a very narrow distance barely within the resolution limit of the electron microprobe.

Since the growth mode has been elucidated, the measured rate constants were used to evaluate the oxygen permeability product $N_O^s D_O$ (N_O^s and D_O are the oxygen solubility at the alloy surface and its diffusivity, respectively) based on the Wagner classical internal oxidation model [10]. The results were included in Table 2. It is evident that this parameter is virtually independent of the aluminum concentration or the presence of the other minority elements despite a conspicuous variation in particle size, shape and distribution (Table 2 and Figure 2). This finding inferred that despite particle continuity, its interface with the alloy matrix does not offer a low resistance path for oxygen transport at this elevated temperature.

This conclusion is in contrast with the behavior at lower temperatures, where a strong dependence of the 'apparent' permeabilities on aluminum content was found [12-14]. Values for a range of alloy compositions, oxidized in 1-atm oxygen, are shown in Figure 3, where it is evident that the difference between the various alloy contents diminished with increasing temperature. Included also in this figure are values of $N_O^s D_O$ in "pure" Ni obtained by extrapolation of plots of $N_O^s D_O$ vs. wt. % Al to 0% Al at the various temperatures [12] and values from independently measured solubilities [15] and diffusivity [16]. The present datum point for 1200 C fits closely with the extrapolated line through the low temperature oxidation data. This enhanced oxygen diffusivity at lower temperatures was probably partially associated with a much finer particle size and consequently a higher precipitate density.

TYPE II MICROSTRUCTURE

Although Al_2O_3 films offer ideal protection against oxidizing environments due to their minute thickening rate,

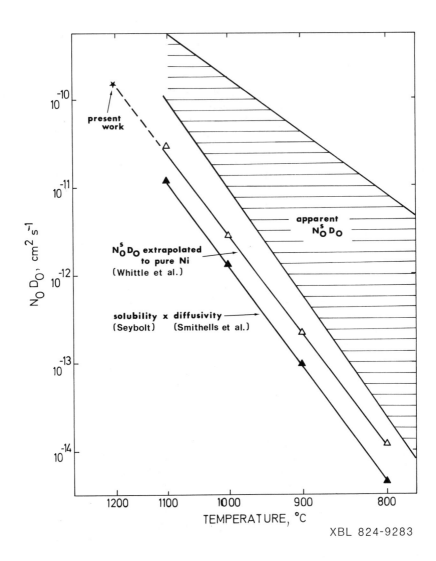

Figure 3. Arrhenius plot of the oxygen permeability product in γ-NiAl solid solution alloys.

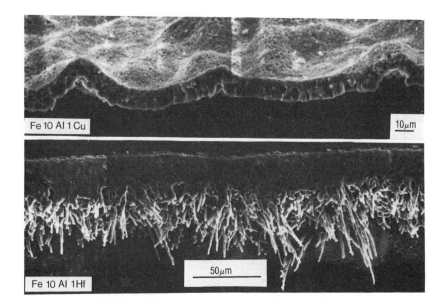

Figure 4. Fracture section of a loose Al_2O_3 scale formed on Fe-10Al-1Cu alloy and a deeply etched section of a tenacious, two-phase scale (Al_2O_3 + HfO_2) formed on Fe-10Al-1Hf alloy.

they tend to grow loosely attached to the alloy or coating developing an excessively wrinkled configuration. A typical observation is included in Figure 4. This stress-induced convoluted growth, which is frequently observed on alloys without additional oxygen-active elements, leads to profuse spallation upon cooling. A succinct review [17] has discussed the growth characteristics of Al_2O_3 scale on alloys and metallic coatings.

In contrast to this behavior, alloys containing minor quantities of oxygen-active elements, e.g., Hf, form tenacious scales. This remarkable improvement in scale adhesion is attributed to the development of an intricate network of inward-growing oxide pegs which was properly revealed by deep etching (bottom micrograph in Figure 4). The Al_2O_3 scale contained randomly-distributed HfO_2 particles (light phase). It is evident that the pegs are an integral part of

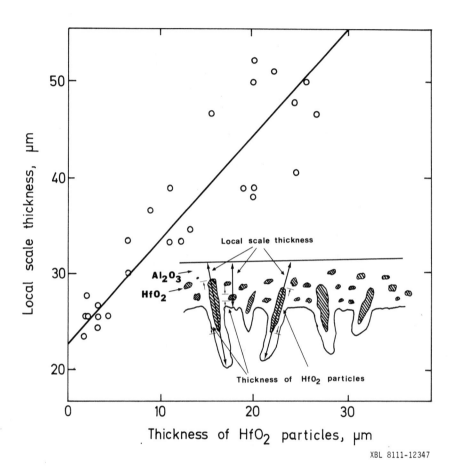

Figure 5. Variation of the scale thickness with the linear density of the HfO_2 particles.

the scale, and unlike Type I, have a similar composition, although richer in Hf. Actually, detailed examination of fractured sections revealed that a peg consisted of a HfO_2 stringer completely enveloped by the Al_2O_3 layer.

Figure 5 demonstrates a direct linear relationship between the scale thickness measured parallel to the local growth direction at random positions, as shown in the inset, and the linear density of the HfO_2 precipitates. The scatter in the data is due to the difficulty in determining the true particle length based on a two-dimensional section.

The isothermal and cyclic oxidation behavior of Al_2O_3-former alloys are being investigated [18]. During relatively short ($\sim$50 hr) isothermal (1200 C) gravimetric tests, approximate parabolic kinetics were obeyed. The rate constants for the hafnium-containing alloys were larger and dependent on their concentrations. In cyclic tests, during which the temperature was varied sinusoidally between 800 C and 1200 C, the addition-free alloy exhibited a transition to accelerated linear kinetics associated with microcracking of the Al_2O_3 scale. The amelioration of scale adhesion, conferred by the hafnium addition, conduced a superior performance under similar conditions.

In order to interpret the oxidation behavior of the Al_2O_3-former alloys, the transport properties of this phase should be considered. Tracer diffusion measurements indicated that both Al and oxygen are mobile in Al_2O_3 single crystals and polycrystalline sintered compacts [19-21]. However, oxygen diffusion was significantly enhanced by grain size reduction [19]. Inert marker observations [22] and O^{18} diffusion profile measurements [23] demonstrated that oxygen transport was predominant in Al_2O_3 scales grown on various alloys containing minor quantities of oxygen-active elements. It should be noted that such scales consistently exhibit a much finer grain size [17]. The incorporation of the HfO_2 dispersoids, as well as unreacted Fe particles in the Al_2O_3 scale, are consistent with the above conclusion [24].

Since the diffusion of oxygen in the highly anion-deficient HfO_2 phase is expected to be several orders of magnitude greater than in Al_2O_3, the dispersoids would act as "short-circuit" paths for oxygen transport leading to preferential localized thickening in the vicinity of these particles. Although this is a relatively rapid transport mechanism, the growth of the HfO_2 particles is limited by the supply of Hf to the alloy/scale interface. The dependence of the oxidation rate on the minority element content is consistent with this proposal.

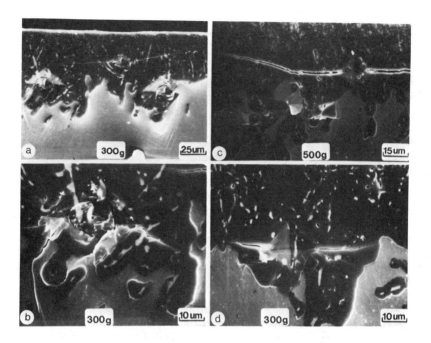

Figure 6. Crack propagation in an adherent Al_2O_3 scale due to microindentation at various loads.

Microindentation and Fracture Behavior of Adherent Scales

   A microindentation technique was employed to study the fracture aspects of adherent scales and assess the adhesion strength originating from the addition of oxygen-active elements. Figure 6 illustrates the salient fracture characteristics of the scale formed on the Fe-10Al-1Hf alloy after being subjected to microindentation at various loads: (1) crack deflection by the HfO_2 particles at small loads; (2) preferential crack propagation in a direction approximately parallel to the outer scale surface, inferring anisotropic stress distribution; and (3) flaw initiation/propagation within the scale at a much smaller load than at the tortuous interface due to the blocking effect of the alloy protrusions adjoining the pegs.

CONCLUSIONS

The role of minority oxygen-active elements in the development of a tortuous alloy/scale interface, emanating from profuse inward oxide peg growth, and the ensuing amelioration of scale adhesion were discussed.

Criteria and Mechanisms of Peg Growth

The conditions and growth mechanisms of the two identified distinct types of pegs are summarized in the following. The general criteria outlined herein would be applicable, in principle, to various alloy systems.

Type I

The growth of this type, illustrated in Figures 1 and 2, is controlled by solute and/or oxygen diffusion in the alloy. It can be interpreted by the classical internal oxidation theory [10] at high temperatures; and, with certain modifications to incorporate the effect of enhanced diffusion of the mobile species along the particle/alloy interface at lower temperatures. In addition to satisfying the necessary conditions for internal oxidation, the precipitates should develop as a continuous "phase" capable of pinning the external scale, rather than isolated particles.

Type II

A typical morphology was illustrated in Figure 4. Its development can be rationalized by "short-circuit" diffusion in a host scale containing a dispersed phase. To develop such a tortuous interface in a predominantly anion-conducting scale, the two phases, which can coexist under local thermodynamic equilibrium, must be at least partially insoluble and non-uniformly distributed in each other, preferably non-interacting and most importantly, their anion transport rate must be significantly different.

Influence of Peg Growth on Scale Adhesion

The efficacy of surface roughness on the development of a highly tortuous interface to improve the adhesion between two dissimilar materials has been demonstrated in various practical applications: chemical etching of a two-phase polymer (e.g., ABS, acrylonite-butadiene-styrene) prior to deposition of thin metallic films; or, graphite fibres prior to incorporation in reinforced plastic or rubber composites [25]; adhesive bonding to metals through the formation of an extremely porous cellular anodic oxide film [26]; and plasma or ion etching of metallic substrates prior to deposition of ceramic coatings [27].

Development of this configuration imparts a greater adhesion strength between the scale and alloy for the following reasons: (1) providing intimate contact between the two phases; (2) relief of thermal as well as athermal stresses. The intermediate mixed alloy/oxide layer would gradually accommodate the mismatch in the specific volumes and thermal expansion coefficients; and (3) improved fracture toughness due to (a) the obvious longer crack propagation path, (b) large interfacial contact area associated with micro-dendritic lateral growth; and, (c) energy dissipation since stress relief is achieved by plastic deformation of the alloy protrusions adjoining the pegs, rather than fracture of the brittle scale.

REFERENCES

1. J. Stringer, Met. Rev., Vol. 11, pp. 113-128 (1966).

2. B. Lustman, Trans. Metall. Soc. AIME, Vol. 188, pp. 995-996 (1950).

3. H. Pfeiffer, Werkstoffe Korros., Mannheim, Vol. 8, pp. 574-579 (1957).

4. J. E. McDonald and J. G. Eberhart, Trans. Metall. Soc. AIME, Vol. 233, pp. 512-517 (1965).

5. J. E. Antill and K. A. Peakall, J. Iron Steel Inst., Vol. 205, pp. 1136-1142 (1967).

6. J. Stringer, B. A. Wilcox and R. I. Jaffee, Oxid. Met., Vol. 5, p. 11 (1972).

7. D. P. Whittle and J. Stringer, Phil. Trans. Roy. Soc. (London), Vol. A295, pp. 309-329 (1980).

8. H. Hindam and W. W. Smeltzer, J. Electrochem. Soc., Vol. 127, pp. 1622-1630 (1980).

9. H. Hindam and D. P. Whittle, J. Mat. Sci., Vol. 18, pp. 1389-1404 (1983).

10. C. Wagner, Z. Electrochem., Vol. 63, p. 772 (1959).

11. R. A. Rapp, Corros.-NACE, Vol. 21, pp. 382-401 (1965).

12. D. P. Whittle, et al., Phil. Mag. A, Vol. 46, pp. 931-949 (1982).

13. F. H. Stott, et al., Oxid. Met., Vol. 18, pp. 127-146 (1982).

14. F. H. Stott, et al., Sol. St. Ionics, (In press).

15. A. U. Seybolt, quoted in Metals Reference Book, Editor C. J. Smithells, 5th ed., Butterworths (1976).

16. C. J. Smithells and C. E. Ransley, Proc. Roy. Soc., Vol. A155, p. 195 (1936).

17. H. Hindam and D. P. Whittle, Oxid. Met., (In press).

18. H. Hindam et al., To be published.

19. Y. Oishi and W. D. Kingery, J. Chem. Phys., Vol. 33, p. 380 (1960).

20. A. E. Paladino and W. D. Kingery, ibid, Vol. 37, p. 957 (1962).

21. D. J. Reed and B. J. Wuensch, J. Amer. Ceram. Soc., Vol. 63, p. 88 (1980).

22. J. K. Tien and F. S. Pettit, Met. Trans., Vol. 3, p. 1587 (1972).

23. K. P. R. Reddy, J. L. Smialek and A. R. Cooper, Oxid. Met., Vol. 17, pp. 429-449 (1982).

24. H. Hindam and D. P. Whittle, J. Electrochem. Soc., Vol. 129, pp. 1147-1149 (1982).

25. S. S. Schwartz and S. H. Goodman, Plastic Materials and Processes, Van Nostrand Reinhold (1982).

26. J. D. Venables et al., Appl. Surf. Sci., Vol. 3, pp. 88-98 (1979).

27. M. A. Bayne, et al., Proc. of the 1st Conf. on Advanced Materials for Alternative Fuel Capable Directly Fired Heat Engines, Editors: J. W. Fairbanks and J. Stringer, U.S. DOE and EPRI, pp. 629-657 (1979).

Microstructure-Property Relationships

Chairpersons: R.J. Gray, P.M. French, M.R. Krishnadev, and J.H. Richardson

METALLOGRAPHIC APPLICATIONS IN INTERFACIAL FREE ENERGY STUDIES

T. A. Roth*

ABSTRACT

The use of metallographic techniques in the determination and interpretation of experimental data and to observe changes in the surface and grain boundary free energies of pure metals and alloys is discussed.

Using the zero-creep method for the study of the interfacial free energies, the determination of absolute free energy values relies greatly on the use of several metallographic techniques. These include numerous examinations and measurements of the surface of the test specimens using light microscopy, measurements of dihedral angles of thermally-etched grain boundary grooves formed at the intersection of grain boundaries with the surface using scanning electron microscopy, and the determination of the distribution of alloying elements in the specimen using scanning electron microscopy and energy-dispersive x-ray analysis.

INTRODUCTION

The interfacial free energies of pure metals and alloys play an important role in numerous properties, characteristics, and phenomena associated with the materials. The importance of the interfacial free energy is readily seen in casting, brazing and welding, sintering, surface modifications,

---

* Department of Chemical Engineering, Kansas State University, Manhattan, Kansas 66506 USA.

such as carburizing and plating, oxidation and corrosion, and fracture, to list but a few areas of general interest. The list can obviously be extended to include most, if not all, cases where surfaces are brought together to form a larger or more complex object. At the other extreme, where objects lose their usefulness, in many cases due to the formation of two new surfaces by some failure mechanism, the significance of the interfacial free energies is equally apparent.

The interfacial free energies of solid pure metals and, more significantly, those of the commercially important alloys should, therefore, find considerable application in the design of metal components and systems, in the evaluation of their continued usefulness prior to possible replacement or abandonment, and in the analyses of the unfortunate cases of failure of the components or systems.

The interfacial free energies being considered in this work are: (1) the free energy at the interface of the solid metal surface and the gaseous environment surrounding the material, i.e., the surface free energy, and (2) the free energy at the interface separating adjacent grains within the solid metal, i.e., the grain boundary free energy.

The surface and grain boundary free energies may be reliably determined experimentally by simultaneously using the zero-creep technique for small diameter wires and measurements of the dihedral angle at the base of thermally-etched grain boundary grooves, as described below.

EXPERIMENTAL PROCEDURE

Although this paper is primarily concerned with considering the use of metallographic techniques in the determination and interpretation of the interfacial free energy values, a brief review of the general experimental procedure is appropriate and will provide a convenient method for introducing and beginning the discussion of where the various metallographic techniques are applied in the general scheme of the studies. More detail regarding the experimental procedure, as well as the results for specific materials studied by this investigator are presented elsewhere [1,2] and need not be repeated here.

The materials tested in the studies were in the form of fine-diameter wire. Test specimens were cut from the spool of wire, a loop tied at one end of the wire was used to hang the specimens in the furnace, and a weight of the same material as the test specimen was tied to the bottom of the wire. During the creep anneal, the added weight causes the wire to elongate, while the surface forces cause the wire to shorten. The purpose of the zero-creep test is to determine the tensile load that just balances the opposing surface forces and, therefore, results in neither elongation nor contraction of the wire. Estimates of the balance load for zero-creep, made from data for the surface tension value for the liquid metal, were used when available to determine the amount of weight to be suspended from the wires. Knots were tied about 5-cm apart with several knots being tied in each wire to serve as gauge markers. Then, the wires were put into the furnace for an initial anneal, at or near the test temperature, in order to soften and straighten them prior to the initial measurement of the distance between gauge marks. The time of the initial anneal varied with the different materials and was experimentally determined to be adequate to bring the metals into the steady-state creep range.

Testing of the wires was carried out in a controlled atmosphere within a sealed furnace tube. After the wires had cooled from the initial annealing temperature, they were removed from the furnace and the gauge lengths of the several specimens per wire were carefully measured. The knots tied into the wires were found to sinter at the elevated temperatures used in the creep experiments and did not take part in the elongation or shrinkage of the wires. In most cases, the initial anneal was adequate to allow the grains to grow to the apparent equilibrium size and shape and the wires, when examined under a microscope, displayed the desired bamboo structure in which the wire is made up of a series of grains, the grain boundaries of which are perpendicular to the axis of the wire and the width of each grain equals the diameter of the wire, as shown in Figure 1.

Gauge length measurements were made on each wire before and after the creep anneals using a filar eyepiece micrometer microscope mounted for vertical measurement and capable of measuring to ± 0.000127-cm (0.0005-in.). The optics of this microscope resulted in a magnification of 40X being used during the various length and change-in-length measurements.

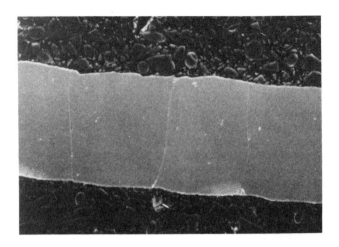

Figure 1. Typical appearance of fine wire specimens after creep showing the thermally-etched grain boundary grooves and bamboo structure. Nominal diameter of the wire is 0.0127 cm.

The grains were generally observed to have retained the bamboo structure developed during the initial anneal and no apparent grain growth was noted. The number of grains per unit length (n/l) was counted for each final gauge length measurement and the average value of (n/l) was determined. The strain for each portion of the wire was plotted against the effective weight, determined by cutting the wire at the mid-point of the gauge length and weighing the wire and weight below this point, as shown in Figure 2. The balance load for zero-creep, W_0, was determined graphically from the least-squares line of the plot of load versus strain, as shown in Figure 3.

The relationship between the surface free energy, F_s, the grain boundary free energy, γ_{gb}, and the load for zero-creep is given by the expression

$$W_0 = \pi r [F_s - \gamma_{gb} r(n/l)] \qquad (1)$$

where r is the radius of the wire.

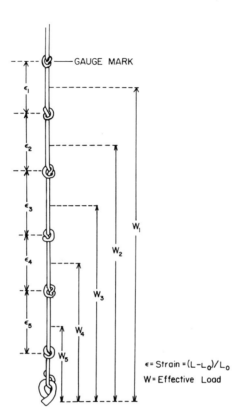

Figure 2. Sketch of wire specimen showing how the effective loads, W, and strains, ε, were determined.

Smith [3] has shown that after a sufficiently long anneal, the dihedral angle, Ω_s, at the base of the grain boundary groove is characteristic of the surfaces and grain boundaries involved, and that the size of the angle is determined by the equilibrium configuration of the surface and grain boundary free energies. At equilibrium, assuming that the surface free energy is independent of crystallographic orientation, the surface and grain boundary free energies are related by the expression

$$\gamma_{gb} = 2F_s \cos(\Omega_s/2) \qquad (2)$$

as shown in Figure 4.

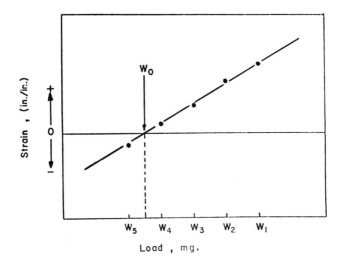

Figure 3. Typical experimentally determined load versus strain plot used to find W_0, the balance load for zero-creep.

If the mean dihedral angle, Ω_s, is obtained by accurate observations of the groove angle at the intersection of the grain boundary with the free surfaces, substitution of Eqn. (2) for γ_{gb} into Eqn. (1) results in a single expression for the average surface free energy,

$$F_s = W_0 / \{\pi r[1 - 2(n/1)r \cos(\Omega_s/2)]\} \qquad (3)$$

The grain boundary groove angles were determined by scanning electron microscopy. Examination of the wires under the microscope showed that at 150-200X the grain boundary groove angles were defined clearly enough for direct measurement. The photographs of the groove angles obtained with the scanning electron microscope were enlarged prior to angle measurement. Numerous groove angles were photographed and measured for each wire. The median value for each test was taken to be the desired groove angle for that test.

Substitution of the experimental parameters into Eqns. (3) and (2) gave the absolute values of the surface and grain boundary free energies of the wire specimens for each creep test.

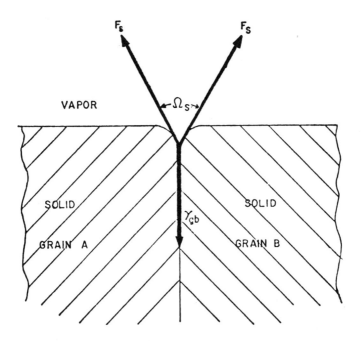

Figure 4. Relationship between surface free energy, F_S, and grain boundary free energy, γ_{gb}, at equilibrium.

DISCUSSION OF METALLOGRAPHIC APPLICATIONS

With the above overview of the experimental procedure in mind, the specifics of the application of the various metallographic techniques should now be readily apparent.

Initial examination of the material under investigation, following the initial anneal, involved the use of the low-power (40X) of the measuring microscope, not only to make initial gauge-length measurement, but also to see that the gauge marker knots had sintered, that the surface of the wires had not oxidized, and that the metal had recrystallized with the desired bamboo structure being formed. Spontaneous thermal-etching should have occurred during the initial heating of the material and no chemical etching is required in the procedure. Following the creep anneal, the measuring microscope was used again to determine the change in distance between the gauge markers, to measure the radius of the wire, and to count the number of grains per unit length.

An improvement to this procedure, awaiting adequate funding, involves the use of radiography to measure the change in length between the gauge markers while the test materials are at the creep temperature. Using a series of radiographs of the gauge length to determine the change in length, the creep rate of the material under the conditions of test could be accurately determined as well. As currently performed, measurements of the distance between gauge markers are done outside of the furnace, at room temperature. Unless considerably more time can be expended per test condition, determination of creep rates is not practical under current practice due to the slow heating and cooling rates of the furnace.

Scanning electron microscopy was used to examine the thermally-etched grooves that were formed during the anneal at the intersection of the grain boundary with the surface wire. Direct angle measurements were made from enlarged photographic prints and the average dihedral angle was determined from many separate angle measurements made on the wire specimens of each individual testing condition. The angle measurements of the grain boundary grooves on the wires composed of the metals of earlier studies, prior to the availability of scanning electron microscopes, were made from enlargement of micrographs obtained using a metallograph. Figure 5 is an example of such earlier data.

As the types and capabilities of the analytical instrumentation have expanded, more detailed, as well as a widening range of data has become available for the characterization of interfaces. Most recent investigations of interfacial free energies have included the examination of the concentration and/or distribution of foreign elements in the base material. The ability to determine the distribution and concentration of impurity or alloying elements is of great value in attempting to explain such phenomena as unexpected variations in measured surface and grain boundary free energies, as related to the subject being considered in this paper, and in confirming suspected cases of segregation of elements to the surface and/or grain boundaries, which can have a significant effect on the interfacial free energies as well. This investigator has examined the distribution of the alloying elements in two titanium alloys using scanning electron microscopy and energy-dispersive x-ray analysis.

Interfacial Free Energy / 289

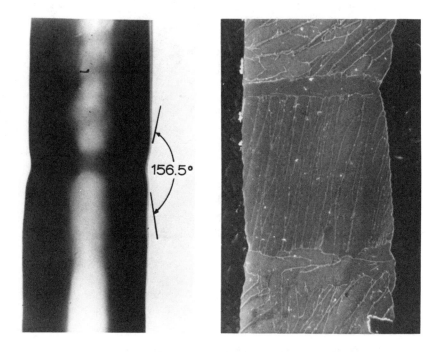

Figure 5. (Left) Measurement of angle at thermally-etched grain boundary groove.

Figure 6. (Right) Scanning electron micrograph of Ti-6Al-4V wire specimen showing alpha-beta microstructure. Etched with Kroll's etch.

Figure 6 shows a segment of a Ti-6Al-4V wire that was mounted, ground, polished, and etched with Kroll's reagent. Figures 7 and 8 compare the distribution of the alloying elements, as well as that of the impurity elements, in the alpha-beta alloy as obtained from a point to point x-ray scan of the wire. Figure 7 records the analysis of a point in the hcp alpha phase while Figure 8 is that for a point in the bcc beta phase. As should be expected, the alpha stabilizing aluminum is seen to be mainly found in the alpha phase region and the beta stabilizing alloy element vanadium, as well as the beta stabilizing impurity elements iron and silicon, are found to be concentrated in the narrower beta regions of the microstructure.

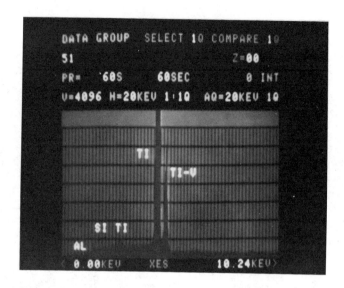

Figure 7. Distribution of elements in the alpha phase of Ti-6Al-4V specimen as determined by energy dispersive x-ray analysis.

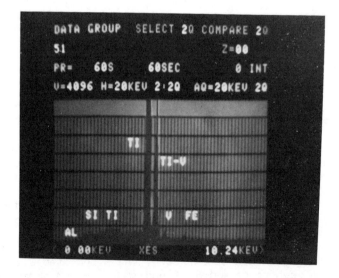

Figure 8. Distribution of elements in the beta phase of Ti-6Al-4V specimen as determined by energy dispersive x-ray analysis.

Analytical techniques and instrumentation available for determining the distribution and concentration of the elements in this and related research have been extensively reviewed and discussed in the literature [4-6], to which the interested reader is referred.

CONCLUSIONS

The use of metallographic techniques in one investigator's approach to the study of interfacial free energies has been considered and discussed. Greater use of the most recent analytical instrumentation and techniques available for the examination and characterization of interfaces will undoubtedly provide more detail and a broader scope to results of future investigations.

REFERENCES

1. T. A. Roth, "The Surface and Grain Boundary Energies of Iron, Cobalt and Nickel," Materials Science and Engineering, Vol. 18, pp. 183-192 (1975).

2. T. A. Roth and P. Suppayak, "The Surface and Grain Boundary Free Energies of Pure Titanium and the Titanium Alloy Ti-6Al-4V," Materials Science and Engineering, Vol. 35, pp. 187-196 (1978).

3. C. S. Smith, "Grains, Phases, and Interfaces: An Interpretation of Microstructure," Transactions AIME, Vol. 175, pp. 15-52 (1948).

4. L. E. Murr, Interfacial Phenomena in Metals and Alloys, Addison-Wesley Publishing Company, Reading, Massachusetts (1975).

5. J. M. Rigsbee, "A Technical Assessment of Electron Optical Systems," Journal of Metals, Vol. 33, No. 3, pp. 13-19 (1981).

6. R. Kossowsky, "Designing an Analytical Microscopy Laboratory," Journal of Metals, Vol. 35, No. 3, pp. 47-53 (1983).

RELATIONSHIPS BETWEEN KNOOP AND SCRATCH
MICRO-INDENTATION HARDNESS AND
IMPLICATIONS FOR ABRASIVE WEAR

Peter J. Blau*

ABSTRACT

Micro-indentation hardness test methods are an important tool for the evaluation of thin metallic layers and coatings. Both quasi-static and scratch indentation methods are currently in wide use. An investigation was conducted on samples of Cu, Fe, Sn, Cd, Ni, Co, AISI 1010 steel, AISI 52100 steel, CDA 638 bronze, CDA 688 bronze, and Nitinol to study the relationships between scratch and Knoop micro-indentation hardness numbers. Indenter loads between 0.0098 and 0.196 N (10-200 g) were used. A standard Knoop indenter was used for vertical testing and a 90 degree cone was used for scratch testing on a commercial scratch testing machine. Correlations between vertical and scratch hardness numbers varied with the testpiece material and the applied load. Microstructural features of the scratches were studied to analyze the cause of these variations. The implications of these variations for abrasive wear/microhardness number correlations are discussed.

INTRODUCTION

Micro-indentation tests of various types have been developed over the years to measure what is commonly referred to as "hardness". Understandings of hardness are for the

---

* Metallurgy Division, National Bureau of Standards, Washington, DC 20234 USA.

most part incomplete, and it seems probable that hardness is, in fact, only a broad class of surface mechanical behavior of materials, not an intrinsic single property. For example, the definition "resistance to penetration" is inadequate because it fails to address three central questions which often arise during testing practice:

(1) penetration by what?
(2) penetration of what?
(3) penetration under what circumstances?

Question (1) relates to the geometry and scale of the indenter as well as to its composition (i.e., the "hard" indenter itself could deform during a test depending on the testpiece). Question (2) indicates that some materials, such as soft polymers or ductile, low-melting point metals, may respond physically much differently than structural and refractory metals under the same indenter geometry and loading rates. In addition (3) specifically addresses the problems of indenter application rate (e.g., rebound, dwell time, stress relaxation, elastic recovery) and environmental sensitivity (e.g., indentation-produced cracks could be dependent upon the ambient conditions of temperature and humidity in some materials). Question (3) is also a central concern in the present study which addresses the problem of how relative material "hardnesses" can differ if criteria are based on sliding (scratch) tests versus quasi-static (vertical) micro-indentation test methods. Because relative quasi-static hardness values have been a central feature of certain models of abrasive wear (e.g., Ref. [1,2]), correlations between abrasion resistance and hardness ratios may be changed if scratch hardness ratios (more akin to the abrasion process) were substituted.

In recent years scratch tests have gained increasing importance in the evaluation of thin coatings and platings (e.g., Ref. [3-5]). Furthermore, detailed research into the nature of microconstituents in composite and polyphase materials have necessitated testing mechanical response of tiny volumes of material (e.g., Ref. [6-8]). Glassy or amorphous metals research has also found a need to use micro-indentation techniques, especially where samples are thin and have little volume (e.g., Ref. [9,10]). In view of these and other burgeoning avenues of research, it was decided to begin an investigation of the relationships between the scratch and quasi-static response of several pure metals and alloys for two standard micro-indentation hardness measurement methods: the Knoop micro-indentation test and the 90 degree cone scratch test.

The question of converting values in one hardness scale to values in another often arises. This is not a simple problem, especially when dealing with micro-indentation hardness values obtained using various indenters, and with the light loads one might need for thin samples or small test areas.

Table 1 shows the geometrical bases for computing the areas (e.g., in micrometer^2) of micro-indentations for various hardness scales using measurements of residual (i.e., indenter removed) dimensions. From the magnitudes of the geometrical constants, it is plain to see that the same numerical values for "microhardness" using different scales would require quite different dimensions of the impressions, and precision of the measurement requirements would vary depending on the chosen scale.

Table 1. Equations for Microhardness Number Calculation

Name of Test (Fig. at bottom)	Type of Area Term	Geometric Constant (Note 1)	Dimension Size (micrometer) to obtain 100 kg/mm^2 at 100 g load
Vickers (a)	Facet Area	1854.4	d = 43.06
Knoop (b)	Projected area	14229.0	D = 119.3
PAH (note 2) (b)	Projected area	2000.0	D = 119.3, d=16.77
Brinell (c)	Projected area	1273.0	D_b = 35.68

Notes:

(1) Microhardness = constant x $\dfrac{\text{load (g)}}{\text{dimension (micrometer}^2)}$

(2) Projected area hardness using Knoop indentation. In the last column, D/d assumed to be 7.114/1.0 as for ideal indentation. PAH = 2000 (Load/(D·d))

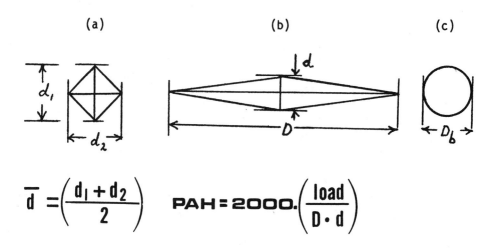

Not only could the geometrical volumes of material affected during indentation be quite different for different scales, but the stress patterns below the indenters could vary widely. Both the lateral and normal (i.e., to the test surface) dimensions could be much different, thereby producing a different proportion of near-surface and bulk material properties to be samples. When one considers both the differences in stress distributions below variously shaped indenters [e.g., Ref. 11 to 13], and the differences in elastic and plastic response to a given stress distribution between one material and another, it is necessary that hardness scale interconversions should be done for each material separately.

Throughout the vast engineering literature involving abrasive wear resistance of metals, hardness has been used frequently. From the time of Kurschov [1], hardness ratios have been used in attempting to correlate relative abrasion resistances among metals. The current investigation, which relates vertical to scratch hardness values for several scales, raises the following question: If hardness scale interconversions are material dependent, which scale(s) would be most likely to correlate best with relative abrasion resistance? A brief moment's reflection should indicate that scratch hardness, being in many respects akin to the abrasion process, should correlate best. In fact, many investigators have used what amounts to a scratch hardness tester to study the mechanistic details of abrasion [e.g., Ref. 14 to 16]. The current investigation of relationships between quasi-static and scratch micro-indentation hardness numbers can therefore be considered as one aid in relating quasi-static micro-indentation hardness numbers with the mechanisms of metal abrasion. The three-fold purpose of this investigation was then:

1. To obtain a correlation between Knoop micro-hardness numbers (KHN) and scratch widths on a range of metals and alloys at relatively light applied loads (10 to 100 g)

2. To examine lateral elastic shape recovery effects of Knoop indentations to see whether improved correlations with scratch widths might be obtained; and,

3. To indicate how abrasive wear models based on relative quasi-static hardness might be affected if scratch test data, rather than vertical micro-indentation test results, would be used

EXPERIMENTAL PROCEDURE

Metals and alloys were tested using commercial micro-indentation hardness testers. Each machine was periodically calibrated by the manufacturer and by using coupons of a narrow range of microhardness numbers. Ocular measurement factors were periodically checked with an etched stage micrometer scale. The scratch tester geometry is shown in Figure 1. It too was a commercial machine. The speed of rotation was between 0.09 and 0.15 cm/s depending on scratch distance from the center of rotation. Despite the conical tip geometry of the scratch indenter, all measured scratch widths were such that only the hemispherical portion of the indenter (radius 76.2 micrometer) was in sliding contact with the testpiece.

Compositions, treatments, and average grain intercept lengths (including annealing twins) of the metals and alloys are given in Table 2. Samples were all metallographically polished by mechanical methods except for the Cu which was electropolished in a 50% (aq.) phosphoric acid solution. Care was taken to polish sufficiently to remove residual damage from each preceding step in reaching the final-polished condition. Microscopical inspection of as-polished surfaces displayed no apparent sample preparation artifacts in the areas chosen for testing.

Table 2. Metals and Alloys Used in this Investigation

Metal or Alloy	Composition (Weight Percent)	Avg. Grain Size** Micrometers	Condition
Cu	99.99+ (OFHC)	503	Vacuum-annealed flat coupon
Fe	99.99+	95.0	As-rec'd rod., polycrystalline
Ni	99.97+	44.2	As-rec'd rod., polycrystalline
Co	99.99+	53.9	As-rec'd rod., finely twinned.
Mo	99.95	8.1	As-rec'd rod., grain size is width of long, narrow grains
Cd	99.999	275.1	As-rec'd rod., polycrystalline
Sn	99.999+	353.6	As-rec'd rod., polycrystalline
638 Bronze*	94Cu, 3Al, 1.8Si, 0.4Co, 0.5Zn, 0.1Ni 0.05Fe	3.5	Mach'd from plate, annealed
688 Bronze*	73.5Cu, 22.7Zn, 3.4Al, 0.4Co, 0.05Pb, 0.05Fe	9.4	Mach'd from plate, annealed
1010 Steel*	0.10C, 0.45Mn, 0.04 max P, 0.05 max S, bal. Fe	8.5	Annealed flat strip
52100 Steel*	1.0C, 1.45Cr, 0.35Mn, 0.28Si, 0.025 max S and P, bal. Fe	5-10	Ball bearing cross-section. Fine, sphero. carbides

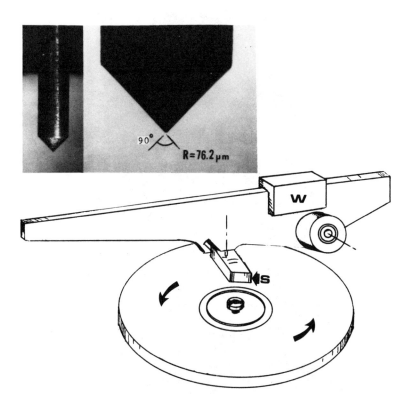

Figure 1. Scratch tester configuration. Insets show indenter tip geometry.

Table 2. (Continued)

Nitinol*   55Ni, 45Ti   Plate, annealed
                                         (500 C, 1 hr),
                                         Transition
                                         temp = 52 C

---

*   Nominal Compositions
** Includes annealing twins, when present

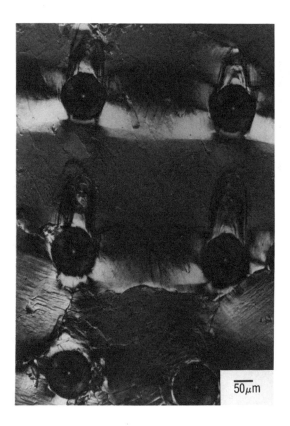

Figure 2. Indentations of the scratch tester tip with 100 g load on polished, unetched Sn. Original 89X.

Knoop indentations were made at least one long diagonal length away from scratches at the same locations as scratch widths were measured. The long diagonals were perpendicular to the scratching direction.

RESULTS

Before considering the difference between the KHN and scratch hardnesses of the various metals and alloys, scratch and quasi-static tests were performed using the scratch testing indenter. The results for six pure metals are shown in Table 3.

Figure 3. Indentations of the scratch tester tip with 100 g load on polished, unetched Cd. Original 89X.

The metals are listed in order of increasing PAH. There is no apparent correlation between the quasi-static impression diameter-to-scratch width ratio with PAH micro-indentation hardness number. Figures 2, 3 and 4 typify the variation in the lateral extent and nature of deformation in quasi-static indents on Sn, Cd, and Mo, respectively. From these preliminary

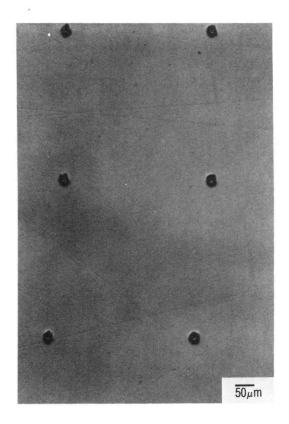

Figure 4. Indentations of the scratch tester tip with 100 g load on polished, unetched Mo. Original 89X.

results, it is apparent that even with the same indenter and load, the scratch width and microindentation hardness are not in the same proportion from one metal to the next. Therefore, when comparing results for differently-shaped indenters, such as the Knoop and the hemispherically-tipped cone indenters used here, one might expect even greater differences between scratch width and quasi-static hardness values.

Table 3. Relationship Among Projected Area Hardness (PAH), Scratch Width (W), and Indent Diameter (D*) Using the Same Indenter and Load on Several Metals

Metal	PAH (100 g) (kg/mm^2)	W (micrometer)	D* (micrometer)	D*/W
Sn	7.1	130.4	140.0 (1)	1.11
Cd	23.1	71.2	110.0 (1)	1.54
Ni	98.2	43.4	40.5 (2)	0.93
Fe	198.0	32.1	34.6 (2)	1.08
Mo	224.0	27.5	31.9 (2)	1.16
Co	250.0	27.5	31.1 (2)	1.13

Notes:

(1) Measured on an SEM photomicrograph. Average of 6 indents.
(2) Measured area, A, on an image analyzer. $D^* = (1.273A)^{1/2}$

Table 4 summarizes the micro-indentation data for the seven metals and five alloys used in this study. Each value is the average of at least five tests. The impression shape recovery index R is also provided to indicate why the correlation between KHN and PAH varies. The more different R is from 1.0, the greater the disagreement of the micro-hardness values.

Table 4. Summary of Scratch Width and Quasi-Static Micro-Indentation Hardness Data

Metal or Alloy	Load (g)	Avg. Scratch Width (micrometer)	Avg. KHN (kg/mm^2)	Avg. R*	Avg. PAH (kg/mm^2)
Sn	10	35.0	6.8	.93	6.3
	25	49.6	7.2	.91	6.6
	50	87.8	7.2	.95	6.9
	100	130.4	7.1	.96	7.1

Table 4. (Continued)

Metal or Alloy	Load (g)	Avg. Scratch Width (micrometer)	Avg. KHN (kg/mm^2)	Avg. R*	Avg. PAH (kg/mm^2)
Cd	10	29.5	20.3	.85	17.3
	25	38.1	21.0	.89	18.7
	50	61.5	22.8	.96	22.0
	100	71.2	23.1	.99	23.1
Cu	10	22.5	36.3	1.32	47.8
	50	45.1	34.1	1.19	40.6
	100	56.7	27.3	1.14	31.0
Ni	10	16.6	125	.82	102
	25	24.0	108	.81	87.2
	50	33.2	106	.87	92.1
	100	43.4	116	.86	98.2
Fe	10	11.6	285	.74	212
	25	17.4	298	.77	228
	50	25.5	298	.72	213
	100	32.1	281	.71	198
Co	10	11.3	324	.70	227
	25	16.4	294	.84	247
	50	21.4	341	.81	275
	100	27.5	279	.90	250
Mo	10	11.3	391	.87	341
	25	14.6	339	.81	275
	50	18.8	301	.76	228
	100	27.5	281	.80	224
1010 Steel	10	15.7	143	.85	122
	50	33.6	135	.90	114
	100	48.6	123	.80	105
688 Bronze	10	13.9	149	.85	127
	25	19.6	159	.78	124
	50	28.1	148	.90	133
	100	42.5	151	.80	120
638 Bronze	10	13.5	150	.81	121
	25	18.8	174	.89	155
	50	25.5	190	.75	142
	100	38.1	160	.84	135

Table 4. (Continued)

Metal or Alloy	Load (g)	Avg. Scratch Width (micrometer)	Avg. KHN (kg/mm^2)	Avg. R*	Avg. PAH (kg/mm^2)
Nitinol	10	9.7	237	1.57	371
	25	16.4	176	1.67	293
	50	22.8	137	2.30	320
	100	25.8	120	2.55	306
52100 Steel	25	-	1031	0.89	933
	50	13.4	1024	1.01	1031
	100	17.1	956	0.95	908

* R is the impression elastic shape recovery index.

Materials deformation due to indentation varied from grain-to-grain as the data in Table 5 (Fig. 5) demonstrate for the large-grained Cu. This orientation dependence of single point abrasion behavior has been long recognized (e.g., Ref.[16]); however, as Table 5 aptly demonstrates, inattention to this effect can lead to errors in scratch to vertical microhardness conversions, particularly for relatively large-grained or single-crystal materials. Figure 6 (upper right) also demonstrates this effect in Fe where the scratch width changes as a grain boundary was crossed by the slider at the center of the field of view.

Table 5. Grain-to-Grain and Twin-to-Twin Variations in the Response of Cu To Micro-Indentations and Scratches

Grain	Location	Load (g)	Scratch Width (micrometers)	KHN (kg/mm^2)	PAH (kg/mm^2)
A	Twin 1	10	22.9	32.9	45.5
	Twin 2		22.7	32.8	43.1
B	Twin 1		22.1	38.7	56.4
	Twin 2		23.8	36.5	44.1
	Twin 3		21.2	40.4	49.7

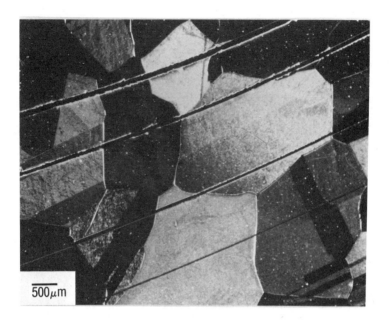

Figure 5. Scratches on the large grained surface of Cu with loads of 200, 100, 50, and 10 g (top to bottom). Original 25X.

Table 5. (Continued)

Grain	Location	Load (g)	Scratch Width (micrometers)	KHN (kg/mm^2)	PAH (kg/mm^2)
C		50	44.7	33.4	40.1
C			39.5	30.0	32.8
E	Twin 1		50.4	36.9	48.7
	Twin 2		45.6	35.9	41.0
F	Twin 1	100	53.2	26.1	32.5
	Twin 2		60.7	31.1	33.7
	Twin 3		55.1	27.5	31.2
			57.9	24.5	26.7

Knoop Versus Scratch Hardness / 307

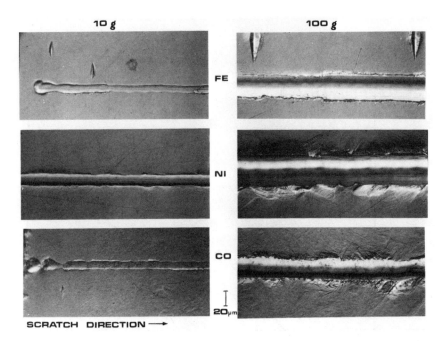

Figure 6. Effects of load on the indenter on scratch appearance on Fe, Ni, and Co.

Results of this study must be viewed only as approximate conversions in the context of demonstrating microhardness relationships for a given set of samples. Samples of similar compositions and microstructural characteristics may show similar behavior; however, in critical applications there is no better substitute than performing one's own correlations on specimens with the same composition, fabrication, and thermomechanical histories as those to be evaluated for the given application.

DISCUSSION

As results for the same indenter geometry used vertically and as a scratching tool showed (Table 3), the different mechanisms of deformation behavior from one material to another

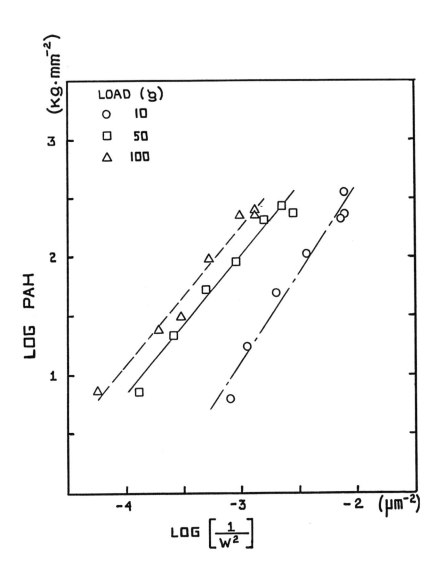

(a)

Figure 7. Relationship between scratch width, W, and projected area hardness (PAH) obtained from Knoop indentations on pure metals (a) and alloys (b).

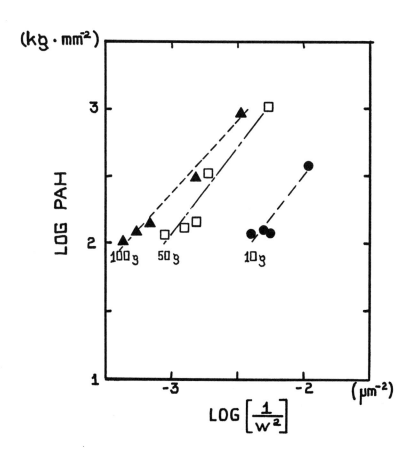

(b)

Figure 7. (Continued)-- caption at left)

affected the quasi-static impression diameter-to-scratch width ratios. With a horizontally translating indenter, the mechanisms of micromachining, ploughing, sliding friction, and others could all influence the resultant scratch characteristics. Add to this the variable of different indenter geometry and the problem's complexity is greatly compounded. The present study has not addressed this plethora of issues, but has focussed more modestly upon the data correlation itself and on its potential usefulness in analyzing abrasion behavior of metals.

Elastic shape recovery ratios, R, in Table 4, indicate that, in general, the indentations failed to achieve the same aspect ratio possessed by the tip of the Knoop indenter (i.e., R=1.0). Therefore, the KHN values were not felt to accurately reflect the true-projected areas of the indentations. Data for KHN was presented mainly because this test is widely used, and because numerical values would compare with those calculated using standardized methodology.

Projected area hardness, PAH, calculated using two-diagonal methods, was felt to be a more accurate scale to use in correlations with scratch width data in view of the above observation that R was not in general equal to 1.0.

Correlations between log (PAH) and log ($1/W^2$) were quite good for pure metals and reasonably good for the alloys at three loads, as Figures 7(a) and (b) show. This permitted estimates to be made for either scratch width or PAH given the other quantity. The good correlations are somewhat surprising in light of the differences between the crystal structures and deformational properties among the seven metals tested. Reasons for such good correlations are not clear, especially when one considers that different crystal structures, stacking-fault energies, twinning energies, indenter geometries, and sliding friction coefficients could conceivably be involved in any fundamental interpretation of the reasons for the correlation.

What is of more immediate interest here are the implications with respect to abrasion. Given the measured PAH, one could estimate the width (and, therefore, the approximate depth) of scratches produced at the same load by a hemispherical indenter of the type used herein. Such information would be useful in selecting coating thicknesses for experimental investigations, and it suggests that for other scratch indenter geometries, more representative of given service conditions, reasonably good correlations might similarly be obtained.

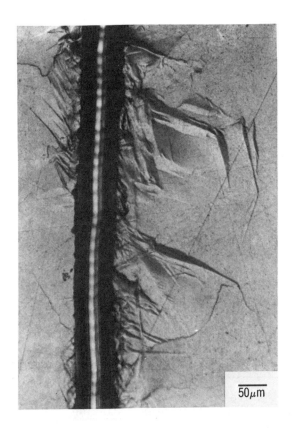

Figure 8. Gross lateral twinning damage flanking a scratch on Cd (100 g indenter load). Original 189X.

Of course, more than scratch depth alone is at issue here. Material immediately below the penetration area is deformed during the abrasion process, and the depth to which damage might extend is dependent on the materials and microstructures comprising the body which is being abraded. The subsurface stress distributions and the response of given materials to these distributions have to be considered. For analogy with lateral (surface) damage extent, consider the differences between Figures 5 and 8 in terms of the apparent zone of visibly deformed metal flanking the scratches. Nevertheless, the data in Table 4, in conjunction with additional information on the depth of disturbed material, may provide another tool for estimating the abrasion behavior of a host of metallic materials.

SUMMARY

The correlation between Knoop, modified Knoop (PAH), and scratch micro-indentation hardness numbers of twelve metals and alloys was observed to be load and material dependent. Using the scratch test indenter, the ratio of quasi-static impression diameter-to-scratch width varied among the seven pure metals tested demonstrating that scratch test data derive from different sets of microstructural properties than quasi-static data and cannot be simply explained in terms of indenter geometry. Correlations obtained between the quasi-static and scratch data may help to link abrasive wear studies to microhardness numbers in a more quantitative manner.

REFERENCES

1. M. M. Krushchov, "Resistance of Metals to Wear by Abrasion, as Related to Hardness," Proc. Conf. on Lub. and Wear, London, pp. 655-659 (1957).

2. K.H.Z. Gahr, "Relation Between Abrasive Wear Rate and the Microstructure of Metals," Proc. Conf. on Wear of Materials, Dearborn, MI, pp. 266-274 (1979).

3. C.C. Lo, "Microindentation and Microscratch Hardness Tests for Thin Electroplates," Plating, pp. 247-250, (March 1973).

4. P. Benjamin and C. Weaver, "Measurement of Adhesion of Thin Films," Proc. Roy. Soc., Ser. A, Vol. 254, pp. 163-176 (1960).

5. J. Ahn, K.L. Mittal, and R.H. MacQueen, "Hardness and Adhesion of Filmed Structures as Determined by the Scratch Technique," ASTM Special Tech. Pub. 640, pp. 134-157 (1978).

6. N.W. Thibault and H.L. Nyquist, "The Measured Knoop Hardness of Hard Substances and Factors Affecting its Determination," Trans. ASM, Vol. 38, pp. 271-348 (1947).

7. H. Buckle, "Progress in Micro-indentation Hardness Testing," Metal Rev., Vol. 4, pp. 49-99 (1959).

8. A.G. Atkins, "Topics in Indentation Hardness," Metal Sci., Vol. 16, pp. 127-137 (1982).

9. M. Rosen, et al., "Crystallization Kinetics of Amorphous Alloys by Ultrasonic Measurements," Proc. Rapid Solidification Processing Conference, Nat. Bur. Standards (1982, in press).

10. R.L. Freed and J.B. Vander Sande, "The Effects of Devitrification on the Mechanical Properties of $Cu_{46}Zr_{54}$ Metallic Glass," Met. Trans., Vol. 10A, p. 1621 (1979).

11. D. Tabor, The Hardness of Metals; Oxford Press (1951).

12. T.O. Mulhearn, "The Deformation of Metals by Vickers-Type Pyramidal Indenters," J. Mech. Phys. Solids, Vol. 7, pp. 85-96 (1959).

13. C.A. Brooks and B. Moxley, "A Pentagonal Indenter for Hardness Measurements," J. of Phys. E, Sci. Instru., Vol. 8, p. 458 (1975).

14. A.A. Torrance, "A New Approach to the Mechanics of Abrasion," Wear, Vol. 67, pp. 233-257 (1981).

15. L. Ahman and A. Oberg, "Mechanisms of Micro-Abrasion -- In-Situ Studies in SEM," Proc. Wear of Materials, Reston, VA, pp. 112-120 (1983).

16. R.P. Steijn, "Friction and Wear of Single Crystals," Mech. of Solid Friction, Elsevier Pub. Co., pp. 48-66 (1964).

THE EFFECTS OF AGING BETWEEN 704 AND 871 C ON
MICROSTRUCTURAL CHANGES IN INCONEL X-750

A. K. Sinha*, M. G. Hebsur*,
and
J. J. Moore*

ABSTRACT

High strength age-hardenable Ni-Cr-Fe alloy, Inconel X-750, is used extensively in both PWR and BWR as bolts, pins, spring components, etc. However, this alloy has experienced some intergranular stress corrosion cracking (IGSCC) in service. It has been reported that the SCC resistance of Inconel X-750 can be improved by thermal treatment in the carbide precipitation regime of 704-871 C. Therefore, the microstructural changes that occur when Inconel X-750 is thermally treated between 704 and 871 C for up to 200 h following a 2 h anneal at 1075 C have been studied in order to provide better understanding of the factors controlling the IGSCC resistance of thermally-treated Inconel X-750. Grain boundary chemistry was studied using electron and Auger microscopy.

INTRODUCTION

Inconel X-750 is a high strength, age-hardenable Ni-Cr-Fe alloy possessing good oxidation and corrosion resistance [1,2,3]. As a result of these properties, it is extensively used in both BWR and PWR installations as bolts, pins, and spring components [4,5]. A major problem encountered with the utilization of

---

* Mineral Resources Research Center, University of Minnesota, Minneapolis, MN USA.

these reactor components in the conventionally (i.e., triple) heat-treated condition is the occurrence of intergranular stress corrosion cracking (IGSCC). However, recent research in Japan has suggested that a single aging treatment after a high temperature solution anneal has resulted in improved IGSCC resistance due to improved carbide and intermetallic morphology [6].

The present work, therefore, was aimed at establishing the most appropriate aging sequence for this alloy which will produce a suitable precipitation morphology and distribution of secondary carbides and intermetallics. In this respect, this investigation examined the carbide morphology, grain-boundary Cr-depletion, grain boundary chemistry, and hardness profiles associated with a solution annealing treatment at 1075 C for 2 h followed by a single aging treatment between 704 and 871 C of Inconel X-750. Optical and scanning electron microscopy (SEM) and scanning Auger microanalysis (SAM) were used in the present study. Fractographs of single aged and cathodically hydrogen charged specimens were also observed with the SAM to determine the grain boundary chemistry and segregation of trace elements.

EXPERIMENTAL PROCEDURE

Electroslag-refined Inconel X-750 alloy was obtained from Huntington Alloy Products Division of the International Nickel Company in the form of 1.5-inch diameter rolled rod. The chemical composition of this alloy is given in Table 1. All specimens cut from the rod were subjected to various heat treatment cycles comprising solution treatment at 1075 C for 2 h, water quenching, followed by aging at 704, 760 and 871 C between 2 and 200 h followed by air cooling. This aging temperature range was chosen because of its coincidence with the nose of TTT curve for many Inconel alloys, the completion of all precipitation processes within a reasonable time at such temperatures and the possibility of grain boundary segregation of trace elements [7], e.g., S, P, B.

Table 1. Chemical Composition (wt. %)

Elements	C	Mn	Fe	S	Si	Cu
	0.06	0.19	7.82	0.001	0.40	0.37
Elements	Ni	Cr	Al	Ti	Nb+Ta	
	72.02	14.82	0.85	2.49	0.98	

Metallographic specimens were etched in an acid mixture of 80-ml HNO_3 and 3-ml HF for 1.5 min. Grain boundary segregation of chromium was determined by measuring the ratio of peak-to-peak heights of Cr and Ni using SAM on differently heat treated samples. Energy-dispersion X-ray analysis (EDXA) with the SEM and SAM point analyses were taken in order to determine the primary (MC-type) and secondary ($M_{23}C_6$-type) carbides, present within the grain and along the grain boundaries of the heat treated samples. In addition, small, round, notched fracture specimens 1.25-inch long x 0.145-inch diameter with an included angle of 60 degrees at 0.5-inch from one side were prepared for each heat treatment regime [8]. These were subjected to hydrogen charging [9] for 6-8 h at 62-85 C and were fractured in situ within the vacuum of the SAM sample changer in order to obtain an intergranular fracture area which could be examined for elucidating grain boundary chemistry and elemental segregation. Microanalyses were taken at different points within the grain and across the grain boundary using the SAM.

RESULTS

Figure 1 shows the aging curves which indicate that aging at 704, 760 and 871 C, each followed by air cooling, results in a peak hardness at 100 h, 10 h and 2 h, respectively. Figure 2 shows the optical microstructures of specimens aged for 2 and 200 h at 704, 760 and 871 C.

Aging at 871 C for as little as 2 h, initiated dissolution of the grain boundary Cr-rich $M_{23}C_6$ carbide with complete dissolution of these carbides being effected after 100 h, eventually resulting in secondary carbide-free zones after aging at 871 C for 200 h (Figure 2f). Figure 3 shows the

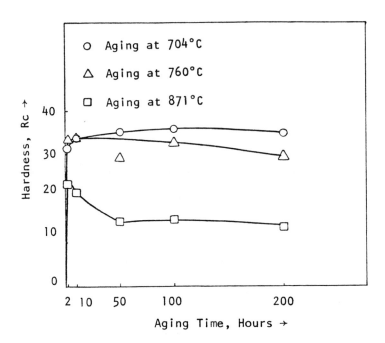

Figure 1. Hardness versus aging-time curve.

continuous precipitation of $M_{23}C_6$ type carbide by SAM for specimens aged for 2 h at 704 and 760 C. These grain boundary carbides started to break up and became discrete on aging after 50 h at 704 and 760 C (Figure 4).

Chromium depletion at the grain boundary was found to be a maximum after 10 h at 704 C and 50 h at 760 C (Figure 5). Grain boundary segregation of chlorine and possibly boron was observed for specimens aged at 704 C for 100 h and at 760 C for 10 h (Figures 6a and b); this was also accompanied by sulfur segregation for specimens aged at 704 C for 200 h (Figure 6c) and also for specimens aged at 871 C for 10 h (Figure 6d).

Figure 7 illustrates the EDXA analysis of a MC-type carbide particle. The darker central portion corresponded to TiC and the outer area (white background) was analyzed as a (Ti,Nb)C composite carbide.

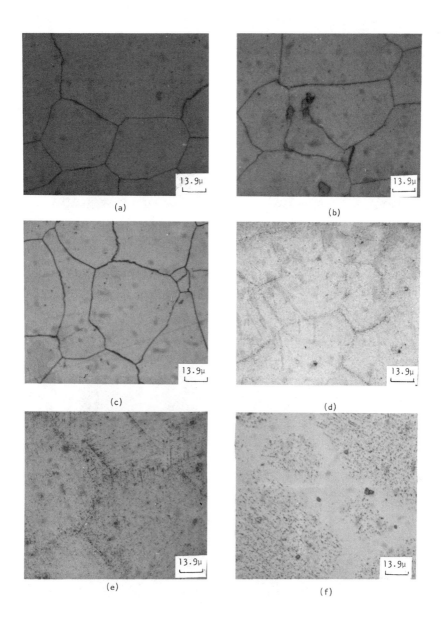

Figure 2. Inconel X-750 solution annealed, water quenched and aged at (a) 704 C, 2 h; (b) 704 C, 200 h; (c) 760 C, 2 h; (d) 760 C, 200 h; (e) 871 C, 2 h; and (f) 871 C, 200 h.

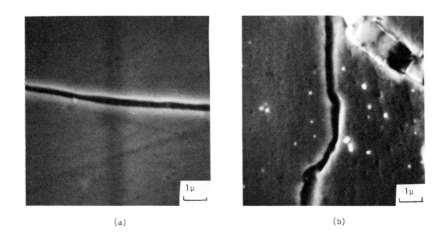

Figure 3. Solution annealed at 1075 C for 2 h, water quenched and aged for 2 h at (a) 704 C and (b) 760 C -- SAM microscopy.

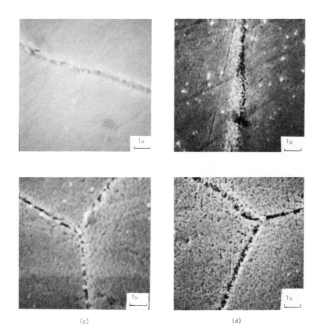

Figure 4. Inconel X-750 solution annealed, water quenched, aged at (a) 704 C for 50 h; (b) 704 C for 200 h; (c) 760 C for 50 h; and (d) 760 C for 100 h -- SAM microscopy.

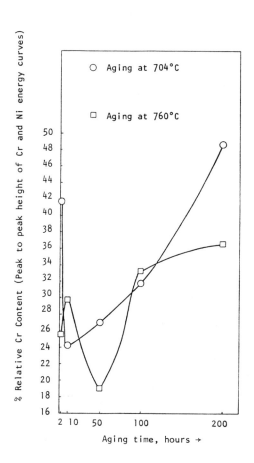

Figure 5. % relative Cr content (peak-to-peak height of Cr and Ni energy curves) versus aging time.

A fractograph of the specimen aged at 871 C for 50 h and subsequently hydrogen charged clearly demonstrates the sulfur and phosphorus segregation at the grain boundary present between the two dimple rupture areas (Figure 8).

DISCUSSION

Maximum peak hardness values are obtained as a result of the strengthening effects of the coherent intermetallic

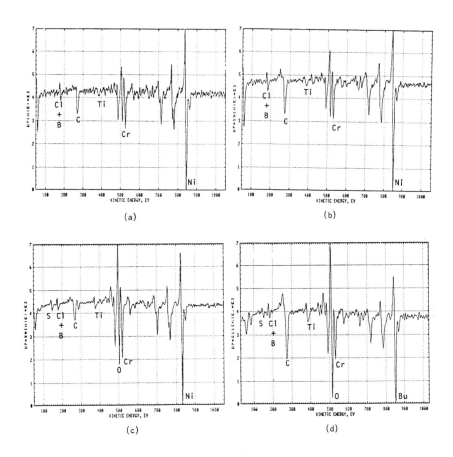

Figure 6. (SAM) point analyses at grain boundary of specimen aged at (a) 704 C for 100 h; (b) 760 C for 10 h; (c) 704 C for 200 h; and (d) 871 C for 10 h.

precipitates. The formation and growth of incoherent intermetallic precipitates causes reduction in the hardness values beyond the peak points. Since the solution temperature of Cr-rich $M_{23}C_6$-type carbides is about 871 C [1], air cooling from this temperature causes the dissolution of this carbide and finally a secondary carbide-free zone is established throughout the matrix after aging for 200 h at 871 C (Figure 2f).

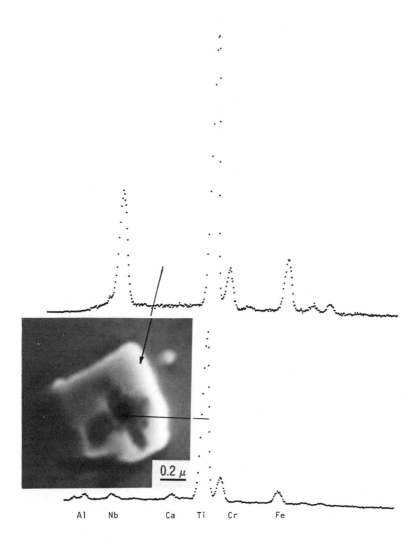

Figure 7. Solution annealed at 1075 C for 2 h, water quenched and aged at 704 C for 2 h -- SEM microstructure and analysis.

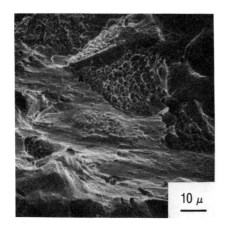 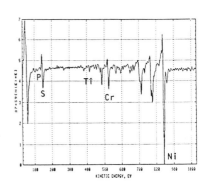

(a) (b)

Figure 8. (a) Fractograph of specimen aged at 871 C for 50 h followed by cathodic hydrogen charging, and (b) point analysis at grain boundary A. SAM fractography and analysis.

The continuous grain boundary $M_{23}C_6$-type carbides, which are present at the lower aging times, begin to break up and become discrete after 50 h aging at 704 and 760 C. As time progresses beyond 50 h, Cr replenishment of the grain boundary occurs and grain boundary carbides are subjected to restructuring to form more discrete particles.

The stages in the formation of the composite MC-type carbide comprise initial formation of TiC with subsequent enveloping by NbC (see Figure 8). This is probably due to more negative free energy change of formation of TiC compared with that of NbC at these aging temperatures [11].

A problem is encountered in interpreting chlorine and boron peaks since the Auger electron energy of LMM transition (181 eV) for chlorine is very close to the KLL transition for boron (179 eV). The identification of chlorine is often aided by peak shape [10] in that chlorine has a more symmetrical upper and lower peak compared with boron. Therefore, from the peak shape in Figure 6, it is evident that chlorine is present. However, bulk chemical analysis of the material using an inductively-coupled plasma (ICP) technique indicated a trace level of less than 100-ppm of boron. Therefore, it is apparent that boron is also segregated with chlorine but this needs to be confirmed.

The Cl (and B) segregation has been observed on the grain boundary for specimens aged at 704 C for 100 h and 200 h; at 760 C for 10 h and at 871 C for 10 h. In addition, sulfur has been found as a segregant on grain boundaries in the samples aged at 704 C for 200 h and 871 C for 10 h.

Sulfur and phosphorus grain boundary segregation was evident on the fractured (after cathodic hydrogen charging) surface between the two dimple rupture areas on specimens aged for 50 h at 871 C. It may be noted that grain boundary segregation, like the Cr-depletion at the grain boundary, is a function of thermal treatment. The grain boundary segregation of sulfur and phosphorus is known to adversely affect the stress corrosion cracking of nickel alloys [10]. Segregation of chlorine and boron may also have a similar effect but this needs to be verified.

The chromium profiles determined by peak-to-peak heights in the SAM (Figure 5) for aging at 704 C for different times were similar to those obtained for Inconel 600 by researchers at MIT [10]. The maximum depletion was reached after 10 h at 704 C and after 50 h at 760 C aging temperatures. Beyond this time, Cr diffused to the grain boundary region accompanied by an increase in width (and hence volume) of the depleted zone. However, the degree of Cr depletion at the grain boundary, being a function of thermal treatment, is controlled by alloy composition (especially C and Cr), diffusivity of Cr in the alloy, grain size and availability of carbon at the grain boundary.

The predominance of MC-type carbides on the fracture surfaces indicates that they are responsible for initiating the fracture. Use of a higher solution annealing temperature than 1075 C should be investigated in order to control the distribution of these primary carbides.

CONCLUSIONS

The microstructural analyses described herein allowed the following conclusions to be drawn with respect to single-stage aging of Inconel X-750 between 704 and 871 C.

1. Inconel X-750 alloy exhibits the maximum hardness in the air cooled samples at 704 C for 100 h, at 760 C for 10 h, and at 871 C for 2 h aging time. Rapid overaging at 871 C is associated with grain boundary $M_{23}C_6$ carbide dissolution, finally resulting in a complete secondary carbide-free zone after aging for 200 h.

2. MC-type carbides within the large dimples appear to initiate the fracture. A predominance of large dimples containing these MC-type carbides was observed on the fracture surfaces of specimens aged at 871 C compared with those aged at 760 and 704 C.

3. Breaking up of the continuous grain boundary $M_{23}C_6$ carbides into discontinuous and discrete particles started after 50 h aging at 704 and 760 C.

4. Metallographic study showed that grain-boundary segregation of chlorine and boron was present in specimens aged at 704 C for 100 h, and 760 C for 10 h. These elements were accompanied by grain boundary segregation of sulfur for specimens aged at 704 C for 200 h and at 871 C for 10 h.

5. Sulfur and phosphorus grain-boundary segregation was also found to occur on the fractured surface of specimens aged at 871 C for 50 h.

6. The Cr content in the grain boundary started to decrease after 2 h aging at 704 C and began to increase after 10 h aging at 704 C. Aging at 760 C resulted in an increase in grain boundary Cr after 2 h which decreased after 10 h and finally increased after 50 h aging.

REFERENCES

1. Inconel Alloy X-750, 4th edition, Huntington Alloys, Inc. (1979).

2. J. R. Kattus, Aerospace Maerials Handbook, Vol. 4, Code 4105, Mechanical Properties Data Center, Belfour, Stulen, Inc., pp. 1-28 (Revised June 1981).

3. E. L. Raymond, Met. Trans. of AIME, Vol. 239, pp. 1415-1422 (Sept. 1967).

4. W. J. Mills, Fractography and Materials Science, ASTM 733, pp. 98-114 (1981).

5. G. Boisde, et al., Effects of Environment on Material Properties in Nuclear Systems, The Institute of Civil Engineers, London, Int'l Conference on Corrosion, Longon, pp. 111-118 (1971).

6. S. Hattori, et al, Hitachi Research Laboratory, Hitachi Ltd.

7. T. R. Mags and R. S. Aspden, Effects of Thermal History and Microstructure on the SCC Behavior of Inconel X-750 in Aqueous Systems at 343-360 C, Westinghouse Research and Development Center.

8. Instruction Manual on Sample Breaker, PHI Model 10-520/15-520/15-530, Physical Electronics Industries, Inc., p. 13 (March 1977).

9. G. P. Airey, Metallography, Vol. 13, pp. 21-41 (1980).

10. R. M. Latanision and R. M. Pelloux, N.P. Research Project 1163-3, Massachusetts Institute of Technology (July 1980).

11. O. Kubachewski, et al., Metallurgical Thermochemistry (Pergamon Press), pp. 426-428 (1967).

CHARACTERIZATION OF NONMETALLIC INCLUSIONS IN STEEL

M. T. Shehata*, V. Moore*,
D. E. Parsons* and J. D. Boyd*

ABSTRACT

The nonmetallic inclusions in two AISI/SAE 1146 free-machining steels are characterized by optical microscopy, image analysis and electron microprobe analysis. A calcium deoxidation practice results in a sharply reduced number of oxide-based inclusions and a predominance of $2CaO \cdot Al_2O_3 \cdot SiO_2$ (gehlenite)-type mixed oxide inclusions. This correlated with reduced tool flank wear rates in machinability tests with high-speed steel and carbide tools, compared with the rate for Al-deoxidized steel where the oxide-type inclusions are predominantly Al_2O_3. Addition of Nb to the Ca-deoxidized steel for grain refinement, results in some Fe(Nb)-NbC eutectic phase at the center of the bar section, but this does not affect the machinability tests which were confined to the bar surface.

The results suggest that further improvements in machinability can be expected with further reduction in the number of oxide-type inclusions and a preponderance of anorthite ($CaO \cdot Al_2O_3 \cdot 2SiO_2$)-type inclusions.

INTRODUCTION

The machinability of steels is controlled primarily by the composition, shape, volume fraction and size distribution

---

* CANMET, Physical Metallurgy Research Labs,
  Ottawa, Ontario, Canada K1A 0G1.

of nonmetallic inclusions [1]. Large sulfide or oxide inclusions promote chip formation by determining the initiation strain and the scale of the local fracture events that occur during machining. At low cutting speeds, the machining process is characterized by the formation of a built-up edge, controlled by inclusions (e.g., MnS) that are plastic at the chip/tool interface temperature [2]. An important factor is the relative plasticity of the inclusion and matrix phases. At the high speeds (and temperatures) attained in machining with carbide tools, oxide inclusions appear to strongly influence tool wear [1,3]. Calcium-treated steels containing $CaO-Al_2O_3-SiO_2$ inclusion phases can produce longer tool life than conventional aluminum-killed steels where the predominant oxide-type inclusion phase is Al_2O_3. However, the composition of the mixed-oxide inclusions in the Ca-treated steels can be critical. For example, it is reported that anorthitic inclusions ($CaO \cdot Al_2O_3 \cdot 2SiO_2$) produce less tool wear, whereas gehlenitic inclusions ($2CaO \cdot Al_2O_3 \cdot SiO_2$) result in a similar tool life to that obtained with Al_2O_3 inclusions [4]. The effect of the different types of oxide inclusions on machinability is related to their softening point; the lower the softening point, the better the machinability.

The problem, then, is to develop a Ca-deoxidation practice which results in the appropriate type and dispersion of oxide inclusions for superior machinability, while maintaining the required transverse- and through-thickness ductility and toughness. To evaluate the results of the differences in deoxidation practice, quantitative metallography and microanalysis are required to characterize the composition and dispersion of inclusions.

In the present work, two commercial heats of fine-grained, continuously-cast, AISI/SAE 1146 grade free-machining steel were studied to determine the effects of Ca-deoxidation practice on machinability. One heat was prepared using standard aluminum-deoxidation practice, and the other employing a modified calcium-deoxidation practice with niobium additions for grain refinement. The characterization of the inclusions in the two steels is reported in this paper. The complete machinability evaluation will be reported later [5].

Table 1. Compositions of Steels, Wt. Pct.
(Spectrochemical Analysis)

Steel	C	Mn	S	P	Si	Al	Ca	Nb
C	0.39/0.42	0.83/0.87	0.080/0.095	0.015	0.26/0.27	0.013/0.019	NA	0.005/0.012
M	0.43/0.49	0.92/0.98	0.090/0.105	0.015/0.017	0.034/0.036	0.003/0.005	0.003/0.007*	0.040/0.049
AISI/SAE 1146	0.42/0.49	0.70/1.00	0.08/0.13	0.04 max	–	–	–	–

* Wet chemical analysis

NA -- Not analyzed

EXPERIMENTAL

Analyses of the two steels investigated are given in Table 1, along with the AISI/SAE 1146 specification. The control steel (C) was deoxidized with Al. The modified steel (M) was deoxidized with Ca, and Nb was added for grain refinement. Both heats were continuously cast as 150-mm billets. Steel C was supplied as normalized, 47.6-mm diameter rolled bar, and steel M as normalized, 50.8-mm diameter bar. The composition ranges shown in Table 1 for steels C and M represent the center-to-surface variation for each element.

The nonmetallic inclusions in the two steels were characterized by optical microscopy, electron microprobe analysis and automatic image analysis on longitudinal sections taken at the mid-radius of the bars. The general distribution of inclusions was first determined by optical microscopy, and representative inclusions were selected for microprobe analysis. Microanalysis was carried out on a CAMECA CAMEBAX MICRO electron microprobe. The elements present in each type of inclusion were first determined by energy-dispersive analysis, and then quantitative analyses were made by the wavelength-dispersive method. A QUANTIMET 900 automatic image analysis system was used with optical microscopy to determine the volume fraction, size distribution, and shape of the different types of inclusions in the two steels. For each measurement, approximately 200 fields, covering about 50-mm^2, and at least 1000 individual inclusion particles, were analyzed.

332 / Microstructural Science, Volume 12

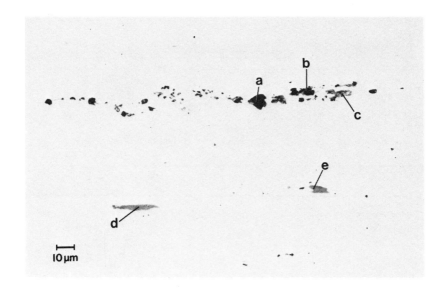

Figure 1. Steel C. Inclusion type: (a) Al_2O_3, (b) Al_2O_3/MnS, (c) Fe/MnS, (d) FeS/MnS and (e) FeS/MnS.

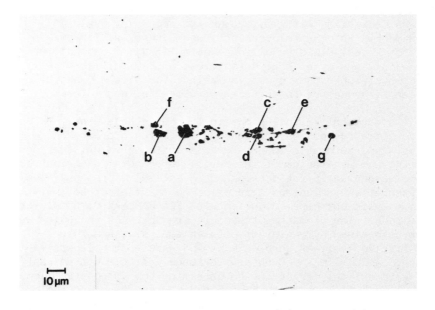

Figure 2. Steel C. Inclusion type: (a) Al_2O_3, (b) Al_2O_3, (c) Al_2O_3/MnS, (d) Al_2O_3/MnS, (e) Al_2O_3/MnS, (f) Al_2O_3/Fe and (g) Al_2O_3/Fe.

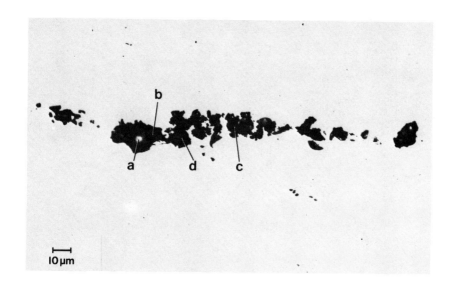

Figure 3. Steel M. Inclusion type: (a) CaS, (b) C_2AS, (c) C_2AS and (d) C_2AS. Note: C:CaO, S:SiO$_2$, A:Al$_2$O$_3$; subscripts refer to the number of each compound. C_2AS is gehlenite, CAS_2 is anorthite.

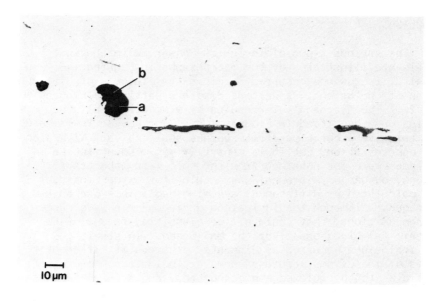

Figure 4. Steel M. Inclusion type: (a) CaS and (b) C_2AS.

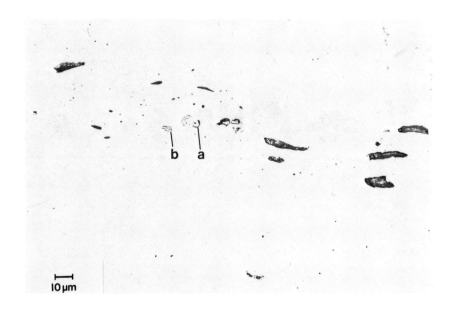

Figure 5. Steel M. Inclusion type: (a) and (b) Fe(Nb)-NbC.

RESULTS

The various types of inclusions observed in the two steels are illustrated by the micrographs in Figures 1-6. In steel C, inclusions occurred as elongated clusters of different types of inclusions, as well as discrete particles (Figures 1 and 2). The microanalysis results by element for all the inclusions identified in Figures 1 and 2 are given in Table 2. The relative accuracies of microprobe analyses vary with element and concentration, but it is generally ±5%. Quantitative analyses were not obtained from the soft x-ray spectra for carbon, oxygen or nitrogen, but a value for oxygen in steel C was calculated by difference, assuming negligible carbon and nitrogen. Although the inclusions varied over a wide range of compositions, they could all be classified as one of two types: (a) Al_2O_3-based, or (b) MnS-based. In steel M, the inclusions occurred as elongated clusters of different types of inclusions, equiaxed complex inclusions, discrete particles (both equiaxed and elongated), and a eutectic phase at the center of the bar section (Figures 3-6). The microanalysis results by element for all identified inclusions in

Table 2. Electron Microprobe Analysis of Inclusions by Element (Wt. Pct.) for Steel C (ref. Fig. 1 and 2)

Inclusion	Fe	Mn	Ti	Si	Al	Mg	Ca	S	O*	Type
1.a	1.38	0.78	0.90	2.18	41.32	0.66	5.35	0.17	47.06	Al_2O_3
1.b	15.54	28.60	0.05	0.09	27.06	0.09	0.76	18.51	9.3	Al_2O_3/MnS
1.c	82.62	8.70	0.10	0.31	0.60	0.04	0.14	6.59	0.91	Fe/MnS
1.d	10.60	52.63	-	-	-	-	0.07	34.52	2.18	FeS/MnS
1.e	11.90	52.17	-	-	-	-	N/D	33.77	2.15	FeS/MnS
2.a	9.06	1.02	0.99	0.63	43.17	0.48	5.20	0.12	39.32	Al_2O_3
2.b	1.58	11.01	N/D	0.02	45.12	0.05	0.40	7.16	34.66	Al_2O_3
2.c	36.50	20.07	N/D	0.08	24.32	0.08	0.24	12.98	5.72	Al_2O_3/MnS
2.d	17.69	17.99	0.01	0.08	29.47	0.16	1.40	11.49	21.71	Al_2O_3/MnS
2.e	18.12	20.81	0.05	0.10	26.29	0.18	2.74	12.93	18.78	Al_2O_3/MnS
2.f	14.75	0.61	0.39	0.25	39.64	0.41	3.92	0.34	39.69	Al_2O_3/Fe
2.g	25.3	7.65	0.42	0.17	29.86	0.30	1.61	5.12	52.33	Al_2O_3/Fe

* Oxygen calculated by difference
N/D -- Not detected

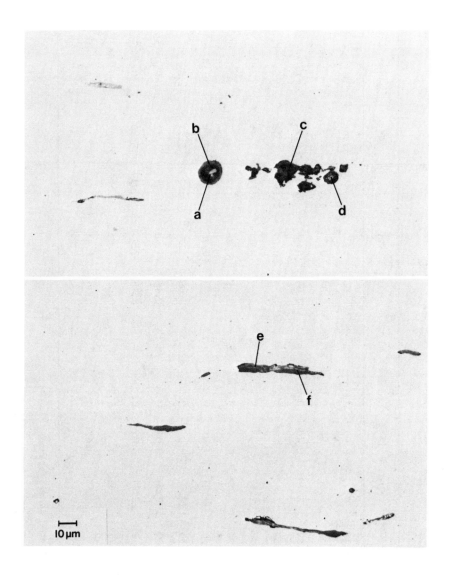

Figure 6. Steel M. Inclusion type: (a) CaS, (b) CaS/Fe, (c) CaS, (d) mixed oxide, (e) MnS and (f) MnS.

Figures 3-6 are given in Table 3. Four basic inclusion types were identified in steel M: (a) $CaO-Al_2O_3-SiO_2$-based, (b) CaS-based, (c) MnS-based and (d) an Fe-Nb solid solution/NbC eutectic (note that the latter phase was verified as being a

Table 3. Electron Microprobe Analysis of Inclusions by Element (Wt. Pct.) for Steel M (Ref. Fig. 3-6)

Inclusion	Fe	Mn	Ti	Si	Al	Mg	Ca	S	Nb	Type
3.a	2.10	0.90	N/D	0.01	0.03	N/D	54.44	41.96	0.05	CaS
3.b	3.49	0.37	0.17	10.61	15.14	8.54	22.21	0.68	N/D	C_2AS*
3.c	1.51	0.42	0.26	11.80	12.55	1.68	28.33	0.33	N/D	C_2AS*
3.d	1.44	0.54	0.48	10.66	14.20	1.29	29.04	1.44	N/D	C_2AS*
4.a	1.72	0.90	N/D	0.02	0.06	N/D	54.54	41.59	N/D	CaS
4.b	1.08	0.26	0.24	12.44	11.37	2.47	28.27	0.58	N/D	C_2AS*
5.a	5.11	0.03	-	-	-	-	0.01	N/D	85.77	Fe(Nb)-NbC
5.b	26.57	0.10	-	-	-	-	N/D	N/D	64.17	Fe(Nb)-NbC
6.a	2.47	1.02	N/D	0.01	0.04	N/D	53.73	41.03	0.01	CaS
6.b	33.04	3.89	N/D	0.12	0.02	N/D	39.33	34.61	0.06	CaS/Fe
6.c	6.83	3.17	0.03	0.15	0.63	N/D	47.67	37.90	0.02	CaS
6.d	21.26	0.32	0.54	2.19	31.61	14.99	1.40	0.06	0.12	Mixed oxide
6.e	3.11	57.60	-	-	-	-	0.09	36.20	0.09	MnS
6.f	10.59	51.87	-	-	-	-	0.07	33.32	0.14	MnS

* C_2AS = $2CaO \cdot Al_2O_3 \cdot SiO_2$ (gehlenite)
Fe(Nb)-NbC: Fe-Nb solid solution/NbC eutectic
N/D -- Not detected

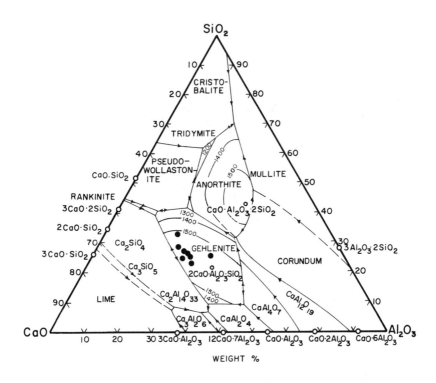

Figure 7. CaO-Al$_2$O$_3$-SiO$_2$ phase diagram [7] with compositions of steel M oxide inclusions shown (•).

carbide rather than a carbonitride by the absence of nitrogen in the soft x-ray spectra). The elemental compositions of the mixed-oxide-type inclusions expressed in terms of the weight percent contents of their respective oxides CaO, Al$_2$O$_3$, SiO$_2$, MgO, and FeO, and normalized to 100% are given in Table 4. The weight percent contents of CaO, Al$_2$O$_3$ and SiO$_2$, normalized to 100% are also given in Table 4, and plotted on the CaO-Al$_2$O$_3$-SiO$_2$ equilibrium phase diagram in Figure 7. Their location on the diagram indicates that the mixed-oxide-type inclusions have composition corresponding to gehlenite (2CaO·Al$_2$O$_3$·SiO$_2$).

Image analysis of optical microscope images is based on grey-level contrast between different phases or features. For steel C, it was possible to distinguish the two types of inclusions in this way (Figure 8), and to measure separately the dispersion parameters of the oxide-type (black) and the sulfide-type (grey). Table 5 gives the measurements of volume

Table 4. Compositions of Mixed-Oxide Inclusions in Steel M
(Wt. Pct. of Individual Oxides)

Inclusion	(CaO + Al$_2$O$_3$ + SiO$_2$ + MgO + FeO) = 100					(CaO + Al$_2$O$_3$ + SiO$_2$) = 100		
	CaO	Al$_2$O$_3$	SiO$_2$	MgO	FeO	CaO	Al$_2$O$_3$	SiO$_2$
3.b	31.53	29.15	20.31	14.43	4.58	38.94	35.99	25.07
3.c	43.94	26.26	24.56	3.09	2.15	46.37	27.72	25.91
3.d	44.44	29.32	21.88	2.34	2.02	46.47	30.65	22.88
4.b	39.40	25.30	22.60	10.81	1.89	45.13	28.98	25.89
*	40.68	19.37	29.42	8.30	2.23	45.47	21.65	32.88
*	43.16	28.00	24.30	2.71	1.83	45.22	29.33	25.45
*	44.02	23.89	25.99	4.56	1.55	46.88	25.44	27.68
*	45.16	23.01	26.83	3.09	1.92	47.54	24.22	28.24

* Inclusion analysis not reported in Table 3.

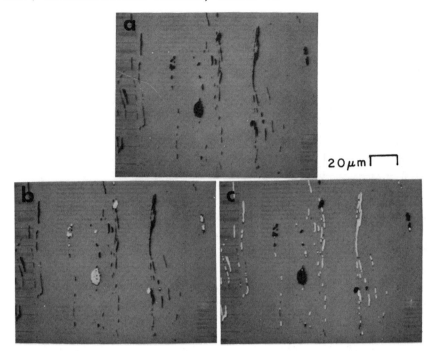

Figure 8. Steel C. (a) optical image, (b) detected image - oxide, (c) detected image - sulfides.

percent (equal to the area percent), number of inclusions/mm^2, mean length (in the rolling direction), mean width (transverse to the rolling direction), and mean shape factor (mean length/width ratio of individual inclusions). For steel M, the CaS-type inclusions could not be distinguished from the oxide-type by grey-level contrast. Hence, these two types of inclusions were counted together and the results reported as "oxides" (Table 5). The Fe(Nb)/NbC eutectic was eliminated from measurements by grey-level contrast and size discrimination. Thus, "sulfides" in Table 5 pertains only to MnS-type inclusions. The size distributions of the "oxides" in both steels are illustrated in Figure 9 and Table 6. The volume percent pearlite was also measured, and was found to be 55 $\pm$ 5% in steel C and 75 $\pm$ 5% in steel M.

DISCUSSION

This metallographic study has shown that there are important differences in the type and distribution of

Table 5. Results of Quantitative Metallography

Steel	Type of Inclusions	Vol. Pct.	No/mm^2	Mean Length, micrometer	Mean Width, micrometer	Shape Factor
C	Oxides	0.038 ± 0.003	24.5 ± 3.08	4.6 ± 0.4	3.6 ± 0.4	1.28 ± 0.05
C	Sulfides	0.81 ± 0.08	324.4 ± 36.4	10.6 ± 0.8	3.3 ± 0.3	3.00 ± 0.24
M	Oxides	0.028 ± 0.003	12.2 ± 1.2	5.3 ± 0.04	4.5 ± 0.4	1.23 ± 0.05
M	Sulfides	0.77 ± 0.06	231.5 ± 22.1	12.3 ± 0.05	3.2 ± 0.4	3.70 ± 0.23

± Indicates the value of the standard error of the mean.

Table 6. Size Distribution of Oxides, (No./cm^2)

Length, micrometer	Steel C	Steel M
0.7 - 2	612	244
2 - 4	710	244
4 - 6	564	301
6 - 8	270	207
8 - 10	125	114
10 - 12	54	53
12 - 14	37	17
14 - 16	20	13
16 - 18	15	12
18 - 20	5	--
>20	40	17
>30	5	3

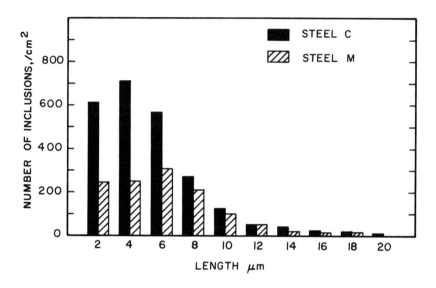

Figure 9. Distribution of length of oxide inclusions in both steels.

inclusions between the Ca-Nb-treated steel (M) and the standard Al-deoxidized steel (C):

1. The modified steel contains inclusions which can be classified as being based on $2CaO \cdot Al_2O_3 \cdot SiO_2$ (gehlenite), CaS, or MnS, whereas the inclusions in the control steel are based on Al_2O_3 and MnS.

2. The modified steel also contains an (Fe-Nb)-NbC eutectic at the center of the bar section. This eutectic has been observed to form to some degree in cast Nb-bearing steels at all Nb concentrations and cooling rates [6], although in continuously cast billets, it tends to concentrate in the equiaxed zone at the center. Calculations indicate that the eutectic should not occur at all in the current bar steels with Nb contents <0.035 wt. pct. [5] but, in any event, its presence at the center of the bar should have no effect on the machinability tests which were confined to the bar surface.

3. The modified steel contains far fewer oxide-based inclusions than the control steel, although the number of sulfide-based inclusions were similar in both (Table 5).

The machinability tests [5], with both high-speed steel (HSS) tools (30-70 m/min cutting speeds) and carbide tools (100-150 m/min), gave tool flank wear rates for the modified steel that were about one-half those for the control steel. This is attributed to the decreased number of oxide inclusions in the modified steel, and the replacement of Al_2O_3-based inclusions by $2CaO \cdot Al_2O_3 \cdot SiO_2$ (gehlenite)-based inclusions. Although it is not possible to separate the effects of these two factors, it has been shown elsewhere that gehlenitic inclusions give the same tool wear as Al_2O_3-based inclusions [4]. This suggests that the smaller number of oxide-based inclusions is responsible for most of the observed reduction in tool flank wear in machining the modified steel, and that further improvements in machinability ought to be obtained by replacing the gehlenite-type inclusions with the softer anorthite-type ($CaO \cdot Al_2O_3 \cdot 2SiO_2$) inclusions.

The carbon content of the modified steel (M) is higher than that of the control steel (C). Thus, steel M has a higher hardness (245 VPN vs. 220 VPN) and a higher volume pct. pearlite (75% vs. 55%). These differences affect the relative machinability of the two steels in a manner not related to the differences in inclusion content. Specifically, the tangential cutting force and the crater rates on HSS tools are higher for the modified steel, which are attributable to its higher strength.

CONCLUSIONS

It is demonstrated how quantitative metallography and microanalysis may be employed to characterize nonmetallic inclusions in steel. It is shown that replacing conventional Al-deoxidation practice for making AISI/SAE 1146 free-machining steel with a Ca-deoxidation practice results in significant changes to the type and distribution of inclusions. The inclusion content correlates closely with the results of machinability tests using high speed steel and carbide tools. The results suggest that further improvements in machinability can be expected with further reduction in the number of oxide inclusions and a preponderance of anorthitic-type ($CaO \cdot Al_2O_3 \cdot 2SiO_2$) inclusions.

ACKNOWLEDGEMENTS

The authors gratefully acknowledge the contributions of D. A. R. Kay and S. V. Subramanian to the preparation of this paper, and B. R. Casault and D. A. Munro for technical assistance.

REFERENCES

1. S. V. Subramanian and D. A. R. Kay, "Inclusion Engineering," International Symposium on Physical Chemistry of Iron and Steel Making, McMaster University Press, Hamilton, Ontario, pp. 30-40 (1982).

2. R. Milovic and J. Wallbank, "The Machining of Low Carbon Free Cutting Steels With High Speed Tools," The Machinability of Engineering Materials, American Society for Metals, Metals Park, pp. 23-41 (1983).

3. J. Fombarlet, "Improvements in the Machinability of Engineering Steels Through Modification of Oxide Inclusions," ibid, pp. 366-382 (1983).

4. T. Fujiwara, et al., "Influence of Calcium Inclusions on Machinability With HSS Tools," International Symposium on Influence of Metallurgy on Machinability of Steel, Iron and Steel Inst. of Japan, Tokyo, pp. 129-138 (1977).

5. D. A. R. Kay, S. V. Subramanian and J. Tlusty, McMaster University, unpublished research.

6. V. K. Heikkinen and R. H. Packwood, "On the Occurrence of Fe-NbC Eutectic in Niobium-Bearing Mild Steel," Scand. J. Metallurgy, Vol. 6, pp. 170-175 (1977).

7. A. Muan and E. F. Osborn, Phase Equilibria Among Oxides In Steelmaking, Addison Wesley Publishing Co. Inc. (1965).

MICROSTRUCTURAL ANALYSIS FOR SERIES 300 STAINLESS STEEL SHEET WELDS AND TENSILE SAMPLES

R. J. Gray*,
R. K. Holbert, Jr.** and T. H. Thrasher**

ABSTRACT

The hot cracking susceptibility of some AISI 300 series stainless steel welds can be minimized by the presence of a small amount of delta ferrite in the microstructure. The identification and location of the delta ferrite in the microstructure of rapidly cooled gas tungsten arc welds of 0.25 mm thick stainless steel sheet require a highly sensitive procedure. A similar challenge is involved in the resolution of strain-induced martensite in the same material. Since martensite significantly affects the mechanical properties, the resolution and identification of the martensite in various stages of tensile specimen preparation is important. The detection of martensite in the as-received sheet, and in the tensile specimen before and after the tensile tests, is vital in the final analysis of the tensile data.

Magnetic etching, which involves the use of an iron colloid solution in a controlled magnetic field, meets many of the demands for resolving these ferromagnetic features in the microstructure. The technique

---

* Metals and Ceramics Division, Oak Ridge National Laboratory, Union Carbide Corporation, Nuclear Division, Oak Ridge, Tennessee 37831 USA.
** Development Division, Oak Ridge Y-12 Plant, Oak Ridge, Tennessee 37831 USA.

produces an analog pattern with the colloid that is observed in-situ with the conventionally etched microstructure and offers high resolution analysis of the ferromagnetic delta ferrite and martensite in the paramagnetic austenitic matrix. The technique, which can be applied up to the upper limit of the magnification of the optical microscope, is described and demonstrated pictorially.

INTRODUCTION

Normally, the structure of wrought or annealed 300 series stainless steels is primarily austenitic. Delta ferrite, the high temperature, body-centered-cubic (bcc) structure of iron, can be retained in the microstructure of some austenitic stainless steels during the rapid solidification experienced in most fusion welding processes. Furthermore, a martensitic transformation can occur in some 300 series stainless steels when they are plastically deformed at low temperature.

Neither the delta ferrite, nor the strain-induced martensite, in thin sheet samples of the austenitic stainless steels can be easily analyzed. However, the magnetic etching technique has been a useful metallographic tool to evaluate these structures in 32-mm thick plate samples [1,2,3].

A major problem experienced in welding some austenitic stainless steels is the formation of hot cracks in and around the fusion zone during solidification. Hot cracking is caused by an accumulation of impurities during the weld solidification generally at the grain or substructure boundaries. The impurities form a low solidus temperature phase which weakens the boundaries. The thermal stresses occurring during solidification are sufficient to cause hot tearing [4,5].

Experience [6] has shown that small amounts of delta ferrite retained in the fusion zone upon solidification can inhibit hot cracking. The impurities in the fusion zone may collect at the boundaries if the weld solidifies as primary austenite (austenite solidifying first). However, if some of the high temperature delta ferrite phase is retained, then the impurities tend to be more evenly distributed throughout the ferrite [7]. If the type of solidification is other than primary delta ferrite, the hot cracking susceptibility of the stainless steel is increased [8,9]. Thus, the hot cracking potential of the fusion zone is dependent not only on the amount of delta ferrite but also its morphology. The morphology and amount of the delta ferrite are determined by the solidification rate, thermal gradients and the chemical content (the amount of ferrite formers and austenite stabilizers) of the weld metal [8].

From the chemical composition of the base material, the amount of delta ferrite formed in thick section welds can be estimated by the Schaeffler Diagram [10], or the DeLong Diagram [11]. Unfortunately, the amount of delta ferrite retained in welds on thin stainless steel sheet (<2-mm thick) is less than that predicted by the diagrams because of the rapid solidification rate. Furthermore, since delta ferrite is ferromagnetic, its amount in most stainless steel welds can usually be measured by instruments, such as the Magne Gage, Severn Gage, and Ferritescope, which measure the strength of its response to an applied magnetic field. However, the volume of influence of the magnetic field exceeds the dimensions of thin sheet resulting in an inaccurate measurement in the fusion zone in thin materials.

Metallographic examination is the only reliable method for determining the ferrite content of thin stainless steel sheet. The main problem with this approach is that it is a highly sensitive technique. The very fine structure of the delta ferrite is usually difficult to etch well enough to resolve. The magnetic etching technique must be confined to a qualitative evaluation due to the accumulation (halo effect) of the colloid particles around a ferromagnetic phase. Not only can the technique provide comparative information about the amount of delta ferrite in a stainless steel weld but the morphology of the delta ferrite and the type of solidification can also be defined.

Strain-Induced Martensite

Martensite can be induced in an austenitic stainless steel by strains experienced when the material is plastically deformed. This strain-induced martensitic transformation will occur at a temperature, Md, above the martensite start temperature (Ms). No amount of deformation will force the martensitic transformation to occur if the austenite is above the Md temperature [12].

The types of strain-induced martensite that can form in 300 series stainless steels with carbon levels below 0.08 wt % are alpha and epsilon martensite. Alpha martensite is a ferromagnetic, body-centered-cubic (bcc) structure; whereas the epsilon martensite is a nonferromagnetic, hexagonal-closed-packed (hcp) structure [13]. The amount of alpha martensite normally increases with increasing strain in austenitic stainless steel. The amount of epsilon martensite will increase with plastic strain until it reaches a maximum, and then it will gradually decrease due to its transformation to alpha martensite [14].

Strain-induced martensite has a dramatic effect on the mechanical properties of austenitic stainless steels. The presence of martensite can produce significant changes in the tensile and notched-tensile strength behavior [15,16], in the flow stress [13,14], in the strain hardening rate [14,17], in the fracture toughness [18], in the low-cycle fatigue strength [19], and in the yield strength [20].

The formation of alpha martensite in austenitic stainless steel can be influenced by the chemical content of the metal, the amount of strain, and the temperature of the material [21]. To a lesser extent, the strain-induced martensitic transformation is also dependent on the strain rate [22], hydrogen content [23], and grain size [21].

The Md temperature is dependent on the chemical content of the austenitic stainless steel. The stability of the austenite with respect to the formation of alpha martensite has been described using the Md30 temperature since the Md temperature is very difficult to determine experimentally [24]. The Md30 temperature is the temperature at which 50 vol. % of the austenite transforms to martensite at 30% true strain. Several formulas for predicting the Md30 temperature have been derived empirically. The equations are listed in Table 1. A compositional increase in these elements in the equations has been experimentally proven to decrease the Md30 temperature [22,25,26].

Table 1. The Equations for Predicting an Md30 Temperature, C.

Angel [25]:	Md30 = 413.0 - 462.0(%C + %N) - 9.2(%Si) - 8.1(%Mn) - 13.7(%Cr) - 9.5(%Ni) - 18.5(%Mo).
Nohara et al. [22]:	Md30 = 551.0 - 462.0(%C + %N) - 9.2(%Si) - 8.1(%Mn) - 13.7(%Cr) - 29.0(%Ni + - %Cu) - 18.5(%Mo) - 68.00(%Nb).
Sjoberg [26]:	Md30 = 608.0 - 515.0(%C) - 7.8(%Si) - 12.0(%Mn) - 13.0(%Cr) - 34.0(%Ni) - 6.5(%Mo) - 821.0(%N).

There are several methods to characterize strain-induced martensite. Both types of strain-induced martensite can be thoroughly defined by x-ray diffraction. The use of the transmission electron microscope allows the identification of the type of martensite [27]. Magnetic etching is a good evaluation tool for alpha martensite but not epsilon martensite since it is not ferromagnetic.

The magnetic etching technique has some advantages over x-ray diffraction for the resolution of alpha martensite. The conventional etching that precedes magnetic etching reveals the grain boundaries and slip lines, whereas a conventional etch is not used in x-ray diffraction. Therefore, alpha martensite formation at grain boundaries or slip within grains cannot be distinguished with x-ray diffraction because of the lack of locational references. X-ray diffraction does have the capability of measuring the precise content of martensite, and the magnetic etching does not. But x-ray diffraction's accuracy is limited when there is 10 vol. % or less of martensite [28]. Due to the halo effect, magnetic etching is not a precise quantitative measurement of a ferromagnetic phase, but it is a good qualitative evaluation tool. Magnetic etching is less expensive and easier to prepare and conduct than x-ray diffraction. Finally, more area of a single sample can be examined more conveniently with magnetic etching than with x-ray diffraction.

EXPERIMENTAL PROCEDURE

Material

The samples examined for delta ferrite and the strain-induced martensite and their respective chemical compositions are listed in Tables 2 and 3, respectively. All the samples are 0.25-mm thick sheet and, except for the high purity Type 316 stainless steel (sample 5), they are typical "off the shelf" products available from commercial vendors.

Table 2. Stainless Steel Sheet Samples

Sample Number	AISI Type	Origin
1	316	Commercially obtained sample used in Lambert Weld Test.
2	304L	Commercially obtained sample used in Lambert Weld Test as a standard.
3	304L	Commercially obtained sample used in tensile tests.
4	316	Commercially obtained sample used in tensile tests.
5	316	Sample with high purity chemistry used in tensile tests.

Table 3. Chemical Analysis of Samples (wt %)

Sample Number	C	Mn	Cr	Ni	Si	Mo	P	S
1	0.056	1.51	17.12	11.40	0.57	2.68	0.032	0.002
2	0.013	1.37	18.02	8.36	0.51	0.16	0.033	0.007
3	0.021	1.37	18.18	9.18	0.61	0.31	0.039	0.002
4	0.056	1.62	17.12	10.56	0.74	2.02	0.022	0.008
5	0.040	0.03	17.19	13.16	0.11	2.17	0.005	*

* Sulfur below limit of detection.

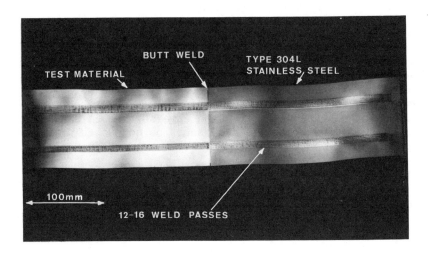

Figure 1. A typical Lambert weld test sample.

Lambert Weld Test

The objective of the Lambert Weld Test [29] is to define qualitatively the hot cracking susceptibility of thin stainless steel sheet. The test coupon is composed of two 100-mm wide by 200-mm long sheets butt welded together to produce a coupon that is 100-mm wide and 400-mm long as shown in Figure 1. One of the sheet samples is an AISI Type 304L stainless steel which has been tested previously with the Lambert Test and no through cracks were found. This AISI Type 304L sample is used as a standard to indicate problems or deficiencies in the test procedure or preparation, such as power supply malfunctions, dirty materials, worn fixturing, faulty shielding gas, etc. The other sheet sample in the test coupon is the material that is to be evaluated.

The test procedure consists of bead-on-sheet weld passes conducted on the sample with an autogenous, gas-tungsten-arc (GTA) welding process along the length of the test coupon. A series of these weld passes are placed side-by-side to form a weld pad. Normally, 12 to 16 weld passes will compose a weld pad. The procedure includes two weld pads per test coupon.

After welding, a dye penetrant is applied to the back side of both of the weld pads on each coupon, and a developer is sprayed on the top side. The coupon is allowed to set for a few minutes to allow the dye penetrant to absorb into any cracks present. Any through cracks which propagate through the fusion zone can now be detected on the top surface of a weld pad. Surface cracking can be found by applying a fluorescent dye penetrant and the developer to the same surface of a weld pad. Since both dye penetrant checks cannot be conducted of the same weld pad, the type of cracking of interest must be decided before examination.

Once a cracked area of the weld is located, the dye penetrant is cleaned from the surface of the sample. Then, the portion of the weld metal containing the crack is sectioned and mounted for examination as shown in Figure 2.

Tensile Test

The purpose of the tensile test is to determine the hydrogen cracking propensity of stainless steel sheet samples strained in a hydrogen atmosphere at low temperature. Magnetic etching was used to determine the relative amount of strain-induced martensite transformed in these samples and to gain an understanding of the morphology and the initiation of the transformation.

The tensile test samples were stamped from a 0.25-mm (0.010-in.) thick sheet. As shown in Figure 3, the sample gauge area is 50.8-mm long x 12.7-mm wide. The specimens were annealed at 1050 C for 5-min in a vacuum furnace before testing.

The test samples were then fixtured in a specially prepared chamber on a screw driven universal tensile test machine. The chamber was filled with hydrogen at a pressure of 0.34 MPa and the sample's temperature was decreased to -54 C. Finally, the samples were loaded to 30% true strain with a cross head speed of 25.4 mm/min.

After straining, the samples were sectioned and mounted for metallographic evaluation using magnetic etching. Examinations were conducted in the middle of the gauge area of the tensile sample, and on an as-annealed sample.

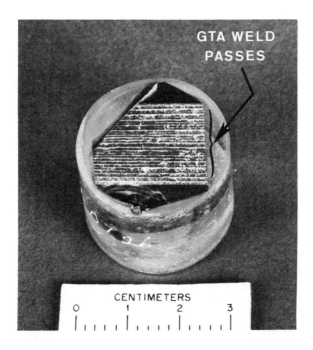

Figure 2. Mounted Lambert weld test sample in preparation for metallographic evaluation.

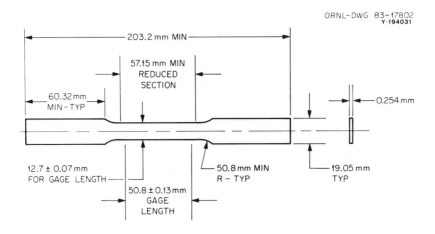

Figure 3. Drawing of stamped sheet specimen used in tensile test.

Magnetic Etching for Both Lambert and Tensile Tests

The magnetic etching technique begins with a delicate specimen preparation [30]. A portion of the 0.25-mm thick sheet with the dimensions of approximately 1- to 2-cm wide x 3-cm long is oriented to examine the flat, horizontal surface. Great care must be exercised to keep the area that is to be examined as flat as possible. To reinforce the thin sheet, the specimen is pressure glued to a 1-cm wide x 1- to 2-cm long x 0.2-cm thick hard rolled aluminum plate. After the glue is set, the ends of the 0.25-mm thick sheet are bent 90 degrees around the back of the aluminum plate forming a flat 1- to 2-cm square portion of the sample for analysis. An electrical contact is made to one of the bent ends of the thin sheet specimen and the entire assembly is mounted in epoxy resin. The added support of the non-magnetic aluminum plate to the thin stainless steel allows the sample to be ground and polished and not buckle. Minimum grinding is done using 400 to 600 grit abrasive sheets. The objective is to have the final plane of examination about midway through the thickness. The specimen is polished in two stages on 30.48-cm (12-in.) diameter Syntron Vibratory units with Linde A (0.3 micrometer) alumina on adhesive backed Texmet* followed by 1/2 micrometer diamond on nylon cloth. The specimen is then electropolished for 10 to 20 s in 10% perchloric acid in glacial acetic acid at 1.5 amps/sq.cm. This step is essential to remove the deformed surface metal that was generated during the final mechanical polishing. The electropolished surface is etched in a conventional manner to reveal the grain structure. The etchant used on both AISI Type 304L stainless steel specimens (samples 2 and 3) was glyceregia (1 part nitric acid, 3 parts hydrochloric acid, and 4 parts glycerol). The AISI Type 316 stainless steel specimens (samples 1, 4, and 5) were electrolytically etched in 10% nitric acid in water at 0.05 amp/sq.cm.

After the specimen preparation is completed, a partial drop (0.0005 cm^3) of colloid (Ferrofluid)+ is dispensed on the specimen surface. As shown in Figure 4, a clean biological type No. 1 cover glass is positioned on the colloid to form a thin layer of Ferrofluid between the cover glass and specimen.

---

* A trademark of Buehler Ltd., 41 Waukegan Road, P.O. Box 1, Lake Buff, Illinois 60044.
+ A trademark of Ferrofluidics Corp., 40 Simon Street, Nashua, New Hampshire 03061.

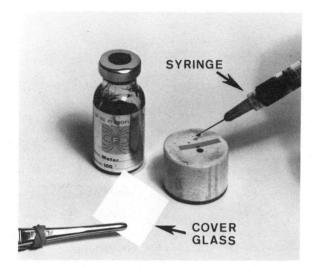

Figure 4. Application of iron colloid to specimen. The cover glass is positioned over the colloid to form a thin fluid layer between the glass and specimen surface.

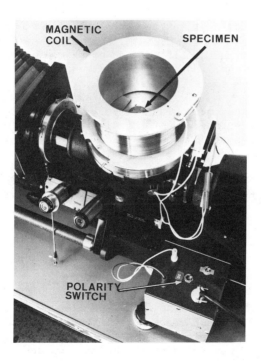

Figure 5. Magnetic etching coil with specimen on the metallograph.

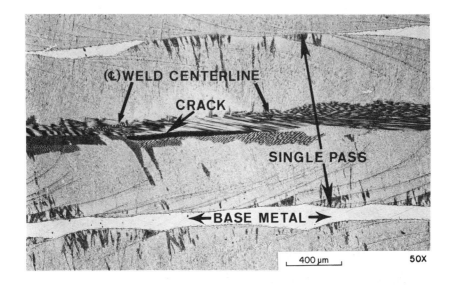

Figure 6. Weld pass of the Lambert Test on the commercial Type 316 stainless steel, 0.25 mm thick sheet sample. A single weld pass is identified; the crack can be seen in the weld centerline.

Cleanliness of the cover glass is very important so the colloid will wet the under surface of the glass. The colloid layer must be thin and transparent enough to see the etched microstructure of the specimen through the colloid. The mounted specimen is then placed under a microscope for examination and a magnetic field is applied as shown in Figure 5. The Ferrofluid will be attracted to the areas above a ferromagnetic phase; thus, the phase will be darkened.

RESULTS

Delta Ferrite

The fusion zones of the AISI Type 316 coupon (sample 1) are more susceptible to hot cracking than the weld pads on the AISI Type 304L coupon (sample 2) as determined by the dye penetrant checks on the Lambert weld test samples. Figure 6 shows that

Stainless Steel Sheet Welds / 357

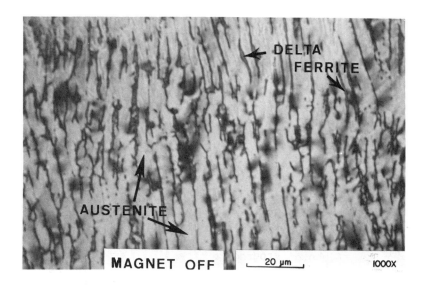

(a)

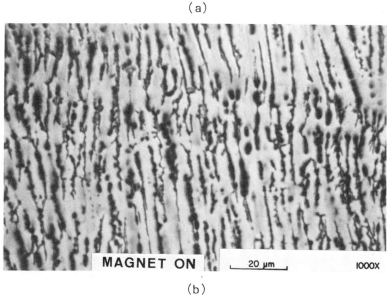

(b)

Figure 7. Magnetic etch of fusion zone in weld on commercial Type 316 stainless steel, 0.25 mm thick sheet. Microstructures are between the weld centerline and the fusion line. The paramagnetic austenite and ferromagnetic delta ferrite are identified. Note the colloid is mildly attracted to the delta ferrite in (a) with the magnet off; but the same field of view with the magnetic on in (b) shows the colloid is highly attracted.

the cracking in the AISI Type 316 stainless steel welded sheet occurs mainly in the weld centerline. Even though the weld passes were placed side-by-side, note that there is a small portion of unmelted base metal between the passes which does not extend through the thickness of the sheet.

The area of the weld in Figure 7 is located between the weld centerline and the fusion line (edge of the weld pass). This region has a duplex structure of austenite and a small amount of delta ferrite. The delta ferrite in the austenitic matrix is darkened by the dark analog pattern the colloid forms over this ferromagnetic phase when the magnetic field is applied. The solidification mode of the delta ferrite has been classified as a Type 1 (vermicular) morphology by David [9]. The morphology does show evidence of the shape instability of the delta ferrite due to the exposure to high temperatures for short times resulting from the subsequent weld pass. A photomicrograph of the AISI Type 304L specimen (sample 2) in Figure 8 shows a similar welded region as Figure 7 located between the fusion line and the centerline. The morphology of the delta ferrite in the AISI Type 304L fusion zone is similar to that of the AISI Type 316 specimen; however, there is more delta ferrite retained in sample 2 than in sample 1.

The type of solidification in the welded region around the crack in sample 1 (Type 316) is different than the primary delta ferrite solidification occurring in the outer fusion zone. Primary austenite solidification, as shown in Figure 9, is predominant with some of the delta ferrite retained at the grain boundaries, i.e., last to freeze. There is less delta ferrite solidifying in the weld centerline than the outer fusion zone in Figure 7. These observations are indicative of a change in solidification rate and the resulting compositional variation. The growth rate of the weld puddle varies from zero at the fusion boundaries to a maximum equal to the weld travel speed at centerline. According to Savage [31], the growth rate, R, of the solidifying weld metal at any given point on the moving solid-liquid interface is as follows:

$$R = V \cos \Theta \qquad (1)$$

where: V = welding velocity, and
                 Θ = angle between the normal to the interface at the given point (i.e., the average growth direction) and the welding travel direction.

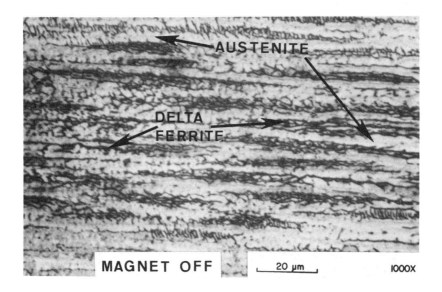

(a)

(b)

Figure 8. Magnetic etch of fusion zone in weld on commercial Type 304L stainless steel, 0.25 mm thick sheet. Microstructures are between the weld centerline and fusion line. The delta ferrite is highly emphasized with the magnetic on in (b).

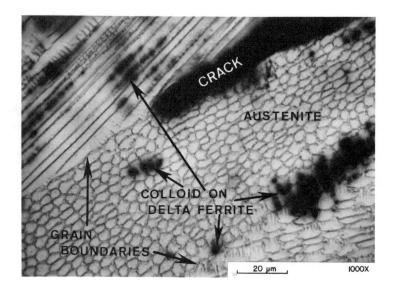

Figure 9. Magnetic etch at the weld centerline on a sample of commercial Type 316 stainless steel, 0.025 mm thick sheet.

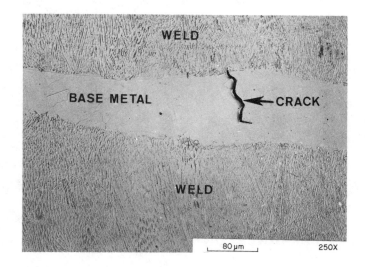

Figure 10. Surface cracks located between the weld passes in Lambert Test sample on commercial Type 304L stainless steel, 0.25 mm thick sheet.

The weld centerline crack shown in Figure 9 is intergranular. This is typical of hot cracking. Another example of hot cracking found in this investigation is shown in Figure 10. These cracks on sample 2 are located between the weld passes in the unmelted base metal. The cracks do not propagate through the welded sheet as the centerline cracks do. Similar cracking has been observed by Lippold and Savage [8] on AISI Type 304L stainless steel.

Strain-Induced Martensite

Using the equations in Table 1, the predicted values of the Md30 temperature are presented in Table 4.

Table 4. The Calculated Md30 Temperatures for the Tensile Test Samples According to the Equations in Table 1

Sample Number	Angel	Md30 (C) Sjoberg	Nohara et al.
3	- 6	- 63	- 47
4	-18	- 64	- 88
5	-58	-188	-177

In this investigation, the formulas were only used to predict the relative degree of the amount of strain-induced martensite introduced into the tensile test specimens after straining at -54 C. These formulas indicate the special chemistry AISI Type 316 stainless steel sheet (sample 5) will have the least amount of strain-induced martensite in the tensile test sample after straining. The commercial AISI Type 316 stainless steel sheet (sample 4) will be the next in order of the minimum amount of martensite and will have less than the AISI Type 304L sample 3.

The results of the metallographic examination employing the magnetic etching technique on samples 3, 4, and 5 are shown in Figures 11, 12, and 13, respectively. The dark areas (ignoring the grain boundaries) are the analog patterns formed

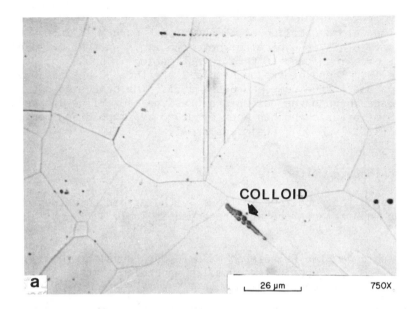

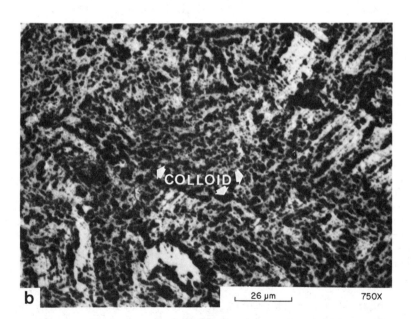

Figure 11. Magnetic etch on tensile test sample of commercial Type 304L stainless steel, 0.25 mm thick sheet. The matrix is paramagnetic austenite. The analog pattern of colloid particles emphasizes the alpha martensite.

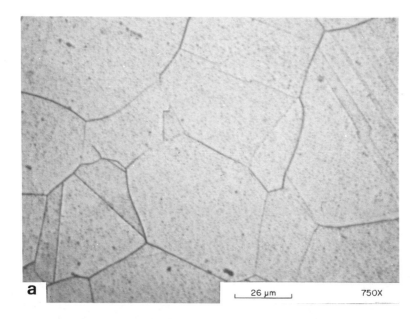

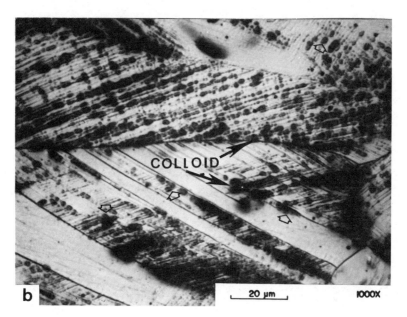

Figure 12. Magnetic etch on tensile test sample of commercial Type 316 stainless steel, 0.25 mm thick sheet. The open arrows in field (b) denote strain-induced, alpha martensite at areas of slip and twinning.

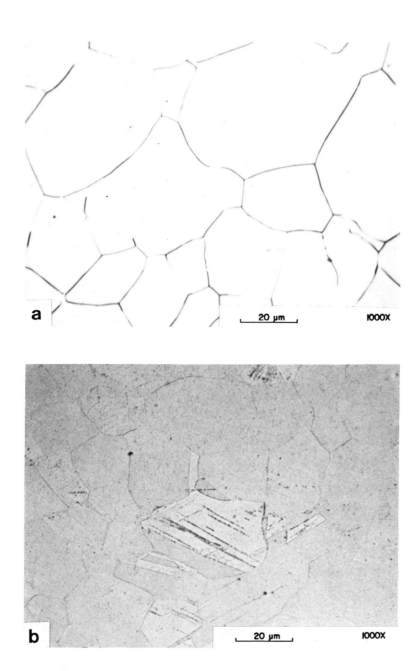

Figure 13. Magnetic etch on tensile test sample of special chemistry Type 316 stainless steel, 0.25 mm thick sheet. There is little or no evidence of strain-induced martensite.

by the 30 nanometer particles over the ferromagnetic phases assumed to be alpha martensite. Part "a" of each of the Figures shows the as-annealed microstructure of the sample. Part "b" shows the microstructure in the gauge area of the tensile specimen after the 30% straining. The magnet was turned on in both parts "a" and "b".

As seen in Figure 11a, there is an isolated ferromagnetic phase in the as-annealed Type 304L specimen (sample 3) which may have been induced into the material before testing; however, no other ferromagnetic phases can be found in this sample. The alpha martensite transformation occurs readily in sample 3 in Figure 11b. In the commercial AISI Type 316 (sample 4) in Figure 12b, there is less martensite formed than in sample 3. There is little or no strain-induced, alpha martensite formed in the high purity AISI Type 316 (sample 5) as indicated in Figure 13b.

Some martensitic transformation products occur at slip and twinning lines denoted in Figure 12b. This indicates possible nucleation sites for the alpha martensite. Furthermore, a photomicrograph taken at the same location in the gauge area at lower magnification in Figure 14 shows that the amount of martensite varies from grain to grain. This inconsistency in the distribution of the strain-induced martensite is probably the result of the variation in the crystallographic orientation of the different grains to the direction of the applied strain. Both of these phenomena are characteristic of the strain-induced martensitic transformations [3,30,31].

CONCLUSIONS

Lambert Weld Test Samples

1. The magnetic etching technique provides a way of examining the morphology and the relative content of delta ferrite retained in a weld on thin stainless steel sheet.

2. There were differences in the content and morphology of the delta ferrite in the weld centerline as opposed to the rest of the fusion zone in the commercial AISI Type 316 stainless steel sample which are probably caused by localized variations in solidification rate and in composition. The microstructure around the crack exhibited less delta ferrite than in the majority of the fusion zone.

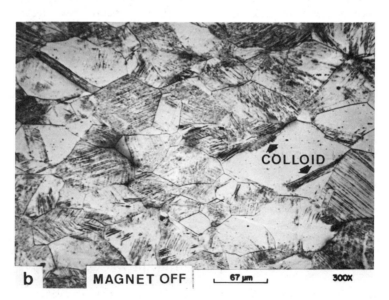

Figure 14. Magnetic etch on tensile test specimen of commercial grade Type 316 stainless steel, 0.25 mm thick sheet in the gauge area of the tensile specimen after the tensile test. The analog pattern of the colloid shows great variations in the amount of strain-induced martensite as related to crystallographic orientation.

3. There was more delta ferrite retained in the microstructure of the fusion zone of the commercial AISI Type 304L stainless steel sample than in the commercial AISI Type 316 sample.

Tensile Tests

1. The distribution and formation of the strain-induced alpha martensite and qualitative stability of austenitic stainless steel sheet can be evaluated with the magnetic etching technique.

2. Magnetic etching shows how different amounts of alpha martensite can exist in a single specimen. This phenomenon may be due to crystallographic orientation.

3. The amount of strain-induced martensitic transformation products is dependent on the chemical content of the material.

ACKNOWLEDGEMENTS

The authors are grateful to C. W. Houck for his assistance in metallography, C. R. Brinkman and G. M. Goodwin for their helpful suggestions in reviewing the report and to L. Johnson for preparing the manuscript.

REFERENCES

1. D. P. Edmonds, D. M. Vandergriff, and R. J. Gray, "Effect of Delta Ferrite Content on the Mechanical Properties of E308-16 Stainless Steel Weld Metal III Supplemental Studies," Properties of Steel Weldments for Elevated Temperature Pressure Containment Applications, The Amer. Soc. of Mech. Engr., New York, NY, pp. 47-61 (1978).

2. D. P. Edmonds and R. J. Gray, "Elevated Temperature Exposure Has Significant Effects on the Structure of Austenitic Stainless Steel Welds," Microstructural Science, Vol. 7, Elsevier North Holland, NY, pp. 345-355 (1979).

3. R. J. Gray, "The Detection of Strain-Induced Martensite in Types 301 and 304 Stainless Steels by Epitaxial Ferromagnetic Etching," Microstructural Science, Vol. 1, American Elsevier Publishing Company, New York, NY, pp. 159-175 (1974).

4. J. C. Borland and R. N. Younger, "Some Aspects of Cracking in Austenitic Steels," British Welding Journal, Vol. 7, pp. 22-60 (1960).

5. F. C. Hull, "Effect of Delta Ferrite on the Hot Cracking of Stainless Steels," Welding Journal, (Research Supplement) Vol. 46, No. 9, pp. 399-409 (1967).

6. W. T. DeLong, "Ferrite in Austenitic Stainless Steel Weld Metal," Welding Journal, (Research Supplement) Vol. 53, No. 7, pp. 273-286 (1974).

7. J. C. Lippold and W. F. Savage, "Solidification of Austenitic Stainless Steel Weldments: Part 2 - The Effect of Alloy Composition on Ferrite Morphology," Welding Journal, Vol. 58, No. 2, pp. 48-59 (1980).

8. J. C. Lippold and W. F. Savage, "Solidification of Austenitic Stainless Steel Weldments: Part 3 - The Effect of Solidification Behavior on Hot Cracking Susceptibility," Welding Journal, Vol. 60, No. 12, pp. 388-396 (1982).

9. S. A. David, "Ferrite Morphology and Variations in Ferrite Content in Austenitic Stainless Steel Welds," Welding Journal, (Research Supplement), Vol. 59, No. 4, pp. 63-71 (1981).

10. A. Schaeffler, "Constitution Diagram for Stainless Steel Weld Metal," Metal Progress, Vol. 56, No. 5, pp. 680-680B (November 1949).

11. W. DeLong, G. Ostrom, and E. Szumachowski, "Measurement and Calculation of Ferrite in Stainless Steel Weld Metal," Welding Journal, (Research Supplement), Vol. 35, No. 11, pp. 526-533 (1956).

12. Paul G. Shewmon, Transformation in Metals, McGraw-Hill, New York, NY, p. 394 (1969).

13. D. T. Read, R. P. Read, and R. E. Schamm, "Low Temperature Deformation of Fe-18Cr-8Ni Steel," <u>Materials Studies for Magnetic Fusion Energy Applications at Low Temperature</u>, pp. 149-172 (1979).

14. V. Seetharaman and R. Krishnan, "Influence of the Martensitic Transformation on the Deformation Behavior of an AISI 316 Stainless Steel at Low Temperature," <u>Journal of Materials Science</u>, Vol. 16, pp. 523-530 (1981).

15. J. F. Breedis and W. D. Robertson, "The Martensitic Transformation in Single Crystals of Iron-Chromium-Nickel Alloys," <u>Acta Metallurgica</u>, Vol. 10, pp. 1077-1088 (1962).

16. A. Rosen, R. Jago, and T. Kuer, "Tensile Properties of Metastable Stainless Steels," <u>Journal of Materials Science</u>, Vol. 7, pp. 870-876 (1972).

17. C. J. Guntner and R. P. Reed, "The Effect of Experimental Variables Including the Martensitic Transformation on the Low-Temperature Mechanical Properties of Austenitic Steels," <u>Trans. of the ASM</u>, Vol. 55, pp. 399-419 (1962).

18. M. A. Meyers, "The Effects of Shock-Loading Temperature and Pulse Duration on the Tensile Response of AISI 304 Stainless Steel," <u>Materials Science and Engineering</u>, Vol. 51, No. 2, pp. 261-263 (1981).

19. W. W. Gerberch, P. L. Hemmings, and V. F. Zackay, "Fracture and Fractography of Metastable Austenites," <u>Metallurgical Transactions</u>, Vol. 2, No. 8, pp. 2243-2253 (1971).

20. G. R. Chanani and S. D. Antolovich, "Low Cycle Fatigue of a High-Strength Metastable Austenitic Steel," <u>Metallurgical Transactions</u>, Vol. 5, No. 1, pp. 217-229 (1974).

21. D. Bhandarkar, V. F. Zackay, and E. R. Parker, "Stability and Mechanical Properties Some Metastable Austenitic Steels," <u>Metallurgical Transactions</u>, Vol. 3, No. 10, pp. 2619-2631 (1972).

22. Kiyohko Nohara, Utaka Ono, and Nobuo Ohaski, "Composition and Grain Size Dependencies of Strain-Induced Martensitic Transformation. I: Metastable Austenitic Stainless Steels," Tetsu-To-Hagane, (Journal of The Iron and Steel Institute of Japan), Vol. 63, No. 5, pp. 772-782 (1977).

23. K. P. Staudhammer, C. E. Frantz, S. S. Hecker, and L. E. Murr, "Effects of Strain Rate on Deformation-Induced Martensite in Type 304 Stainless Steel," Shock Waves and High-Strain-Rate Phenomena in Metals: Concept and Applications, Plenum Press, New York, NY, pp. 91-112 (1981).

24. A. J. West and J. H. Holbrook, "Hydrogen in Austenite Stainless Steels: Effects of Phase Transformation and Stress State," Hydrogen Effects in Metals, ed. I. M. Bernstein and A. W. Thompson, TMS/AIME, Warrendale, PA, pp. 607-618 (1981).

25. Tryggve Angel, "Formation of Martensite in Austenitic Stainless Steels," Journal of the Iron and Steel Institute, Vol. 177, pp. 165-174 (1954).

26. Jorgen Sjoberg, "The Influence of Analysis on Properties of Stainless Spring Steel," Wire, pp. 155-158 (1973).

27. J. M. Rigsbee, "TEM Observations on Hydrogen-Induced Epsilon-HCP Martensite," Metallography, Vol. 11, pp. 493-498 (1978).

28. Private communication with H. L. Yakel, Oak Ridge National Laboratory, Union Carbide Corp., Oak Ridge, TN.

29. J. A. Brooks and F. J. Lambert, Jr., "The Effects of Phosphorus, Sulfur and Ferrite Content on Weld Cracking of Type 309 Stainless Steel," Welding Journal, Vol. 56, No. 5, pp. 139-143 (1978).

30. R. J. Gray, "Magnetic Etching with Ferrofluid," Metallographic Specimen Preparation, Plenum Publishing Corp., New York, NY, pp. 155-177 (1974).

31. W. F. Savage, "Solidification, Segregation, and Weld Imperfections," Welding in the World, Vol. 18, No. 5/6, pp. 89-114 (1980).

32. R. Lagneborg, "The Martensite Transformation in 18% Cr-8% Ni Steels," Acta Metallurgica, Vol. 12, pp. 823-843 (1964).

THE PRECIPITATION BEHAVIOR OF MARTENSITIC Fe-Ni-W ALLOYS

Günter Petzow and Heinrich Hofmann*

ABSTRACT

A metallographic investigation of the precipitation behavior of martensitic Fe-Ni-W alloys was undertaken. Compositions ranging between 5 and 20 wt % W were studied on heating up to 1000 C and by isothermally annealing at 450 C. During the martensite-austenite reaction, metastable Fe_2W precipitates form, leaving the matrix depleted with respect to W. To facilitate the SEM study of the morphology of the fine Fe_2W precipitation the matrix was electrolytically dissolved. Also, during the martensite-austenite reaction a decomposition of austenite into a Ni-rich and a Ni-poor composition has been observed. During isothermal annealing of the alloys a metastable Ni_3W and a W-rich phase form which is isomorphic with W, but has a composition of 10 wt % Fe and 90 wt % W; the diameter of the particles are about 50 nm after annealing 2000 h at 450 C.

INTRODUCTION

Maraging alloys are a class of high-strength Fe-Ni steels with small amounts of additional elements such as Mo, Ti, W, etc., and a very low carbon content. The substitutional age-hardening elements have a great influence on the age-hardening

---

* Max-Planck-Institut für Metallforschung, Institut für Werkstoffwissenschaften, Pulvermetallurgisches Laboratorium, D-7000 Stuttgart 80.

of the Fe-Ni martensite. Several basic characteristics of these maraging type alloys are directly related to the Fe-rich corners of the phase diagrams concerned. In this paper, the equilibrium diagram, as well as the metastable phases in the Fe-rich corner of the Fe-Ni-W-system, were investigated to determine the influence of W.

The isothermal section at 1400 C is shown in Figure 1. In the Fe-rich corner, bcc-Fe(α) and fcc-Fe (γ) solid solutions, the two phase fields ($\alpha+\gamma$); ($\alpha+ \mu-Fe_7W_6$) and ($\gamma+ \mu-Fe_7W_6$) and a three phase field ($\alpha+\gamma + \mu-Fe_7W_6$) have been observed. Samples with compositions corresponding with the shaded area (Figure 1) showed a martensitic microstructure after quenching (Figure 2). In this alloy, lath martensite with a cubic structure (α') has formed [1]. In this investigation, the precipitation behavior and martensite-austenite transformation during heating and isothermal annealing of these martensitic alloys have been studied.

EXPERIMENTAL PROCEDURE

The alloys were prepared from the mixture of pure powders by solid state sintering. Pellets were uniaxially compressed with a pressure of 500 MPa, sintered in pure hydrogen at 1400 C for 20 h and then quenched into water. After quenching, alloys with a Ni content of 15 wt % and 5, 10, 15 and 19 wt % W were further annealed at 450 C for 2330 h. Dilatometric analysis was carried out with a differential dilatometer using alumina as the reference material and a heating rate of 5 K/min.

In order to identify the precipitates within the martensite laths by metallographic and x-ray methods, it was necessary to enrich these phases by electrolytic dissolution of the matrix phase (α') in an acid solution (e.g., 90 ml H_2O, 10 ml HCl (1,19)).

The Vickers hardness measurements were performed at room temperature with a load of 3 N.

RESULTS

Typical dilatometer curves of three different compositions are shown in Figure 3. For the Fe-Ni alloys (no W), the $\alpha' \rightarrow \gamma$ and $\gamma \rightarrow \alpha'$ transformations occur in a single step during heating and subsequent cooling. The $\alpha' \rightarrow \gamma$ transformation is characterized

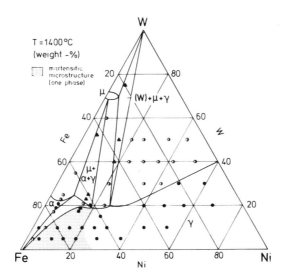

Figure 1. Isothermal section through the Fe-Ni-W system at 1400 C. Sample with compositions corresponding with the dotted area show a martensitic microstructure ($\mu = Fe_7W_6$).
○ one phase; ◐ two phases; △ three phases.

Figure 2. Microstructure of the 70 wt % Fe, 15 wt % Ni, 15 wt % W alloy after quenching from 1400 C. Etchant: 10 ml H_2O, 10 g $K_2S_2O_5$; 4 min.

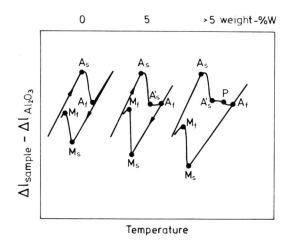

Figure 3. Typical dilatometer curves (schematic).

by the temperatures A_s and A_f, the $\gamma \rightarrow \alpha'$ transformation by the temperatures M_s and M_f. For alloys with a W-content between 0 and 5 wt %, the transformation occurs in two steps, i.e., a third characteristic temperature, A_s' occurs during transformation. Above 5 wt % W an $\alpha' \rightarrow \gamma$ transformation in three steps is found for all martensitic alloys. The characteristic temperatures for this transformation are given by A_s, A_s', P, and A_f. Whilst the $\alpha' \rightarrow \gamma$ transformation is strongly influenced by the W-content, the $\gamma \rightarrow \alpha'$ transformation remains unchanged.

The transformation temperatures for the 11 alloys studied are given in Table 1. All transformation temperatures decrease with increasing Ni-content.

After heating and subsequent cooling, the microstructures of alloys with low W-contents were unchanged, whereas alloys with W-contents >5 wt % exhibited two phases (α'+Fe_2W) or three phases (α'+Fe_2W+γ). Figure 4 shows the microstructure of the 66 wt % Fe, 15 wt % Ni, 19 wt % W alloy. The micrograph depicts only martensite α' (grey) and retained austenite (white); Fe_2W is not visible at light microscopy magnifications. After removing the α'+γ matrix by electrolytic etching, the small needle or plate-like Fe_2W precipitates are visible (Figure 5) at high magnification. The lattice constant of martensite decreases when the Fe_2W phase appears (Figure 6).

Table 1. Results of the Dilatometric Analysis (One Cycle)

Composition (wt %) Fe	Ni	W	Microstructure before heating	Microstructure after heating	A_s	A_s'	P	A_f	M_s	M_f
90	10	0	α'	α'	696	–	–	729	580	503
80	20	0	α'	α'	557	–	–	629	285	175
90	5	5	α'	α'	797	836	–	863	658	565
85	10	5	α'	α'	653	735	–	876	459	345
80	15	5	α'	α'	573	668	–	743	312	175
80	10	10	α'	$\alpha'+Fe_2W$	706	748	–	830	429	320
75	15	10	α'	$\alpha'+Fe_2W$	573	687	724	790	290	175
70	20	10	α'	$\alpha'+Fe_2W+\gamma$	550	620	701	790	130	<RT
70	15	15	α'	$\alpha'+Fe_2W$	575	677	730	780	240	125
67.5	15	17.5	α'	$\alpha'+Fe_2W+\gamma$	580	687	730	770	222	<RT
66	15	19	α'	$\alpha'+Fe_2W+\gamma$	575	623	695	743	72	<RT

Figure 4. (Left) Microstructure with retained austenite of the 66 wt % Fe, 15 wt % Ni, 19 wt % W alloy.

Figure 5. (Right) SEM photograph of Fe$_2$W precipitates after heating and subsequent cooling of a 70 wt % Fe, 15 wt % Ni, 15 wt % W alloy (the α' matrix is electrolytically removed).

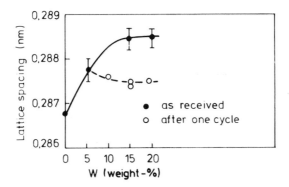

Figure 6. Lattice constants of martensite before and after the dilatometer cycles.

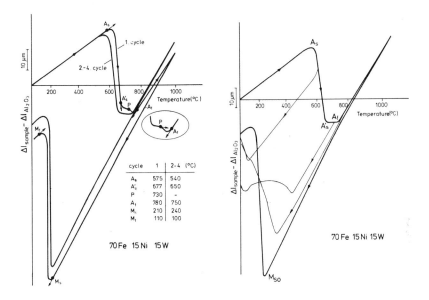

Figure 7. (Left) Dilatometer curves of an alloy of 70 wt % Fe, 15 wt % Ni, 15 wt % W.

Figure 8. (Right) Dilatometer curves of 70 wt % Fe, 15 wt % Ni, 15 wt % W alloy, obtained during cooling from different temperatures ranging from the A_s to slightly above A_f.

Upon subsequent heating cycles (5 K/min from RT to 1000 C) of alloys with >10 wt % W, the transformation at P was not observed and a displacement of A_s, A_s' and A_f towards lower temperatures occurred (Figure 7).

During cooling, directly after heating to different temperatures between A_s and slightly above A_f, a two step martensitic transformation takes place with M_s-temperatures higher and/or only lower than M_{so} (Figure 8) (M_{so} = martensite start temperature measured by cooling from temperature >1000 C). Samples quenched from maximum temperatures ranging between A_s and A_s' showed a decrease of the lower M_s temperature and an increase of the higher M_s-temperature with increasing maximum temperatures. By quenching from temperatures ranging between A_s' and slightly

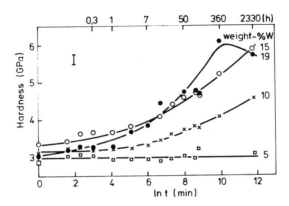

Figure 9. Hardness vs. annealing time at 450 C of Fe-Ni-W alloys with 15 wt % Ni and 5, 10, 15 and 19.5 wt % W, respectively.

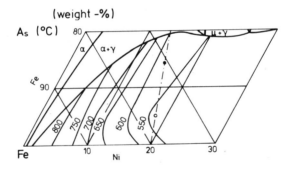

Figure 10. Composition change of α' during precipitation of Fe_2W of the 70 wt % Fe, 15 wt % Ni, 15 wt % W alloy (● starting composition, o final composition). In addition the A_s temperatures and the solubility lines at 1400 C are shown.

above A_f, the two M_s-temperatures tend to M_{so} with increasing maximum temperature.

During isothermal annealing of 15 wt % Ni alloys with 5, 10, 15 and 19 wt % W at $(A_s+M_s)/2$ = 450 C, the hardness increases with annealing time and W content except for the 5 wt % alloy (Figure 9). Metallographic and x-ray investigations show that precipitation of Ni_3W and η-W with approximately 50 nm diameter occurs, where η-W is a solid solution of Fe in W with about 12 wt % Fe.

DISCUSSION

Precipitation of Fe_2W during the $\alpha' \rightarrow \gamma$ transformation leaves the matrix depleted with respect to W. For example, the change of composition in the matrix during this precipitation is shown by the dotted line in Figure 10 for an alloy of 70 wt % Fe, 15 wt % Ni, 15 wt % W (solid point in Figure 10). Figure 10 also shows the A_s-temperature and solubility lines at 1400 C in the Fe-rich corner of the Fe-Ni-W phase diagram. With the knowledge of the A_s-temperature and the lattice spacing after the first cycle (Figures 6 and 7), the final composition of the α' matrix was determined as 77 wt % Fe, 18 wt % Ni, 5 wt % W (open point in Figure 10). Therefore, the dilatometer curves of the 2nd to 4th cycles (Figures 7 and 8) show the transformation behavior of alloys corresponding to this final composition. However, the first cycle shows the transformation behavior of an alloy whose martensite composition changed during the $\alpha' \rightarrow \gamma$ transformation. This observation agrees with the results of Servant et al. [2] on the transformation behavior of an alloy with similar composition (75.4 wt % Fe, 16.4 wt % Ni, 7.2 wt % W).

Investigations of the Fe-Ni-W specimens after annealing times >2000 h [3] show that Fe_2W is a metastable intermetallic phase. Therefore, the observed two phase field ($\alpha'+Fe_2W$), i.e., ($\gamma+Fe_2W$) at higher temperatures, with a solubility of W in α' (γ) of 4-5 wt %, must also be metastable. Because the transformation temperature P was not observed in the second cycle, this temperature might be related to the precipitation of Fe_2W.

The characteristic transformation temperatures A_s, A_f and M_s measured at the second cycle are used to estimate a realization diagram at 5 wt % W through the Fe-Ni-W phase diagram including metastable and stable states (Figure 11). With increasing Ni content the A_s, A_f and M_s temperatures decrease. This behavior is also observed in the binary Fe-Ni system [4]. Figure 11 also shows the austenite compositions during heating, as determined from the changes in M_s-temperatures measured by dilatometric analysis during quenching from different temperatures above A_s (Figure 8). During

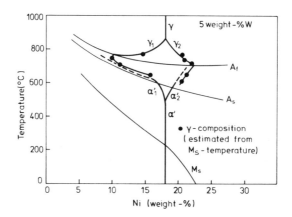

Figure 11. Realization diagram at 5 wt % W through the metastable Fe-Ni-W phase diagram. In addition the decomposition of the austenite $(\gamma \rightarrow \gamma_1 + \gamma_2)$ and the martensite-austenite transformation $(\alpha_1' \rightarrow \gamma_1, \alpha_2' \rightarrow \gamma_2)$ during heating is shown.

heating, a decomposition into a Ni-rich and Ni-poor composition takes place [2]. The Ni-rich martensite transformed at A_s into Ni-rich austenite, whereas the Ni-poor martensite transformed at temperatures between A_s' and A_f into Ni-poor austenite. Therefore, by quenching from temperatures ranging between A_s and A_f, two M_s temperatures were observed, the M_s temperature from the Ni-rich austenite (below M_{so}), and the M_s temperature from the Ni-poor austenite (above M_{so}).

A possible reason for decomposition of austenite during heating is a miscibility gap $((\gamma_1 + \gamma_2), (\alpha_1' + \alpha_2'))$ in this area of the phase diagram. For the realization diagram at 5 wt % W (Figure 11), for <750 C, the miscibility gap ranges from 8 to 23 wt % Ni, whereas for temperatures >900 C a homogeneous γ-solid solution exists. It is likely that the miscibility gap widens with decreasing temperature, but because of the very slow kinetic processes the equilibrium state is not reached. Similarly, a miscibility gap in the binary Fe-Ni system might exist also, as some authors [5-7] have proposed. Also, Inden [8] showed that in systems with ferromagnetic components a miscibility gap near the Curie temperature exists.

The precipitation of intermetallic compounds was also observed during isothermal annealing at 450 C. In accordance with the results, the increase in hardness during isothermal annealing is connected with the precipitation of Ni_3W and η-W, both of which are metastable [3,9]. Since the alloy with 5 wt % W showed no changes in hardness, the content of W in α' must be above 5 wt % W, because there is no change in hardness, and below 10 wt % W, because such an alloy had an increase of hardness. From the thermodynamic point of view, the precipitation of η-W and Ni_3W cannot be understood because the most stable decomposition product observed in the martensite lath of those alloys is Fe_2W [10]. However, nucleation of η-W and Ni_3W might be easier, probably because the lattice constants correspond more closely to that of the matrix.

SUMMARY

Dilatometric and metallographic investigations of martensitic, Fe-rich Fe-Ni-W alloys show a martensite-austenite transformation in two steps (low W-content, between 0 and 10 wt %) or, in three steps (W-content above 10 wt %). During the transformation, the precipitation of metastable Fe_2W takes place, thus depleting the martensite with respect to W. Samples quenched from temperatures between the A_s- and A_f-temperatures exhibit two martensite start temperatures. This behavior can be explained by the decomposition of the austenite to a Ni-rich and Ni-poor composition on heating. It is, therefore, proposed that a miscibility gap must exist ranging from 8 to 23 wt % Ni at approximately 750 C and 5 wt % W. During isothermal annealing at 450 C an increase in hardness was found which corresponds to the precipitation of the metastable phases Ni_3W and η-W in the martensite.

REFERENCES

1. Z. Nishiyama, "Martensitic Transformation," M. E. Fine, M. Mehis and C. M. Wayman (Editors), Academic Press, New York (1978).

2. C. Servant, G. Maeder and P. Lacombe, "Investigation into the Effect of Substituting Mo by W on the Behavior, During Aging, of the Maraging Type Ternary Alloy," Met. Trans., Vol. 10A, pp. 1607-1620 (1978).

3. H. Hofmann, Ph.D. Thesis, TU Berlin (1983).

4. F. W. Jones and W. I. Pumphrey, "Free Energies and Metastable States in the Iron-Nickel and Iron-Manganese Systems," J. Iron Steel Inst., Vol. 163, pp. 121-131 (1949).

5. S. Kachi and H. Asano, "Concentration Fluctuations and Anomalous Properties of the Invar Alloys," J. Phys. Soc. Jap., Vol. 27, pp. 536-541 (1969).

6. Y. Tanji, H. Moriga and Y. Nakagawa, "Anomalous Concentration Dependence of Thermoelectric Power of Fe-Ni (FCC) Alloys at High Temperatures," J. Phys. Soc. Jap., Vol. 45, pp. 1244-1248 (1978).

7. J. R. C. Guimãraes, J. Danon, R. B. Socorzelli and I. S Azuvedo, "Phase Stability in Iron-Nickel Invar Alloys," J. Phys. F. Metal Phys., Vol. 10, pp. L197-L202 (1980).

8. G. Inden, "The Effect of Continuous Transformations on Phase Diagrams," Bulletin of Alloy Phase Diagrams, Vol. 2, pp. 412-422 (1982).

9. E. Hornbogen, "Entmischung von Eisen-Molybdan und Eisen-Wolfram-Mischkristallen, " Z. Metallkde., Vol. 52, pp. 47-56 (1961).

10. A. F. Yedneral, O. P. Zhukov and M. D. Perkas, "Structural Changes in the Martensite During the Aging of Iron-Nickel-Tungsten and Iron-Nickel-Cobalt-Tungsten Alloys," Fiz. Met. Metalloved., Vol. 36, pp. 339-346 (1973).

AN INVESTIGATION OF SHELL AND DETAIL
CRACKING IN RAILROAD RAILS

Ravi Rungta,* Richard C. Rice,*
Richard D. Buchheit* and David Broek**

ABSTRACT

This paper presents a failure analysis case study on railroad rails. The work, performed under the sponsorship of the Department of Transportation, addresses the problem of shell and detail fracture formation in standard rails. Fractographic and metallographic results coupled with hardness and residual stress measurements are presented. These results suggest that the shell fractures form on the plane of maximum residual tensile stresses. The formation of the shells is aided by the presence of defects in the material in these planes of maximum residual stress. The detail fracture forms as a perturbation from the shell crack under cyclic loading and is constrained to develop as an embedded flaw in the early stages of growth because the crack is impeded at the gage side and surface of the rail head by compressive longitudinal stresses.

INTRODUCTION

In a recent Battelle study completed for the Department of Transportation (DOT) [1], it was found that nearly 10 percent of over 10,000 1-mile segments of revenue-service

---

* Battelle-Columbus Laboratories, 505 King Avenue, Columbus, Ohio 43201 USA.
** FractuResearch, 2992 Heatherleaf Way, Columbus, Ohio 43229 USA.

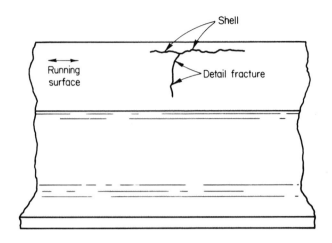

Figure 1. Transition from shelling to detail fracture.

railroad rail contained one or more detail fracture-type defects. These flaws were identified as a result of from 1 to 20 detector car passes over 53 different rail segments within a period of 2-1/2 years. Most of the defects were found in high MGT (million gross tons) and low-weight rail, but some detail fracture-type defects were found in rails with as low as 50 MGT's of service and in nearly all rail sizes.

Comparatively, detail fracture-type rail defects are uncommon (they comprised only about 3-1/2% of some 30,000 defects observed in the above cited study). However, these defects are of great concern in terms of rail reliability and safety. This is true because detail fractures typically initiate and grow within the head of the rail and they may go unidentified until the rail strength is so deteriorated that a sudden failure of the entire rail can occur.

It has been observed that a detail fracture is often associated with a shell fracture. The shell fracture initiates and grows parallel to the running surface. Branching may occur in the shell crack, and for reasons that have not been very well understood, one of these shell cracks may turn into a detail fracture. Figure 1 shows the typical relationship between a shell fracture and a detail fracture.

0.5X

Figure 2. Fracture surface of rail 5-17.

The present study is a part of a DOT-supported effort to better understand and thereby control and/or predict the incidence of shell and detail fractures in railroad rails. To this end, fractographic and metallographic investigations and hardness measurements were conducted to determine the cause of shell and detail crack formation. Although several rails have been investigated, one rail will be discussed in this paper as a typical example.

EXPERIMENTAL RESULTS

Fractography and Metallography

Fractographic and metallographic investigations were conducted on both detail and shell fractures in an attempt to identify the cause of cracking and to establish conditions under which a shell turns into a detail crack. Figure 2 presents the complete section of a rail identified as rail 5

1.5X

Figure 3. Detail fracture in rail 5-1.

containing a detail fracture. The two halves of the fracture surface were identified as 5-17 and 5-1. The flaw was detected in service, at which time the rail was removed and the flaw was broken open at DOT. A slightly magnified view of the other half of the detail fracture is presented in Figure 3. The crack located at the top of the detail fracture indicates the presence of a shell associated with the detail fracture. From the ring-shaped beach marks, it is apparent that the detail fracture initiated at more than one location. Eventually the cracks merged together to form a main detail crack. Region A marked in this figure is suspected to be the origin of the detail fracture. Higher magnification views of this region are shown in Figure 4. The presence of the ridge clearly indicates that two cracks initiated and merged along the ridge. It is also apparent from these figures that the fracture surface has undergone mechanical damage in service. This is very typical of detail fractures investigated so far [2-4]. Typical examples of damage include rubbing and corrosion debris on the fracture surface. Because of the damage, no fatigue striations could be identified on the detail fracture even though this is clearly a fatigue failure. In some cases where the crack propagated through a pearlite colony, pearlite lamellae were visible on the fracture surface and they give the appearance of fatigue striations but, in fact, are not. Such features are very common in pearlitic steels.

Shell and Detail Cracking / 387

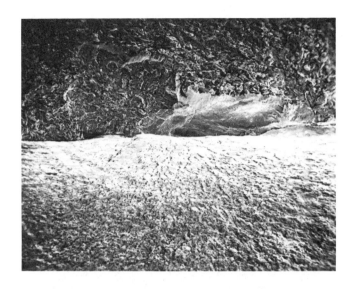

8.5X (a)

130X (b)

Figure 4. Detail crack initiation region in rail 5-1 suggesting merging of two cracks and mechanical damage on fracture surface.

Figure 5. The shell fracture associated with the detail fracture in rail 5-1 (2.5X).

To expose the shell fracture surface, the detail fracture was sectioned along the secondary crack noted in Figure 3. Figure 5 shows the shell and the detail fracture. The top view of the shell is shown in Figure 6. Based on the flow pattern of the material, it is suspected that the shell may have initiated along the line AB in Figure 6 and grown laterally in the rail head. This is supported by a slightly magnified view of the shell in Figure 7. It may be noted that point A in Figures 6 and 7 just about coincides with the point A in Figure 3. A magnified view of region A (looking at the shell) is presented in Figure 8. It is clear that repeated contact of the cracked surface has obliterated any details of the detail fracture crack initiation site.

Energy-dispersive x-ray analysis was conducted along the streak AB to determine why the shell may have initiated around it. Figure 9 presents the results of a scan along the streak. A standard rail contains small amount of silicon and manganese so the detection of these elements is not of concern. The results indicate the presence of aluminum and calcium in

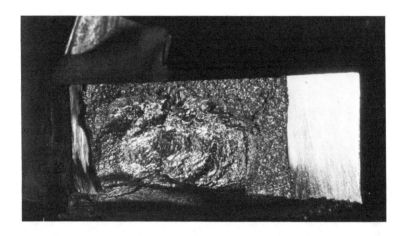

Figure 6. Shell fracture in rail 5-1 (3.75X).

Figure 7. Magnified view of the shell in rail 5-1 (13X).

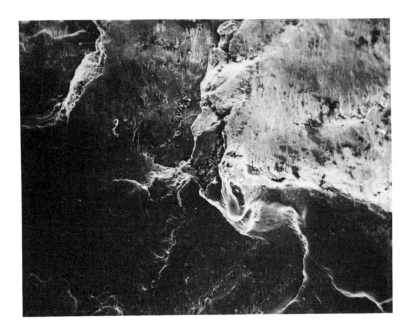

Figure 8. Magnified view of region A in Figure 7, viewing from the top in the same orientation as Figure 7 (85X).

the streak, most likely as oxide inclusions. Aluminum is added for deoxidation while calcium may arise from interactions with the slag to form non-metallic inclusions. Because nonmetallic (oxide) particles are hard and brittle, it is conceivable that shell cracking would initiate around these particles. A similar scan in the base metal away from the streak did not reveal any aluminum or calcium.

For further investigation, the specimen was rotated 90 degrees so that the streak in the shell could be viewed from the side. Figure 10 presents the view close to the site where the shell turned into a detail fracture. An energy-dispersive x-ray analysis inside the crack along the shell again revealed elements that are typically found in nonmetallic inclusions. As Figure 11 shows, in addition to calcium and aluminum, titanium was also detected inside the crack. These results suggest that the presence of nonmetallic inclusions initiated the shell crack in this rail. Similar defects were observed over 30 years ago by Cramer on rail samples tested in the laboratory [5].

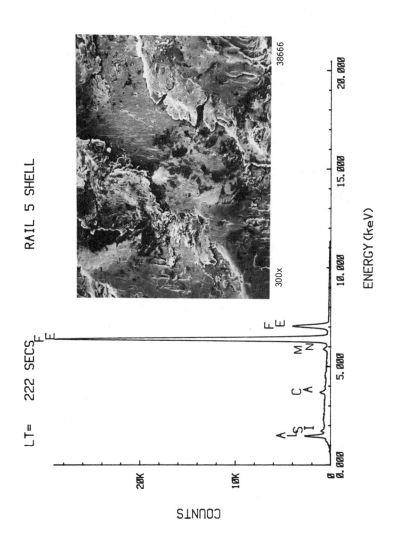

Figure 9. Energy-dispersive x-ray analysis along the streak in rail 5-1 (reduced 30% in reproduction).

Figure 10. A side view of the shell crack in rail 5-1 (75X).

To further examine why the shell crack turned into a detail fracture, a metallographic section was prepared through point A in Figure 1 so that the shell and the detail fracture could be viewed from the gage side. Figure 12 presents the as-polished surface showing the contour of the detail fracture in relation to the shell fracture. It is interesting to note how the shell tilted slightly into the transverse orientation and then quickly turned completely transverse. The figure also shows a secondary crack emanating from the primary shell and terminating in the detail fracture. Energy-dispersive x-ray analysis along the secondary crack again detected the presence of aluminum. At the intersection of the secondary shell with the main crack (point P in Figure 12b), aluminum, magnesium, and calcium were detected. Apparently the shell cracks commonly form at these nonmetallic inclusions (defects) and begin to propagate by linking with other defects in the immediate vicinity.

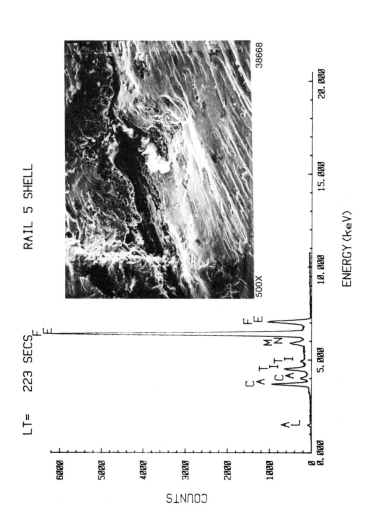

Figure 11. Energy-dispersive x-ray analysis inside the crack in the shell in rail 5-1 (reduced one-third in reproduction).

Shell Surface

Detail Fracture Surface

215X   (a)

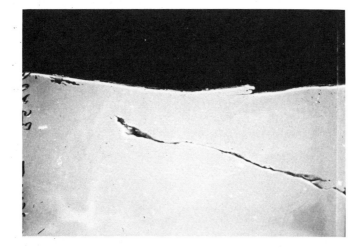

425X   (b)

Figure 12. Metallographic section showing the shell and the detail fracture in rail 5-1.

Figure 13. Martensite layer on the shell fracture surface in rail 5.

Etching the polished surface revealed patches of white layers along the shell fracture surface (Figure 13). Microhardness measurements in the white layer with a 25 g load indicated very high hardness compared to material just below the layer. The measurements are presented in Table 1.

Table 1. Knoop Microhardness Measurements with 25g Load

White Layer Along the Shell	In the Base Metal Moving Away From the Shell
1080	462
1240	361
1330	314
1080	340

Based on the microhardness measurements, it is believed that
the white layer is martensite. Similar martensite formation
has been noted in other rails [2] but it is believed that the
martensite forms after the shell has developed. Formation of
a subsurface crack produces free surfaces which can rub against
each other under the high pressure of a passing wheel. Such
rubbing would concentrate the plastic strain in this region
which could produce martensite in these steels. In the absence
of a crack, there would be no driving force for strain
concentration in a region removed from the running surface.
As will be discussed later, shear strain occurs away from the
region where shells form, and could not account for formation
of martensite. Shear deformation-induced martensite has been
known to form in bearings.

Microhardness Measurements and Relative Shell Position

Microhardness measurements were conducted on the rail
head on the gage side to see if there was any relationship
between the location of the shell and the hardness of the
material. Figure 14 shows the measurements in an orthogonal
coordinate system, with an approximate, partial running surface
profile drawn in for rail 5. The readings were taken with a
400 g load on the Vicker's hardness scale [2]. It is apparent
that the material near the running surface is work hardened.
Work hardening decreases away from the running surface down
into the more moderately stressed core material. Superimposed
on the hardness measurements is the location of the shell
relative to the running surface. Similar relationships between
microhardness measurements and the shell location have been
noted for other rails [4]. The shell typically forms at the
interface of the cold worked material and the soft material
in the bulk of the rail. At such locations large residual
tensile stresses have been observed [6]. These large tensile
residual stresses, combined with the live wheel loads, produce
the cyclic stresses that can cause cracks to initiate and grow.

DISCUSSION

Hypothesis for Shell and Detail Fracture Development

Lack of space prevents a rigorous presentation of the
hypothesis and associated fractographic evidence. Salient
features are presented here to put forward the argument.
The interested reader may refer to the original report for
details [4].

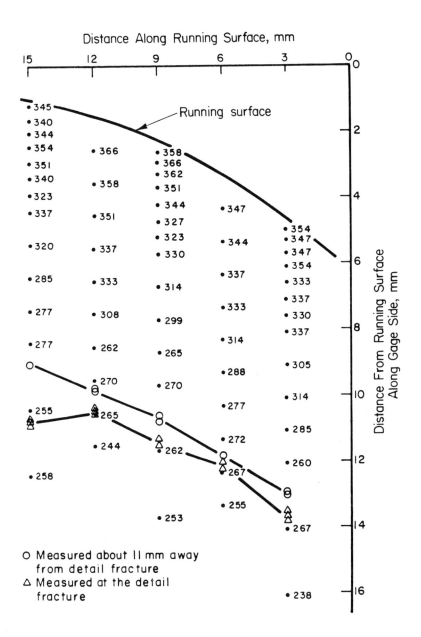

Figure 14. Shell position relative to the running surface in rail 5. Vicker's hardness readings show that shell forms at the interface of cold worked and soft material.

Figure 14 shows that the shell forms just underneath the region of increased hardness, where the hardness has returned to its normal value. This observation leads to an important conclusion which forms the basis for the hypothesis. The increased hardness is due to plastic deformation which, in turn, occurs due to passage of wheels and leads to an attempted expansion of the top layer of the rail. As this expansion is constrained, a residual compressive stress system is built up causing a strain equal and opposite to the plastic expansion, so that the resulting deformation is zero. The constraining action is performed by the substrate material, which as a consequence, develops residual tensile stresses. These tensile stresses are highest just below the plastically deformed top layer. Hence, if the shell forms just below the plastically deformed layer, then it follows that it forms in the region of highest residual tension.

The next logical step was to develop qualitative residual and live stress distributions. The residual stress distribution was obtained from the work of Groom [6] while the live stress distribution was obtained from the work of Johns et al [7,8]. The rail undergoes cyclic stresses because after a wheel has passed, the live load returns to zero, and the stresses are again equal to the residual stresses. Thus, the stress fluctuation between residual stress and residual plus live stress constitutes the cyclic stresses. The cyclic stress distribution so derived is shown in Figure 15 [4]. Although cyclic compressive stresses are of importance to crack initiation, they are not very important for crack growth. Therefore, only the tensile part of the cyclic stresses are shown.

Figure 15 basically presents the hypothesis, which is as follows:

> High cyclic compressive stresses occur in the rail head. These stresses cause fatigue damage and occasional cracks, especially at sites where nonmetallic (oxide) inclusions are present. However, most of these cracks are of little consequence, and will normally show erratic growth, if any. A few cracks which develop in the area of high cyclic tensile stresses (immediately under the plastically deformed layer of material), will grow as a result of the cyclic in-plane tensile stresses. Such a crack develops into a shell.

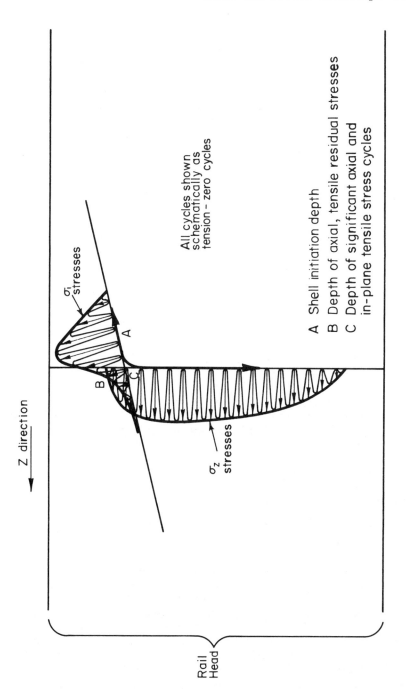

Figure 15. Cyclic stresses on the plane of the detail and on a plane slightly inclined to the shell.

All cracks wander, and show perturbations from their "straight" path. If a shell wanders upward (inclined plane in Figure 15) it enters a region with rapidly decreasing cyclic in-plane tension. Therefore, the perturbation will not persist, and another perturbation will keep the shell approximately in its plane. A perturbation downward also causes the crack to enter a region of rapidly decreasing cyclic in-plane tension. However, a perturbation downward also turns the plane of the crack into a direction of high cyclic tension in the Z-direction. Hence, such a perturbation can persist, turning the crack decisively downward where it encounters increasing cyclic tension in the Z-direction, and the detail crack is formed. Depending upon the relative position of points A, B, and C a shell may or may not form a detail. These positions will, of course, differ from rail-to-rail (not all rails develop fractures and detail cracks).

Inferences Made From the Hypothesis and Fractographic Confirmation

If the hypothesis holds true, then many features of the development of shell and detail cracking can be inferred. Several inferences will be made here, and fractographic evidence will be presented as confirmation.

The first inferences are made on the basis of Figure 16, which shows the distributions of cyclic tension for an evolving shell and detail crack.

Inference 1

Shells will tend to persist on a plane perpendicular to the direction of maximum cyclic tension. This plane is coincident with the plane of maximum in-plane residual tension. From the directions of in-plane residual tension it follows that the angle (Figure 16) of the shell must be from 20-30 degrees (of course varying slightly from rail to rail due to variations in usage, rail wear and consequent residual stresses). The shell should persist in this plane (be relatively straight, instead of curved) because the direction of principal stresses varies only slightly along the plane.

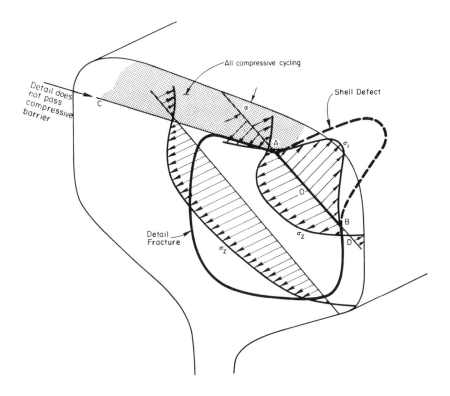

Figure 16. Detail and shell development and associated cyclic stresses.

Confirmation of this is found in Figures 3 and 17. The shell is at an angle of 20-30 degrees with the running surface and is very straight. The same evidence was found in all other rails [2-4].

Inference 2

The shell will spread only from its origin to A and B in Figure 16, because there is no cyclic tension on the plane of the crack beyond A and B. The detail will spread through the head, but it finds a "barrier" along the line A-B above which there is no cyclic tension. As A-B is fairly straight and slightly inclined downward to C, the top of the detail will be straight and probably slightly inclined -- the detail will not enter the top 0.4 to 0.5 inch or so of the rail. By the same token, the detail will probably not progress past D

Figure 17. Fracture surface of rail 15 (2.5X).

(in Figure 16) and, therefore, seldom break out at the gage side (at least until the detail crack becomes sufficiently large to alter the axial residual stress field). Confirmation of these points is obtained from Figures 3 and 17, as well as from all other fractures which have been examined [2-4].

Inference 3

As the detail develops due to a perturbation of the shell, the material above the shell being intact, the detail is at all times anchored (or pinned) to the shell. Thus, as shown in Figure 18, the crack front of the shell at successive positions is anchored to the shell. Consequently, if any growth marks (successive crack front positions) exist, they will show a typical pattern similar to the one shown in Figure 18.

Confirmation of this point is abundant; see Figures 3 and 17, and, in particular, Figure 18.

Figure 18. Fracture surface of rail 1 (2.5X).

Inference 4

The detail forms by a perturbation of the shell. Hence, at the origin of the detail, there should be a gradual turn from the shell into detail. However, once the detail persists it becomes a separate crack (although it is anchored to the shell). Therefore, beyond the initiation site there will generally be a sharp transition from the shell to the (independent) detail. This is shown in Figure 19. There is ample fractographic evidence to support this argument [2-4].

There are other inferences that need further confirmation. Shells and details propagate as a result of high cyclic tension and they grow on principal planes (no shear). The heavy shear deformation and extreme microstructural disarray on planes of shear may sometimes have a crack-like appearance, but the shell normally forms on a lower plane. Fractographic evidence is lacking at the present time to prove this observation. Also, the live stress depends strongly on wheel load and are mostly compressive. The cyclic tensile stress, therefore, is determined primarily by the residual stress and is hardly effected by wheel load. It has also been observed that the residual stress is not affected greatly by wheel loads and

4.25X (a)
Sharp corner, rail No. 4

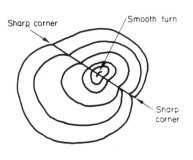

(c)
Schematic of shell and detail transition

(b)
Magnified view of sharp corner, rail No. 4, 64X

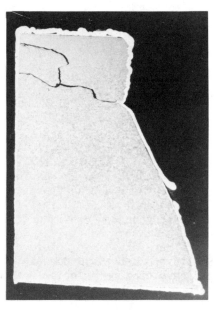

(d)
Smooth transition, rail No. 15, 2.4X

Figure 19. Shell and detail transitions in rail Nos. 4 and 15.

MGT [4,6]. Hence, the crack growth rates should depend mostly on the number of wheel passages, and be weakly dependent upon wheel load and MGT. Confirmation of this should be obtained from the rails tested at the Facility for Accelerated Service Testing (FAST) where actual growth measurements can be made.

CONCLUSIONS

This most recent study of detail fracture growth patterns in railroad rails has led to the formulation of what appears to be a rational and consistent explanation of the cracking process. The hypothesis which has been presented has been confirmed, in large part, by the fractographic evidence presented in this and previous reports.

The hypothesis suggests that shells can propagate in tension and Mode I (crack opening) if they happen to initiate in the zone just below the work hardened surface layer of the rail head. (Although this hypothesis does not specifically address crack initiation, evidence provided in this report suggests that most shells do initiate at oxide inclusions or stringers that happen to be located in this region of high multiaxial stress). Depending on the exact nature of the residual stress field and the alignment of the shell, this horizontal flaw will continue to grow under the influence of substantial longitudinal tensile stress excursions. If stress conditions are right, a downward perturbation in the direction of the shell will cause it to turn gradually into a transverse plane and once again grow in Mode I, but as a detail fracture rather than a shell. The detail fracture will be constrained to develop as an embedded flaw in the early stages of growth because the crack will be impeded at the gage side and surface of the rail head by compressive longitudinal stresses. After the detail flaw grows through a significant portion of the rail head (usually about 35 percent), and the internal tensile residual stresses are redistributed, the flaw will break out on the gage side of the rail. At this point the growth rate of the flaw will accelerate significantly and failure of the rail will be imminent.

ACKNOWLEDGEMENT

The work has been supported by the Department of Transportation under Contract No. DOT-TSC-1708. Many helpful discussions with Dr. Oscar Orringer and Mr. James Morris of the Transportation Systems Center are gratefully acknowledged. The support of Battelle's Columbus Laboratories in preparing the manuscript is appreciated.

REFERENCES

1. G. A. Mack et al., "Statistical Analysis of Burlington Northern and Atchison, Topeka and Santa Fe Rail Failures," Task 5 Final Report to the Department of Transportation from Battelle-Columbus Laboratories, Contract No. DOT TSC-1708 (June 8, 1982).

2. R. Rungta et al., "Post-Service Rail Defect Analysis," First Interim Report to DOT by Battelle-Columbus Laboratories (July 1982).

3. R. C. Rice et al., "Post-Service Rail Defect Analysis," Second Interim Report to DOT by Battelle-Columbus Laboratories (January 1983).

4. R. C. Rice et al., "Post-Service Rail Defect Analysis," Third Interim Report to DOT by Battelle-Columbus Laboratories (July 1983).

5. R. E. Cramer, "Eighth Progress Report of the Shell Rail Studies at the University of Illinois," Proceedings AREA, Vol. 51, pp. 597-607 (1950).

6. J. J. Groom, "Residual Stress Determination in Railroad Rails," Final Report on Task 2 for Contract No. DOT TSC-1426 from Battelle-Columbus Laboratories to the Transportation Systems Center (1979).

7. T. G. Johns et al., "Engineering Analysis of Stresses in Railroad Rail: Phase I," Battelle Report to TSC (1977).

8. T. G. Johns et al., "Engineering Analysis of Stresses in Railroad Rails," Federal Railroad Administration Report, Contract No. DOT/TSC-1038 (1981).

HYDROGEN EMBRITTLEMENT OF AISI 316 STAINLESS STEEL

R. C. Wasielewski*
and
M. R. Louthan, Jr.*

ABSTRACT

Cathodic charging of hydrogen into AISI 316 stainless steel and subsequent aging treatments at room temperature caused extensive surface cracks to develop. Cracking morphology was dependent on the charging current density, time and the metallurgical condition of the sample. Specimens were tested in the high-energy-rate forged, fully annealed, and sensitized conditions. Metallographic observations of non-sensitized samples showed that slip bands were more susceptible to hydrogen embrittlement than were twin boundaries or grain boundaries. Sensitization increased the susceptibility of the steel to hydrogen embrittlement by increasing the ease for grain cracking. These combined results are used to explain the effects of thermomechanical treatments on the hydrogen compatibility of austenitic stainless steels.

INTRODUCTION

The adverse effects of hydrogen on the mechanical behavior of austenitic stainless steels have been recognized for over two decades [1-5]. In spite of this, the technical literature continues to contain references to the resistance

---

* Materials Engineering, Virginia Polytechnic Institute, Blacksburg, VA 24061 USA.

of such alloys to hydrogen embrittlement [6-7]. One reason for such apparently contradictory results is the large influence of thermomechanical treatment on the susceptibility of austenitic lattices to hydrogen induced damage. For example, tests with a series of 21-6-9 stainless steel samples, processed by a variety of techniques and annealed to various strength levels and microstructures, showed that while some specimens were unaffected by exposure to high pressure hydrogen other samples of nearly identical composition underwent very large scale hydrogen induced reductions in ductility [8]. These studies, and those of other investigators [9-11], concluded that hydrogen absorption reduced the strength of grain boundaries, twin boundaries and other interfaces. However, the relative susceptibility of the various metallurgical interfaces to hydrogen-induced cracking in austenitic steels has not been established. This lack of information on the relative importance of metallurgical structure to hydrogen-induced cracking has led some investigators to conclude that if we are to improve our understanding of hydrogen embrittlement of stainless steels a detailed microstructural picture of hydrogen effects on crack propagation must be obtained [12].

Cathodic charging of hydrogen into austenitic steels will cause extensive surface cracking and can promote a variety of phase transformations in the fcc austenite lattice [13-14]. Examination of micrographs of surface cracked, cathodically charged samples, shows that cracking occurs along grain boundaries, twin boundaries and other crystallographic traces. Differences in the extent of surface cracking, resulting from a specific cathodic charging treatment, have been used to rank the relative susceptibility of steels to hydrogen embrittlement [6]. Furthermore, it has been suggested that a lack of cracking in a steel during cathodic charging corresponds to a compatibility of that steel with high pressure gaseous hydrogen. In any event, both electrochemical charging of steels with hydrogen, and plastic deformation of steels in high pressure hydrogen gas, can cause cracking along specific microstructural features. Modification of these microstructural features by thermomechanical processing changes the susceptibility of steel to hydrogen-induced effects [8]. However, the reasons for the changes in susceptibility are relatively obscure partially because very few attempts have been made to rank the relative susceptibility of microstructural defects to hydrogen embrittlement. This paper develops such a ranking based on a metallographic study of surface cracking caused by cathodic charging.

Figure 1. Microstructure of as-forged 316 stainless steel (200X).

MATERIALS AND METHODS

The AISI 316 stainless steel used in this study was supplied by Sandia National Laboratory, Livermore as high-energy-rate forged (HERF) plate stock having the composition given in Table 1. The microstructure of the as-forged material (Figure 1) was typical of other HERF austenitic steels [15] and contained highly elongated grains with deformed annealing twins and wavy slip bands. Sections of the as-forged material were heat treated to vary the microstructure without changing the chemical composition of the material. The heat treatments are summarized in Table 2 and the resulting microstructures are shown in Figure 2.

Table 1. Chemical Composition Range for 316 Stainless Steel

C	Mn	P	S	Si	Cr	Ni	Mo	Fe
.08	2.0	.045	.030	1.0	16.0–18.0	10.0–14.0	2.0–3.0	balance

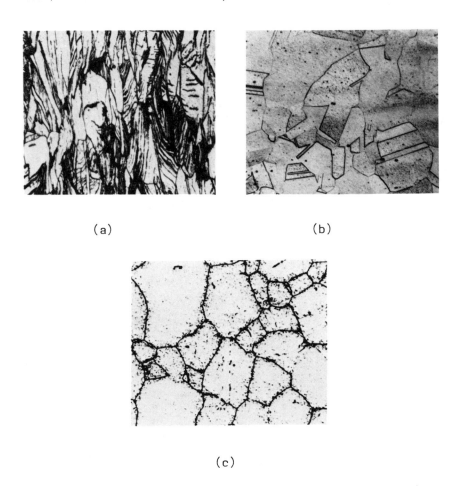

(a) (b)

(c)

Figure 2. Microstructures of heat-treated type 316 stainless steel samples (160X). (a) As-forged and sensitized structure, (b) Recrystallized structure, (c) Recrystallized and sensitized structure.

Sensitization of the HERF structure caused minimal changes in the optical metallographic appearance (Figure 2a) while the fully recrystallized structure contained numerous straight edged annealing twins and relatively clean grain boundaries (Figure 2b). Sensitization of the recrystallized structure caused extensive carbide precipitation at the grain boundaries and a minimal amount of precipitation along annealing twins (Figure 2c).

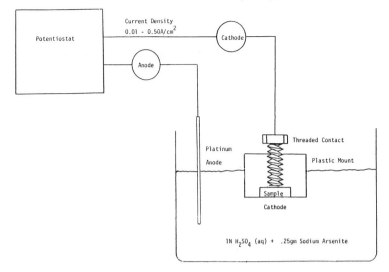

Figure 3. Sketch of apparatus used to cathodically charge metallographic sections of test samples.

Table 2. Heat Treatments for As-Forged Type 316 Stainless Steel

1. As-Forged
   A. No Heat Treatment
2. Recrystallized
   A. 1050 C for 1 hour
   B. Air Cooled to Room Temperature
3. As-Forged and Sensitized
   A. 650 C for 24 hours
   B. Air Cooled to Room Temperature
4. Recrystallized and Sensitized
   A. 1050 C for 1 hour
   B. Then, 650 C for 24 hours
   C. Air Cooled to Room Temperature

Metallographically mounted samples of each microstructure were charged with hydrogen in the apparatus sketched in Figure 3. Each polished (0.05 micron finish) sample was cathodically charged in 0.5 liters of a fresh 1N sulfuric acid solution at 20 C. The charging solutions were poisoned with 0.25 gms. of sodium arsenite. Charging times ranged between 1 and 24 hours. The constant cathodic current density was controlled at selected values between 0.01 and 0.5 amps/cm^2 and the electrolyte was agitated during the charging operation.

The cathodic charging caused surface deformation and subsequent cracking of all samples. Estimates of the susceptibility of the various microstructural features to hydrogen-induced damage were made by optical metallographic examination of charged surfaces.

RESULTS AND DISCUSSION

Cracking Sequence

The cathodic charging process caused cracks to develop in the near surface region of all charged samples. Hydrogen absorption in a stable austenite at room temperature is accompanied by a reversible expansion of the lattice [13]. This expansion, $\gamma \rightarrow \gamma^*$, has been termed a phase transformation and the expanded austenite, γ^*, has even been considered to be a fcc hydride phase [14]. Cathodic charging of alloys similar in stability to AISI 316 stainless steel used in this study also causes the formation of both the α' and ε martensitic phase [10,13,14]. These hydrogen-induced phases probably appear in the sequence $\gamma^* \rightarrow \varepsilon \rightarrow \alpha'$. Thus, when AISI 316 is hydrogen charged, the austenitic lattice is deformed and new phases can be produced. The deformation or expansion of the near surface lattice places that region in compression because of the constraint due to the underlying unexpanded, hydrogen-free metal. With continued charging these compressive stresses exceed the yield strength of the steel and cause the surface to develop ripples along (111) traces (Figure 4). The ripples are regions of large scale, localized, plastic strain. After

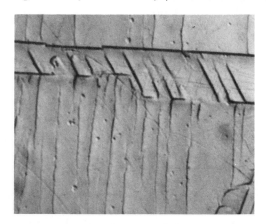

Figure 4. Surface deformation caused by cathodic charging. Note rumples at bottom of photomicrograph. These surface deformations lead to cracks with continued offgassing (750X).

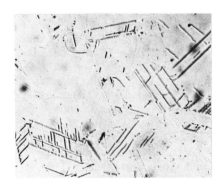 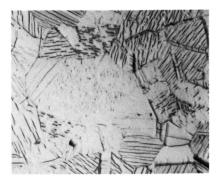

10 min. after charging   5 h after charging

Figure 5. Surface cracking accompanying offgassing of cathodically charged samples of recrystallized type 316 stainless steel. Note orientation dependence of cracking and the lack of cracking in selected grains (200X).

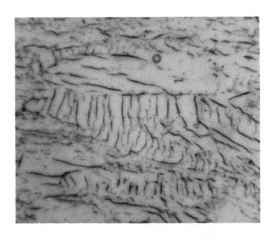

Figure 6. Surface cracking accompanying offgassing of cathodically charged samples of as-forged type 316 stainless steel. Note the tendency for the deformation bands to crack (250X).

completion of the charging sequence, the surface of the samples begins to outgass or lose hydrogen and the expanded $\gamma*$ lattice begins to contract. Because irreversible, plastic compressive strains accompanied charging, this lattice contraction of the austenite places the surface in tension. These tensile stresses can lead to cracks in the charged region. The magnitude of the tensile stresses increases as outgassing continues. Thus, surface cracking becomes more severe with increasing time after charging (Figure 5). Reference to the microstructures shows that in the fully annealed structure, the hydrogen-induced cracks occur on grain boundaries and twin boundaries. Single surface stereographic analysis shows that the intergranular cracks which also occur are along (111) traces. These results with annealed samples are quite similar to the effects of hydrogen charging on the as-received HERF material, where cracks were observed on grain boundaries, slip bands and deformed annealing twins (Figure 6).

This series of tests showed that the cracking sequence for AISI 316 stainless steel coupons, charged with hydrogen while exposed to no externally applied stresses, is:

1) Surface deformation and compressive plastic flow due to the expansion of the lattice during charging (because the diffusivity of hydrogen in austenite is approximately 10^{-12} cm^2/sec at room temperature [16], very steep concentration gradients were obtained and the shape of this gradient contributes significantly to the stress development).

2) Tensile stresses begin to develop in the surface as hydrogen outgassing occurs. These tensile stresses are in a lattice which contains dissolved hydrogen even though the hydrogen content is decreasing.

3) The combination of high tensile stresses and high hydrogen content in the lattice at near surface region causes hydrogen embrittlement. This embrittlement is seen as surface cracking.

4) The surface cracks occur along grain boundaries, annealing twins and slip bands or (111) traces.

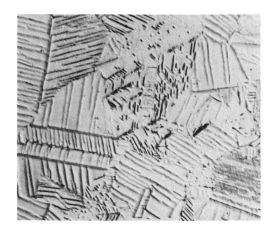

Figure 7. Photomicrograph illustrating initiation, growth and coalescence of surface microcracks in recrystallized type 316 stainless steel. Note small cracks in the center grains, the surface rumpling in several grains and the long cracks in the upper center of photomicrograph (250X).

(a)   (b)

Figure 8. Photomicrographs illustrating crack paths in sensitized type 316 stainless steels. (a) Recrystallized and sensitized. Note tendency for grain boundary cracking (b) As-forged and sensitized. Note slightly increased tendency for grain boundary cracking.

Susceptibility and Microstructure

The experimental studies to establish the cracking sequence for cathodically charged austenitic steel samples showed that the crack path was typically along grain boundaries, twin boundaries and deformation bands. To establish the relative susceptibility of these microstructural features to hydrogen-induced cracking, low cathodic charging current density experiments were performed. The deformation bands in both HERF and the recrystallized samples appeared to be the first crystallographic feature to crack. The cracking occurred by the initiation, growth and coalescence of microcracks, which were only a fraction of a grain diameter (Figure 7). Sensitization heat treatments of the annealed samples changed the relative susceptibility of the microstructural features to hydrogen embrittlement by causing the grain boundaries to be the most susceptible feature to crack formation (Figure 8a). However, sensitization had little effect on the cracking tendency in the HERF structures (Figure 8b).

The observation of preferential cracking along deformation bands and twin boundaries is in agreement with the results of hydrogen embrittlement studies made by tensile testing of austenitic steels in high pressure gaseous hydrogen. The high pressure gas phase studies also showed that hydrogen exposure promoted failure along both twin boundaries and slip bands [8]. Such results lead to the conclusion that hydrogen absorption reduced the strength of the various metallurgical interfaces. The strength reductions are assumed to be dependent on the hydrogen concentration at the interface. Clearly, the observation that the preferred crack path can be changed by heat treatments, such as sensitization, demonstrates the importance of thermomechanical treatments to the compatibility of austenitic steels in hydrogen environments.

A model illustrating the relative importances of interfaces, heat treatment, and hydrogen content in the embrittlement process is sketched in Figure 9. The slopes of hydrogen-induced reductions in interfacial strength are shown at different values because the character of a coherent twin boundary is far different from that of a slip band or grain boundary. Sensitization causes chromium depletion at grain boundaries as well as the precipitation of carbides along the boundary. The combined action of both of these factors lowers the hydrogen compatibility of the interface, and thus promotes hydrogen embrittlement in sensitized lattices. Because sensitization induces a change in relative hydrogen compatibility of the grain boundary, gas phase hydrogen tests with annealed steels in 69 MPa hydrogen show little or no intergranular

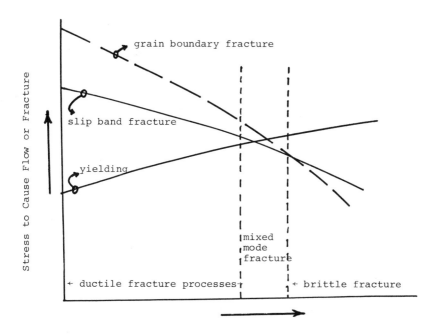

Figure 9. Sketch depicting effects of hydrogen on yield strength and interfacial strength of austenitic steels.

cracking while tests with sensitized steels show large areas with an intergranular fracture morphology. The steep slope for hydrogen-induced reductions in grain boundary strength is sketched in Figure 9 to be consistent with sustained load crack growth studies [17]. These studies show that the grain boundary is the preferred fracture path for samples tested in 172.5 MPa hydrogen but that tensile tests in 69 MPa hydrogen show the preferred path to be twin boundaries and slip bands.

The effects of hydrogen on the various microstructural features, as illustrated in Figure 6, demonstrate the difficulty in determining the compatibility of any particular steel to hydrogen embrittlement. Grain size, hydrogen content, heat treatment, slip band structure, twin morphology and inclusion or particle distribution all affect the relative compatibility of the steel with hydrogen. This electrochemical study has shown that the crack path for hydrogen-induced fracture can be determined by cathodic charging studies, and that these susceptible paths are consistent with the results of tests in gaseous hydrogen. However, the combined results of these experimental charging studies and previous work demonstrate that absolute measures of hydrogen compatibility cannot be made without the inclusion of a relatively complete description of the microstructure and thermomechanical history of the steel of interest.

CONCLUSION

Cathodic charging experiments have shown that hydrogen-induced cracking in AISI 316 stainless steel is a sensitive function of both the hydrogen charging current density and the specimen microstructure. Slip bands and twin boundaries were shown to be the most susceptible microstructural features to hydrogen cracking in non-sensitized structures. Sensitization was shown to increase the susceptibility of grain boundaries to hydrogen embrittlement. These results are consistent with observations made with samples tested in high pressure gaseous hydrogen and thus demonstrate that cathodic charging techniques can be used in testing for hydrogen compatibility in gaseous hydrogen.

REFERENCES

1. R. M. Vennet and G. S. Ansell, Trans. ASM, Vol. 62, p. 1007 (1969).

2. M. L. Holzworth, Corrosion, Vol. 25, p. 107 (1969).

3. M. R. Louthan, Jr., et al., Mater. Sci. Eng., Vol. 10, p. 357 (1972).

4. N. A. Nielsen, Journal of Materials, Vol. 5, p. 821 (1970).

5. M. R. Louthan, Jr. Hydrogen in Metals, ed. I.M. Bernstein and A. W. Thompson, ASM, p. 53 (1974).

6. J. Chene and M. Aucouturier, Hydrogen Effects in Metals, ed. I.M. Bernstein and A. W. Thompson, AIME, p. 583 (1981).

7. Air Force Materials Handbook.

8. A. J. West and M. R. Louthan, Jr., Met. Trans., Vol. 13A, p. 1049 (1982).

9. G. R. Caskey, Jr., Environmental Degradation of Engineering Materials in Hydrogen, Virginia Polytechnic Institute, p. 283 (1981).

10. C. L. Briant, Hydrogen Effects in Metals, ed. I.M. Bernstein and A. W. Thompson, AIME, p. 527 (1981).

11. H. Hanninen, et al., Environmental Degradation of Engineering Materials in Hydrogen, Virginia Polytechnic Institute, p. 347 (1981).

12. C. L. Briant, Environment Degradation of Engineering Materials in Hydrogen, Virginia Polytechnic Institute, p. 3335 (1981).

13. M. L. Holzworth and M. R. Louthan, Jr., Corrosion, Vol. 24, p. 110 (1968).

14. D. Eliezer, Hydrogen Effects in Metals, ed. I.M. Bernstein and A. W. Thompson, AIME, p. 565 (1981).

15. M. R. Louthan, Jr., et al., Corrosion, Vol. 29, p. 108 (1973).

16. M. R. Louthan, Jr. and R. J. Derrick, Corrosion Science, Vol. 15, p. 565 (1975).

17. M. W. Perra, Environmental Degradation of Engineering Materials in Hydrogen, Virginia Polytechnic Institute, p. 321 (1981).

Microstructure-Fracture Relationships

Chairpersons: M. Ryvola and M.H. Rafiee

FRACTOGRAPHY AND METALLOGRAPHY
OF POLYMERS

Ian D. Peggs*

ABSTRACT

The increased use of polymers such as
polyethylene (PE), polyvinyl chloride (PVC) and
nylon in sewer pipe, fasteners, line pipe for
fluids and gases etc., has introduced a new range
of materials to fractography and failure analysis.
Failures in PE and PVC pipes due to fatigue, variable
rate tensile loading, environmental stress cracking
and improper fusion welding are described in terms of
fractographic evidence. While some features are similar
to metallic failures, several are significantly different.

INTRODUCTION

Plastics, more correctly known as polymeric materials
to the materials engineer, continue to replace metals in the
aircraft, automobile, chemical, fluid and gas distribution
industries.

A great deal of effort has been expended on chemical
formulation and manufacturing processes to develop more
chemically inert, higher strength and temperature resistant
polymers. In many instances, they are handled by engineers
and technologists applying conventional metals-related
engineering techniques to these new materials. Not
surprisingly, mechanical failures and other physical

---

* Hanson Materials Engineering (Western) Ltd., 7450-18 St.,
Edmonton, Alberta, Canada.

problems have occurred, the characteristics of which are only recently being collated and evaluated. Studies in the polymer failure analysis field have begun to delineate the differences between polymers and metals in the areas of mechanical properties, microstructure and fracture surface appearance.

Polymer Microstructure

Long-chain macromolecules developed from low-molecular weight monomers, such as ethylene, can be combined to form a polymer in linear, branched and cross-linked interaction. The linear and branched macromolecules are held together by van der Waals forces or hydrogen bonds with an energy of approximately 10 kJ/mol. The branching within the macromolecules and cross-linking between them are effected by chemical bonds with an energy of approximately 400 kJ/mol.

Under the effect of heat, the linear and branched polymers lose their bonding, the chains become mobile with respect to one another, and the material becomes plastic. On cooling the secondary bonding increases again, the material solidifies and crystallizes. Such materials are termed THERMOPLASTIC and include polyethylene (PE), polyvinyl chloride (PVC), polypropylene (PP) and polyamide (PA, nylon).

Polymers that are cross-linked will not become soft and plastic under the application of heat because the chemical bonds are not broken. These materials are THERMOSETS and include phenolics, polyesters and epoxies, most commonly the matrices of reinforced composites.

ELASTOMERS have fewer and generally longer cross-links than thermosets and thus have some flexibility at room temperature which is not increased on heating. They include natural rubber and polyurethane (PUR).

Epoxies, phenolics and polyurethane (PUR) are used as corrosion and abrasion resistant coatings on line-pipe, ship hulls, etc. Extruded polyethylene (PE) is used as a corrosion resistant jacket on line pipe. Moulded and extruded polyethylene (PE) is used for containers and, together with polyvinyl chloride (PVC), for natural gas distribution, water and sewer lines. Nylon is used for fasteners.

Polymers fabricated from one basic molecule are strictly termed homopolymers but if "alloying" occurs with two different molecules a copolymer results.

Linear macromolecules can form regions with a regular lattice-type structure termed crystallites and, because of the complexity of the chain, the degree of crystallization can never exceed about 80%. The remainder of the material remains amorphous. When an individual crystallite grows, it nucleates others which tend to orient themselves in a radial pattern which grows to form a spherulite. The spherulites are resolvable in thin microtome sections using transmission light microscopy. Due to their chemical cross-links, thermosetting polymers are always amorphous.

Mechanical Properties

Unlike metals, the stress/strain curves of polymers do not exhibit true proportional limits and yield points. The curves are extremely dependent on temperature and strain rate due to the visco-elastic nature of polymers. The polymer behaves midway between an elastic solid and viscous fluid. Strains to failure in excess of 1000% at low strain rates are not uncommon. At room temperature, high density PE has a tensile yield strength of approximately 27 MPa (4 ksi).

During deformation, if the loading rate is short compared to the time taken for macromolecular rearrangement to occur due to relaxation or thermal processes, the material is brittle and rigid. At lower loading rates rearrangement can occur and the polymer is tough and flexible. Such differences are clearly seen in Figure 1 where the low-strain rate fracture face is dull, white, and stringy while the high-strain rate face is shiny, smooth, and "conchoidal".

The temperature dependence of mechanical properties is demonstrated by changes in the shear modulus with reference to the glass transition temperature, the temperature at which the mechanical loss factor is a maximum. This is close to the temperature at which the amorphous material solidifies on cooling or softens on heating. Amorphous thermoplastic polymers start to flow about 60 C above the glass transition temperature whereas crystalline thermoplastics flow above the crystallite melting temperature,

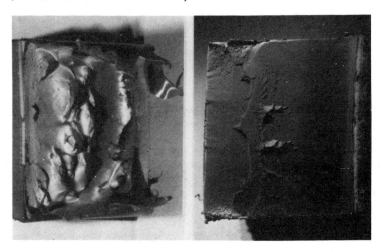

Figure 1. Slow (left) and rapid (right) fracture faces of polyethylene.

perhaps 150 C above the glass transition temperature. Thermosets show very little change in mechanical properties above the softening point until the decomposition temperature is reached.

Like metals, polymers suffer from fatigue failures but with two major differences caused by their visco-elastic nature. Some polymers, thermoplastic and thermosets, such as nylon, PE terephthalate and epoxy, do not show a low stress endurance limit, i.e., the S/N fatigue curve decreases linearly.

Analysis of the stress and strain cycles during fatigue testing show that there is a phase difference between the two, i.e., the strain cycle lags behind the stress cycle. The energy difference represented by this hysteresis is dissipated by heating. This is particularly evident adjacent to defects within the material where local softening, and even melting, can occur due to the relatively low thermal conductivity of polymers. Taken to an extreme, it is thus possible to obtain a ductile fatigue fracture face with no discernible striations. Such hysteretic heating is of less significance in the thermosets than in thermoplastics such as PP, PE and PA.

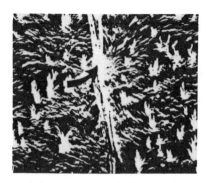

Figure 2. Spherulite, arrowed, nucleating on PE weld interface. Plasma etch.

Metallography

Conventional reflected light microscopy is primarily limited to macrostructural examination of polymers - the existence of cracks, voids, reinforcement orientation, coating adhesion, etc. The use of polarizing filters and angular reflected light are useful. Etching by chemical solvents or by plasma etching will reveal the spherulitic microstructure but these techniques are of most interest when examining features, such as welded joints, in the scanning electron microscopy, Figure 2. Scanning electron microscopy is, of course, invaluable for examining fracture surfaces.

Microstructures of crystalline polymers are best examined in thin (15 micrometer) microtomed sections using transmitted polarized light. In addition to the spherulites, processing flow lines and residual stresses are readily observed.

CASE EXAMPLES

Polyvinyl Chloride

PVC has been largely replaced by PE for small diameter natural gas distribution line pipe due in part to the higher susceptibility of PVC to environmental stress cracking (ESC), a similar phenomenon to stress-corrosion cracking in metals but producing a fatigue-type crack.

A section through the suspected fracture initiation site of a PVC pipe, which had been struck by a spade on the O.D., revealed a flat region normal to the inside surface of the pipe, Figure 3. A cursory inspection of the I.D. had previously shown no obvious defects but more persistent examination under varying lighting conditions showed an extensive array of parallel ESC, Figure 4.

A pipe coupling had been leaking through an improperly cemented joint, Figure 5. Closer examination revealed a brown stain on the surface outside the cemented joint covered by excess cement. Using a combination of polarized light and angular reflected light, Figure 6 was obtained showing the cement, the stained area, which resulted from the chemical contaminant entering the surface ESC cracks, plus extended ESC cracking which had not previously been identified.

Epoxy

Fusion bonded epoxy is becoming the predominant line pipe coating for corrosion protection. A cross-section examined by reflected light microscopy showed an excessive pore volume, Figure 7, but polarized light is required to define the surface defect. Note that the polishing orientation marks at the surface of the coating are not apparent under polarized light. Similar effects are noted at ESC cracks on an extruded PE jacket, Figure 8.

Polyurethane

Visual surface inspection of a 100% solids PUR coating on pipe showed no defects and the thickness was acceptable; but, almost continuous through-defects were indicated with a "holiday" detector. Examination of a section which flaked-off during a bend test revealed porosity that exceeded 75% of the coating thickness which reduced the interfacial bond strength permitting moisture penetration after the slightest abrasion.

Polyethylene

A sewer was crushed prior to installation in a pumped system. Longitudinal cracking eventually occurred along gouges on the O.D. surface. A region of partial cracking was opened up to reveal the fatigue crack shown in Figure 9.

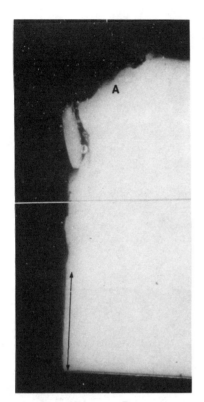

Figure 3.

Figure 4.

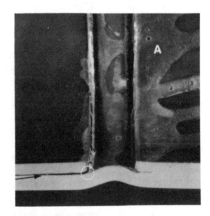

Figure 5.

Figure 3. Fracture initiated by ESC, arrowed, in PVC pipe. Polarized light.

Figure 4. ESC inside PVC pipe.

Figure 5. Cross section of leaking PVC pipe coupling. Excess cement - A.

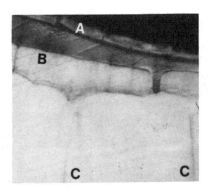

Figure 6. ESC in area circled on Figure 5. A - cement,
B - surface ESC, C - extended ESC.

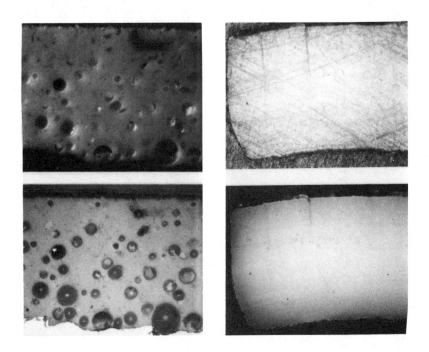

Figure 7. (Left) Epoxy coating with (top) and without polarizing filters. Note curled surface defect at top right.

Figure 8. (Right) ESC in PE with (bottom) and without polarizing filters.

Beach marks were not observed at low magnification. The "ripple" marks around the edge of the fatigue crack resulted from jerkily progressing, brittle, force fractures probably the result of handling after excavation.

Moisture present during the extrusion of a PE pipe jacket produced dimples on the surface of the coating causing failure during an ESCR bend test, Figure 10. An axial section through the coating, examined with combined polarized and angularly reflected light, clearly revealed the deep stress cracks at the dimples, Figure 8. Note that no crack opening is apparent.

Butt fusion welding is the process used to join two pipe sections or a pipe to a stub-end for subsequent mechanical joining. The two surfaces are cleaned, aligned and softened against an interfacial heater plate. The plate is removed and the ends butted together and held until fusion occurs. The extruded material forms a bead on the I.D. and O.D. of the pipe, Figure 11. Occasionally misalignment occurs with consequent reduction in wall thickness and close alignment of I.D. and O.D. notch defects which can initiate early failures.

We have examined several pipe-to-flange welds made between different grades of PE which require temperatures differing by as much as 60 C to achieve identical viscosities for satisfactory fusion. Heater plate temperatures are not variable from side to side so adjustments are made with insulating spacers. In the field, control of such a procedure is difficult and a significant number of welds have failed when subjected to normal service stresses. Failures occur due to overheating causing material decomposition, gas formation, excessive fluidity and dry bonding, Figure 12.

Microtomed sections across the weld show the extrusion and welding flow lines and whether one, or several attempts, have been made to make the weld, Figure 13. This figure shows enclosed defects, dry welds and evidence of misalignment during welding.

Figure 14, using polarized light, shows that failure on the weld line has occurred only in the region of high residual stress. As the residual stress decreases, the fracture moves away from the weld line. In this particular case, a moulded flange was joined to an extruded pipe which, due to anisotropic differences in thermal expansion coefficients, resulted in a joint containing high residual stresses.

 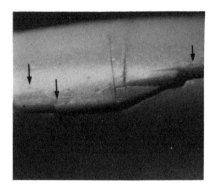

Figure 9. (Left) Fatigue crack in PE pipe wall.

Figure 10. (Right) ESCR test failure at surface dimples on PE coating.

Figure 11. Cross section of pipe to stub PE fusion welds. Poor alignment on right.

Polymers / 433

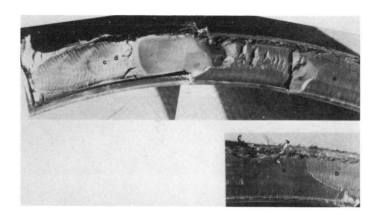

Figure 12. PE pipe weld failure, a) no bonding, b) poor bonding, c) good bonding, d) stepwise crack propagation.

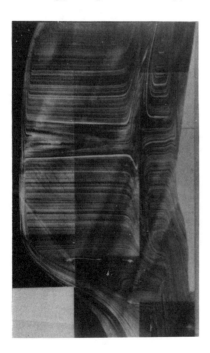

Figure 14. Fracture at PE weld. Weld line is top right to bottom left. Dark band, arrowed in flange, identifies residual stress. Microtome, polarized light.

Figure 13. Microtome of PE fusion weld. Flange black, pipe O.D. (top) pipe I.D. (bottom). First good weld on left. Two poor welds close together on right.

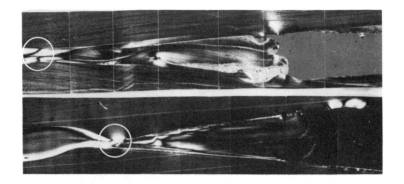

Figure 15. Microtome of extrudate between welded PE liners. Residual stress regions circled. Weld temperatures 400 C (top) and 500 C (bottom), polarized light.

Similar observations are made in welded seams in PE membrane liners for ditches and ponds. Excessively heated welds show much higher residual stresses in and adjacent to the weld, Figure 15, compared to welds made at the correct temperature.

CONCLUSIONS

While metallographic techniques similar to those used for metals can be employed with polymeric materials, the visco-elastic nature of polymers requires refinements in the approach. In reflected light microscopy both polarized light, and particularly variations in off-normal incident light, are of invaluable assistance in defining the extent of environmental stress cracking. Microtomed sections examined with transmitted polarized light reveal flow patterns, dry bonding, crystalline regions, and residual stress distributions.

OBSERVATIONS ON FATIGUE CRACK GROWTH/MICROSTRUCTURE
INTERACTIONS USING ADVANCED TECHNIQUES

Fraser Smith*
and
David W. Hoeppner*

ABSTRACT

Polycrystalline copper was loaded to fatigue failure by sinusoidally alternating uniaxial tension to observe the formation of defects and the propagation of subsequently formed cracks. The fatigue process was carried out in the viewing chamber of a scanning electron microscope (SEM) to enable both high magnification photographs and real-time videotape recordings to be taken. This report is a brief summary of the preliminary findings of these tests. Results of this study show general agreement with previous research of a similar nature that employed standard metallographic techniques. Thus, the coupling of the SEM and the fatigue unit provide a valuable adjunct to other methods of observing the formation of fatigue "damage". The principle advantage of the system used is the high resolution and ability to record the dynamic action of damage formation.

INTRODUCTION

The purpose of this investigation is to study the nature of the surface microstructural response of a model plastic material subjected to cyclic loading and to compare the results obtained with those generated by others [1-7] in the field using different equipment and specimen geometries.

---

* Department of Mechanical Engineering, University of Toronto, Toronto, Ontario, Canada M5S 1A4.

It is to be noted throughout this research that arbitrarily defining a crack as that separation of material that can be detected by the instruments that our current level of technology provides us is inadequate; rather it is the study of the mechanisms involved in the cracking process, from the time the crack becomes visible, that is of interest, regardless of whether the crack was present in the material to begin with or not.

EXPERIMENTAL DETAILS

Polycrystalline copper of nominal grain size ASTM 6.5 was selected due to its historical popularity as an approximation of an ideally plastic material. Metallographic sectioning and etching indicated that the material had been cold rolled and then had undergone a stress relief anneal. Tensile testing determined that the yield and ultimate strengths were 293.8 ± 5.4 [MPa] and 312.5 ± 2.2 [MPa], respectively.

The fatigue specimen geometry used (Figure 1) is non-standard; due to its small section depth, a fine crystal structure is preferred in order that sufficient intergranular constraint exists to provide a representative thickness. Figure 2 shows the specimen in place in the reaction frame along with the SEM viewing chamber. The details of the fatigue test unit are described in a recent paper [8].

In preparation for viewing, specimens were electro-polished and then lightly etched in a solution of 33% HNO_3 and 67% CH_3OH. It is therefore important to realize, when viewing the accompanying micrographs, that the surfaces were initially exceptionally smooth and flat (Figure 3).

OBSERVATIONS AND RESULTS

Specimen Cu #1 was unnotched and, as with all the specimens, it was loaded to 2/3 its yield strength and cycled at an R-ratio of 0.2 at frequencies never in excess of 20-Hz. As expected, the initial cycles led to widespread slip line formation, along with the fracture and extrusion of surface-embedded nonmetallic inclusions. However, after the first 100,000 cycles, strain hardening had caused a saturated hardness for that load span and the surface

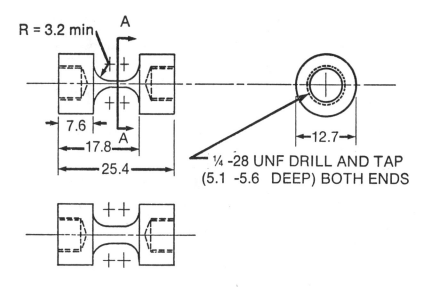

Figure 1. Specimen geometry.

Figure 2. Load frame coupled to SEM.

Figure 3. Specimen Cu #5: Notch area prior to loading.

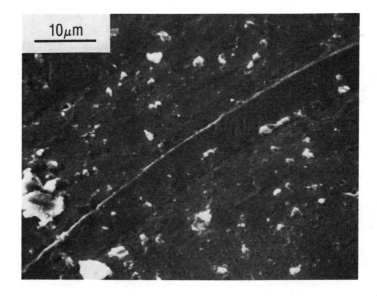

Figure 4. Specimen Cu #1: Twin; became visible gradually due to relief effects produced by mobile slip systems.

features remained constant until the test was discontinued at 600,000 cycles. Figure 4 shows one of many twins that became visible during the course of the test due to relief effects produced by mobile slip systems.

The rest of the specimens tested were notched as shown in Figure 3, enabling constant observation of the area of more probable crack origin. Figure 5 shows the specimen 20,000 cycles later, with a crack that can be easily spotted even though it is still very small. The presence of a crack off to one side of the specimen is very convenient in this research since it enables the observation of many fatigue effects at once. In the crack tip region the traditional "butterfly" type plastic zone is shown, while much farther into the specimen the parent material is cycling in a relatively stress concentration free setting. Therefore, there exists a transitional area which undergoes deformation as a function of its location between these two extremes. Areas experiencing close to far-field stress levels would act somewhat like specimen Cu #1; i.e., reach a saturated hardness level relatively quickly and then be in equilibrium for that stress range. On the other hand, areas close to the crack tip experience comparatively high loads and can conceivably never reach a saturated hardness level before the crack propagates through them. Figure 6 is a good example of a transitional area showing wavy slip, slip bands, intrusion-extrusion pairs and fissures (as a point of reference, the crack is approaching this area from the right hand side).

Figure 7 illustrates one of the inherent advantages of these real-time observational systems; that being, the capability of at least a limited view into a crack while the specimen is maintained under load. They are also of great assistance conceptually, as shown in Figure 8, where the mechanisms of complexing and, to a lesser extent, hindering are graphically embodied to a degree very hard to attain with conventional laboratory metallographic arrangements.

Correlations of these pictorial results with work done by both Wood, Forsyth, and previous research by the second author was excellent. Of interest is that Wood's and Forsyth's use of torsionally-loaded cylindrical copper specimens fatigued in air and periodically observed under an SEM, displayed similar surface microstructural effects to the uniaxially loaded specimens fatigued in vacuo.

440 / Microstructural Science, Volume 12

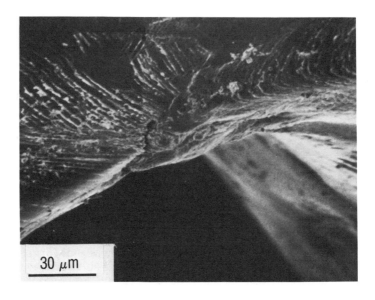

Figure 5. Specimen Cu #5: Crack site is clearly visible.

Figure 6. Transitional area of specimen surface showing fissures, wavy slip and slip bands.

Advanced Techniques / 441

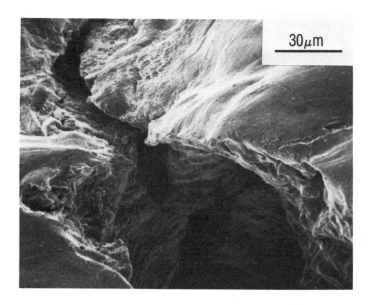

Figure 7. View of crack with specimen under load, close to notch root.

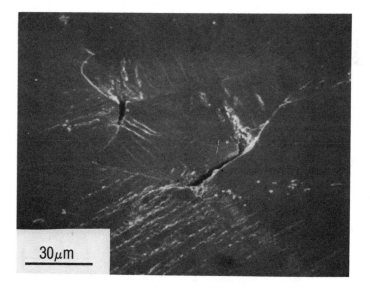

Figure 8. Grain being wrenched from the matrix due to complexing effect.

CONCLUSIONS

From the results of these studies the following conclusions are rendered:

1) Real-time viewing provides both visual and conceptual opportunities hard to come by without similar systems.

2) Results obtained herein are in agreement with previous research work.

Similar research is underway in the structural integrity, fatigue, and fracture research laboratories (SIFFRL) at the University of Toronto.

ACKNOWLEDGEMENTS

Most of the research reported herein is supported by Rolls Royce Aeroengine Division of Derby, England, and the Natural Sciences and Engineering Research Council of Canada (NSERC). The authors are grateful to R.R. and NSERC for support. In addition, we are extremely grateful to David Cameron (Ph.D. student at U.T.) and R. Jeal of R.R. for many stimulating discussions. Portions of this paper are based on the BASc. thesis of Fraser Smith, University of Toronto, 1983.

REFERENCES

1. P. J. Forsyth, The Physical Basis of Metal Fatigue, American Elsevier Publishing Co., New York (1969).

2. J. P. Hirth and J. Lothe, Theory of Dislocations, John Wiley & Sons, Second Ed., Toronto (1982).

3. R. W. K. Honeycombe, The Plastic Deformation of Metals, St. Martin's Press, New York (1968).

4. D. Hull, Introduction to Dislocations, Pergamon Press, Second Ed., Toronto (1975).

5. D. McLean, Grain Boundaries in Metals, London: Oxford University Press (1957).

6. W. A. Wood, "Four Basic Types of Fatigue," Treatise on Materials Science, Vol. 5, Academic Press, New York (1974).

7. S. Missana, A New Apparatus for Making In-Situ Observations of the Fatigue Process in Metals, MASc. thesis, University of Toronto (1983).

8. D. W. Cameron and D. W. Hoeppner, "A Servohydraulic Controlled Load Frame for SEM Fatigue Studies," to be published in International Journal of Fatigue.

FATIGUE CRACK INITIATION IN SHOT-PEENED BLENDED ELEMENTAL Ti-6Al-4V POWDER COMPACTS

D. Eylon*, C. M. Cooke*,
and
F. H. Froes**

ABSTRACT

Fatigue crack initiation mechanisms in shot-peened blended elemental Ti-6Al-4V cold-pressed and sintered powder compacts were investigated. The shot-peening was carried out at two intensity levels in an attempt to improve the fatigue life by closing the surface pores because past work has shown that pores are responsible for initiation of most premature fatigue cracks. The shot-peened specimens demonstrated higher than baseline fatigue strength which, however, was still lower than that of the fully dense prealloyed powder compacts. The fatigue crack initiation morphologies of the baseline and the shot-peened conditions were very similar and resulted from surface or near-surface porosity. Shot-peening did not heal the deformed surface pores, which resulted in only modest improvement in fatigue strength. After stress relief, the shot-peened compacts exhibited fatigue life lower than baseline, associated with multiple fatigue crack initiations. The surface damage caused by shot-peening is considered to be the cause for the multiple surface initiations.

---

* Metcut-Materials Research Group, P.O. Box 33511, Wright-Patterson Air Force Base, OH 45422 USA.
** Air Force Wright Aeronautical Laboratories, Materials Laboratory, Structural Metals Branch, Wright-Patterson Air Force Base, OH 45433 USA.

INTRODUCTION

Titanium alloy powder metallurgy (PM) has advanced to the stage that commercial products are now available for the aerospace and industrial markets [1] with prices and properties competitive with ingot metallurgy (IM) materials [2,3]. Titanium powder metallurgy is divided into two major categories: Prealloyed PM which provides fully dense compacts with properties equivalent to or exceeding those of IM materials; and, blended elemental (BE) pressed and sintered compacts which have good static properties [4], but exhibit somewhat lower fatigue strength; the latter property degradation relating to the residual porosity [5]. The blended elemental compacts were the subject of this work. This method provides low cost titanium alloy components because low cost starting powder (sponge fines) is used along with a low cost process of cold pressing and sintering [2]. The resulting components are close to net-shape which eliminates much of the costly machining expenses [6]. However, the inherent residual porosity limits this technology to non-fatigue critical components due to relatively lower fatigue life [2]. In past work, surface and near-surface porosity was identified as the source of early fatigue cracks [5]. The objective of the present work was to improve the fatigue strength by shot-peening the fatigue specimen surfaces in an attempt to close the surface pores and induce compressive residual stresses on the surface.

EXPERIMENTAL

Cold-pressed and sintered Ti-6Al-4V BE compacts were produced by Imperial Clevite with the chemical composition listed in Table 1. Smooth bar high-cycle fatigue specimens, with a gage length of 5 mm (0.2 in) diameter x 31 mm (1.25 in) long were machined from the compacted bars and processed as listed in Table 2. The shot-peening intensity is in Almen units [7], 6N designated the low intensity and 12N the high intensity shot-peening. Some of the shot-peened specimens of both groups were stress relieved in vacuum to check if the shot-peening effect is the result of structural changes or surface residual stresses.

Table 1. Chemical Composition of Ti-6Al-4V BE Compacts

Element	Al	V	Fe	C	N	O	H	Cl	Ti
wt pct:	6.0	4.1	0.17	0.02	0.007	0.191	0.001	0.14	Bal.

Table 2. Material Conditions

Condition Number	Shot-Peening Almen Intensity	Stress Relief Heat-Treatment	Designation
1	None	None	Baseline
2	0.006N	None	6N (low intensity shot-peen)
3	0.012N	None	12N (high intensity shot-peen)
4	0.006N	620 C (1150 F)/2h	6N plus stress relief
5	0.012N	620 C (1150 F)/2h	12N plus stress relief

Fatigue tests were performed on a Servohydraulic MTS machine at room temperature, ambient air, 5Hz frequency, triangular waveform, and in load control mode.

All fatigue fracture surfaces were examined by scanning electron microscopy (SEM) to study the nature of the crack initiation sites. Metallographic cross sections were made to check if the shot-peening caused any surface microstructural changes, and one specimen was precision sectioned [8,9] through the initiation site on a plane perpendicular to the fracture surface to study the microstructural nature of the initial cracks.

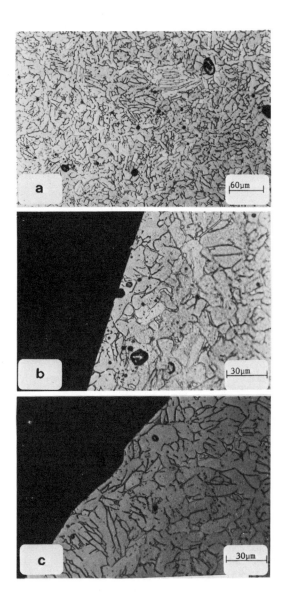

Figure 1. a) Condition 1 (baseline) cold compacted and sintered microstructure. b) Optical micrograph near the surface of a condition 1 fatigue specimen. c) Optical micrograph near the surface of a condition 5 fatigue specimen after high intensity shot-peen plus stress relief treatment.

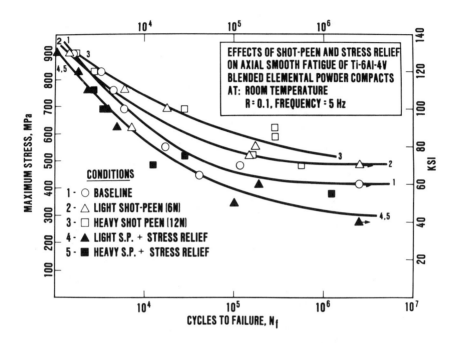

Figure 2. S-N fatigue curves for conditions 1-5.

RESULTS

The microstructure of conditions 1 and 5 are presented in Figures 1a, b, and c. It can be seen that the sintered material (cond. 1) is not fully dense, with a low aspect ratio alpha-plate structure. The shot-peening and the subsequent stress-relief heat treatment of condition 5 did not cause any microstructural change at the specimen surface and only a higher degree of surface roughness is evident (Figure 1c). Some near-surface pores are also visible in this photomicrograph.

The fatigue life vs. maximum stress (S-N) curves for all 5 conditions are shown in Figure 2. The baseline data (cond. 1) is in agreement with the results of previously published data [2,4,5]. The shot-peened specimens (cond. 2 and 3) display higher fatigue life curves except in the high stress (low cycle) regions in which fatigue lives similar to

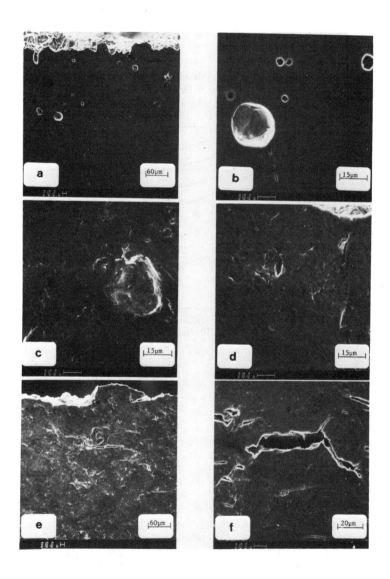

Figure 3. a) Low magnification SEM photomicrograph of condition 1 fatigue sample surface directly below the initiation site. b) High magnification SEM photomicrograph of same area shown in (a). c) SEM photomicrograph of a condition 2 (light shot-peen) fatigue sample surface. d) SEM photomicrograph of a condition 3 (heavy shot-peen) fatigue sample surface. e) SEM photomicrograph of a condition 4 (light shot-peen plus stress relief) fatigue sample surface. f) SEM photomicrograph of a condition 5 (heavy shot-peen plus stress relief) fatigue sample surface.

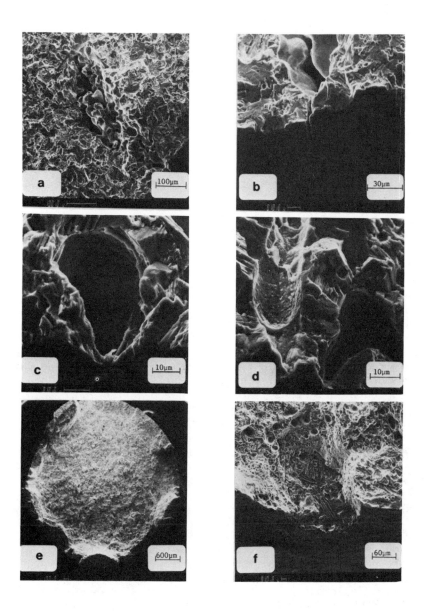

Figure 4. a and b) Typical pore initiations in a fatigue sample. c and d) Typical contaminants in fatigue initiation sites which are also associated with pores. e and f) Multiple initiation at the higher stress (low cycle) test regimes of conditions 4 and 5.

baseline material were measured. The higher intensity shot-peened condition 3 has the highest fatigue strength. The shot-peened and stress-relieved specimens (cond. 4 and 5) display fatigue life lower than the baseline material, especially in the low stress (high cycle) region.

The gage surface features for all 5 conditions, in a location directly below the fatigue crack initiation site, are shown in Figures 3a through f. The fractographic results are shown in Figures 4a through f and the SEM photomicrographs of a condition 5 precision-sectioned specimen are shown in Figures 5a through d.

DISCUSSION

The shot-peened and stress-relieved conditions (no. 4 and 5) have lower fatigue strengths than the baseline material, indicating that the fatigue life improvement of conditions 2 and 3 was primarily achieved by the compressive residual stresses in the deformed surface [10]. The higher fatigue curve of the high intensity shot-peened condition (no. 3) when compared to the lower intensity condition (no. 2) reinforces this conclusion. In this work, no attempt was made to measure the compressive residual surface stresses. Comparison of the photomicrographs in Figures 1b and c also indicates that there were no surface microstructural changes after shot-peening beyond the shallow surface deformation and the partial pore closure. A better view of the specimen surface topography for all 5 conditions is given by Figures 3a through f. The surface porosity of the baseline condition 1 (Figures 3a and b) is partially closed after the light shot-peening (cond. 2 in Figure 3c) and, as expected, only highly deformed large pores are visible on the surface of the high intensity shot-peened material (cond. 3 in Figure 3d). The stress-relief heat treatment seems to cause some morphological changes at the specimen surface, since there is no evidence of deformed surface porosity in conditions 4 and 5 (Figures 3e and f, respectively). Condition 4 (light shot-peen + stress relief in Figure 3e) exhibits limited surface cracking developed as a result of the fatigue loading. Condition 5 (high intensity shot-peen plus stress relief in Figure 3f) developed extensive interlinked surface cracking leading to multiple initiations especially in the high stress (low cycle) region. The nature of this secondary cracking will be discussed later with the help of precision sectioning observations.

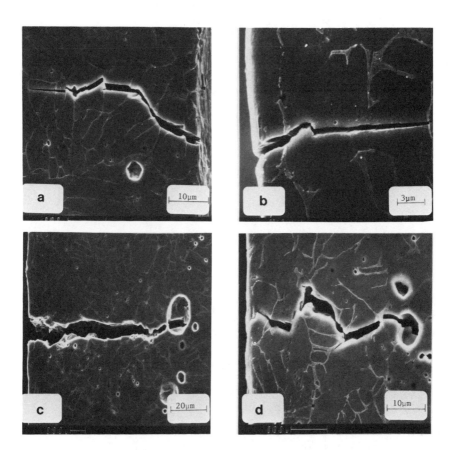

Figure 5. a and b) Precision sectioned condition 5 sample showing surface cracking not related to surface pores. c and d) Precision sectioned condition 5 sample showing surface initiated crack blunting at sub-surface pores.

Results of the fractographic crack initiation study are summarized in Table 3. A typical pore initiation site is shown in Figures 4a and b and contaminants initiation in Figures 4c and d. Most multiple initiations occurred in the higher stress (low cycle) regimes of conditions 4 and 5 and are shown in Figures 4e and f.

Table 3. Nature of Fatigue Crack Initiation in Ti-6Al-4V BE Powder Compacts

Cond.	Number of Specimens Examined	Single Initiation No. (%)	Multiple Initiation No. (%)	Pore* Initiation No. (%)	Contaminant Initiation No. (%)	Microstructural Initiation** No. (%)	Surface Initiation No. (%)	Near Surface*** Initiation No. (%)
1	8	6 (75)	2 (25)	8 (100)		– –	7 (88)	1 (12)
2	8	7 (88)	1 (12)	7 (88)		1 (12)	6 (75)	2 (25)
3	6	6 (100)	–	6 (100)		– –	3 (50)	3 (50)
4	8	4 (50)	4 (50)	8 (100)		– –	7 (88)	1 (12)
5	7	3 (43)	4 (57)	6 (86)	1 (14)	– –	4 (57)	3 (43)
TOTAL	37	26 (70)	11 (30)	35 (94)	1 (3)	1 (3)	27 (73)	10 (27)

* Includes also pores with contaminants.

** No relation to pore or contaminant.

*** 10 to 100 micrometers below surface.

It is evident from the fractographic work (Figures 4a through f) and the resulting Table 3, that conditions 1, 2 and 3 have the same crack initiation behavior, namely initiation at pores (deformed or undeformed) at surface or near-surface locations (Figures 4a and b). The specimen surface topographic observation of these conditions (Figures 3a through d) also indicate that deformed porosity exists even on the surface of the most heavily deformed shot-peened condition 3 (Figure 3d). However, the lack of visible surface porosity in the stress-relieved condition 4 and 5 (Figures 3e and f), which could be the result of localized recrystallization and/or pore healing (from the prior shot-peening), lead to a different type of multiple-surface crack initiation which is not exclusively related to the porosity. The section of a condition 5 specimen with extensive surface cracking (Figures 5a through d) shows that most of those initial cracks are not related to surface or subsurface pores. Those cracks are initiated at surface alpha grains in what appears to be a shear across alpha grains (Figures 5a and b) and some later link to subsurface pores (Figures 5c and d). In this case, the crack tip opening at subsurface pores seems to be quite large which indicates that linkage to those pores can lead to crack blunting and hence to crack growth rate slow-down. The fatigue strength improvement of the shot-peened conditions 2 and 3 could be interpreted on the basis of:

a. Increased compressive residual surface stresses
b. Blunting of the surface pores which leads to lower stress concentration factors

The substantial drop in fatigue strength of conditions 4 and 5 could be explained by:

a. Loss of surface compressive stresses by the stress-relief heat treatment
b. Promoting sheer cracking across alpha grains at the specimen surface

The statistical summary of the crack initiation morphology in Table 3 indicates the following points:

a. All initiation sites for all 5 conditions were either at or near the specimen surface. No failures were initiated deeper than 100 μm below the surface. Surface initiation was observed in 73% of the failures and near surface in 27%.

b. Crack initiation at pores (with or without contaminants) was observed in 94% of the cases. Only 3% initiated at contaminates not associated with a detectable pore, and only 3% at a non-defect related microstructural feature.

c. Most of the multiple initiation cases were in the stress-relieved conditions 4 and 5. Single initiation occurred in 70% of the total cases while multiple initiation only in 30% (mostly in the high stress/low cycle region).

The initiation statistical results are different than those found in fatigue fractures of prealloyed Ti-6Al-4V powders hot isostatically pressed (HIP) to full density [11] which showed most failure to be related to metallic and nonmetallic inclusions with many failures initiating at subsurface locations. However, it should be noted that the full density and the higher level of cleanliness of the prealloyed compacts resulted in higher fatigue strengths, which in some cases were even higher than ingot metallurgy (IM) material [12].

SUMMARY AND CONCLUSIONS

The effect of shot-peening on the fatigue strength and fatigue crack initiation mechanisms in blended elemental (BE) Ti-6Al-4V pressed and sintered powder compacts was investigated. Untreated baseline material, as well as shot-peened and stress-relieved conditions, were also tested to allow a more comprehensive study of the role of residual stresses and surface deformation on crack initiation mechanisms. It was found that:

1. Shot-peening improves the fatigue life of BE compacts with higher intensity shot-peening leading to further improvement.

2. Residual surface stresses and surface pore blunting are considered to be responsible for the fatigue strength improvement of the as shot-peened conditions.

3. Stress relief of shot-peened material leads to a drop in fatigue strength to a level even lower than the untreated baseline material.

4. In both the shot-peened and baseline conditions, fatigue cracks initiated at surface or near-surface pores. The similarity of the crack initiation mechanism is possibly due to the existence of unhealed pores in the shot-peened specimen surface.

5. Shot-peened and stress-relieved conditions exhibited multiple shear related surface cracking. The major failure occurred from shear cracks across alpha grains often combined with large near-surface pores.

6. It is considered that the loss of surface residual stresses, and the surface shear cracking, are responsible for the lower fatigue strength in the shot-peened and stress-relieved conditions.

7. Statistical analysis of the initiation morphology indicated that:

    a) All failures initiated at the specimen surface or very near the surface.
    b) Most failures were associated with pores.
    c) Most multiple initiation cases were in shot-peened and stress-relieved specimens tested at high stress levels.

ACKNOWLEDGEMENTS

The authors wish to thank Dr. P. Eloff, N. G. Lovell, W. A. Houston and R. D. Brodecki for their assistance. Parts of this work were done under USAF Contracts F33615-79-C-5152 and F33615-82-C-5078.

REFERENCES

1. L. Parsons, et al., "Titanium Powder Metallurgy Comes of Age," Metal Progress.

2. F. H. Froes, et al., "Developments in Titanium Powder Metallurgy," J. Metals, Vol. 32, pp. 47-54 (February 1980).

3. D. Eylon, et al., "Titanium Powder Metallurgy for Industrial Applications," to be published in ASTM STP Industrial Applications of Titanium and Zirconium (1983).

4. P. J. Anderson, et al., "Fracture Behavior of Blended Elemental P/M Titanium Alloy," Modern Developments in Powder Metallurgy, MPIF Publications, Vol. 13, pp. 537-549 (1981).

5. F. H. Froes, D. Eylon and Y. Mahajan, "The Effect of Microstructure and Microstructural Integrity on the Mechanical Properties of Ti-6Al-4V PM Products," Modern Developments in Powder Metallurgy, MPIF Publications, Vol. 13, pp. 523-535 (1981).

6. D. Eylon, et al., "Manufacture of Cost-Affordable High Performance Titanium Components for Advanced Air Force Systems," Proceedings of the 12th SAMPE Technical Conference, Seattle, WA, p. 356 (1980).

7. Military Specifications MIL-S-13165B, Amendment 2 (25 June 1979).

8. D. Eylon and W. R. Kerr, "The Fractographic and Metallographic Morphology of Fatigue Initiation Sites," ASTM STP 645, Fractography in Failure Analysis, ASTM, Philadelphia, PA pp. 235-248 (1978).

9. W. R. Kerr, D. Eylon and J. A. Hall, "On the Correlation of Specific Fracture Surface and Metallographic Features by Precision Sectioning in Titanium Alloys," Metallurgical Transaction, Vol. 7A pp. 1477-1480 (1976),

10. W. P. Koster, L. R. Gatto and J. T. Cammett, "Influence of Shot-Peening on Surface Integrity of Some Machined Aerospace Materials," Proceedings of the First International Conference on Shot-Peening, Paris, France, pp. 287-293 (1981).

11. S. W. Schwenker, D. Eylon and F. H. Froes, "Performance Optimization in PM Titanium Components," to be published in the Proceedings of 1983 SAMPE Meeting, Anaheim, CA.

12. P. R. Smith, et al., "Process Control for Mechanical Property Optimization of PM Titanium Compacts, Advanced Processing Methods for Titanium, ed. by D. Hasson, TMS-AIME Publications, pp. 61-77 (1982).

THE EFFECTS OF WELDING-INDUCED RESIDUAL STRESSES AND MICROSTRUCTURAL ALTERATIONS ON THE FATIGUE-CRACK GROWTH BEHAVIOR OF COMMERCIALLY-PURE TITANIUM

David C. Wu*
and
David W. Hoeppner*

ABSTRACT

An investigation was carried out to establish the effects of welding-induced residual stresses and microstructural alterations on the fatigue-crack growth behavior of an electron-beam welded model material, commercially-pure titanium. Fatigue-crack growth tests were performed on base material specimens and welded specimens with the weld in two orientations. The results of the experiments showed that over the range of fatigue-crack growth rates attained, welding-induced residual stresses were a major factor on the crack-growth behavior. Where applicable, the crack-growth data could best be presented as plots of the mode I crack-growth rate versus the effective mode I stress-intensity factor range. Correlated on the basis of the effective mode I stress intensity factor range, the macroscopic mode I crack-growth rates were not noticeably affected by the different microstructures in the neighborhood of the weld although fractographic examination of the fracture surfaces showed definite microstructural influences in the cracking behavior.

---

* Department of Mechanical Engineering, University of Toronto, Toronto, Ontario, Canada M5S 1A4.

INTRODUCTION

It is well known that welding can, and does, introduce physical discontinuities into weld joints. Most of these welding-induced discontinuities act as stress raisers and in the process enhance the nucleation and/or propagation of a crack under alternating stress, i.e., fatigue loading. To all intents and purposes then, the fatigue life of a weld joint can be considered to consist wholly of crack propagation. Although crack propagation in weldments is not wholly mode I, mode I fracture mechanics is still applicable to describe most of this process.

Previous studies of fatigue-crack growth in weld joints have largely been empirical. Differences in the observed crack-growth rate (da/dN) vs. the stress-intensity factor range (ΔK)* type of crack-growth data between welded and unwelded materials have been attributed to either welding-induced microstructural alterations [1] and/or residual stresses [2-6]. Measurement of residual stresses has not been a common practice.

There are two approaches in analyzing fatigue-crack growth in residual stress fields. One is based on the superposition of the applied and the residual stresses. Because of its simplicity, its validity is not often questioned. However, when a crack is propagating in the vicinity of a residual stress field, the residual stresses can be expected to relax in the plastic zone ahead of the crack. In addition, with an advancing crack residual stress re-distribution can also be expected to occur as traction-free surfaces are created.

An alternative approach based on the observation that propagating fatigue cracks have a tendency to close at a tensile load is termed the "crack closure model" [7,8] for fatigue crack growth (FCG). Essentially, this model states that crack propagation can only occur when the crack is open, and that the effective crack driving force may be derived from the effective load range which is the difference between the maximum applied load and the load at which the crack opens.

---

* All K's in this paper refer to mode I K's.

Hence, the observed da/dN may be better correlated with a parameter which represents this effective crack driving force, for example, the effective stress intensity factor range, ΔK_{eff}. It can be expected that a residual stress field would affect the opening of a fatigue crack. Bucci [9] has pointed out the importance of residual stresses and the possible use of crack closure concepts in FCG rate measurements. Nordmark et al [10] have applied the crack closure concept to describe FCG in high deposition rate welds in 5456 aluminum plate.

This study was undertaken in order to further elucidate the influences of welding-induced microstructural alterations and residual stresses on FCG in the vicinity of weld joints by utilizing crack closure concepts. WOL-type specimens, with the weld in two orientations parallel and perpendicular to the initial notch (LW- and TW-, TWR-specimens) as shown in Figure 1, were tested. In order to reduce the complexity of possible microstructural alterations around the weld joint, a model material and an autogenous welding process were used. The material was commercially-pure titanium. The welding process was electron-beam welding. The vacuum environment of welding and the fact that no filler material was required ensured that the chemical composition across the weldment would nominally be the same.

WELD PREPARATION: MICROSTRUCTURAL AND RESIDUAL STRESS CHARACTERIZATION

The weld preparation was that of a square butt. Two-pass electron-beam welds were made by clamping two 1.4-cm thick CP-titanium strips together in a welding jig and making one welding pass from each side.

A macrosection of the resulting weld joint is shown in Figure 2. The base material was composed of equiaxed alpha. The heat-affected zone (HAZ) was composed of acicular alpha near the unaffected base material, and martensitic near the weld metal. The weld metal was composed of martensitic alpha and some retained beta dendrites. Microhardness measurements showed that the weld metal was the hardest, followed by the martensitic HAZ, acicular HAZ, with the base material being the softest.

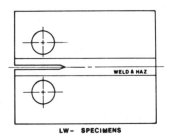

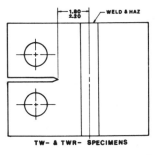

Figure 1. Orientations and locations of welds in standard WOL specimens.

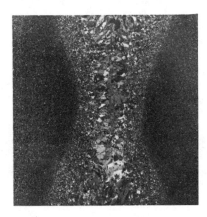

Figure 2. A macrosection of the weld joint. Plate thickness: 1.4 cm.

Initial residual stress characterization in a welded plate was carried out via blind hole drilling and sectioning. Important observations with regard to the measured residual stresses were: 1) Residual stresses did exist in the welded plate. The maximum average longitudinal residual stress (+ 155 MPa) in the weld was approximately 1/3 of the material yield strength. 2) Erratic compressive transverse residual stresses (-20 MPa to -140 MPa) were measured in the weld. 3) The magnitudes of the residual stresses were not equal on the two sides of the plate.

TEST PROCEDURE

The WOL-type fatigue-crack growth specimens were machined from the welded plates. The specimens were then tested in two conditions: as welded and stress-relieved (595 C for 1/2 hr, air cooling). Base material specimens, including one in the stress-relieved condition, were also tested.

Constant-load amplitude testing was carried out in accordance with ASTM standard E647. Test frequency was 30-Hz at a R ratio of +0.1. The crack closure load was determined by taking load (P) vs. crack-opening displacement (COD) readings periodically during testing. The COD readings were obtained from a clip gauge mounted across the mouth of the specimen. The crack-closure load was taken to be the load corresponding to the intersection of the two straight line segments on the P vs. COD traces.

TEST RESULTS

Typical base material FCG results are shown in Figure 3. The difference between the FCG data plotted as a function of the applied ΔK and the effective ΔK was not great.

Longitudinal Weld Specimens

Difficulties were encountered during testing of the as-welded specimens in terms of uneven crack growth on the two surfaces of the specimen and a tendency for the crack to deviate out of the weld. These difficulties were not completely unexpected as hole-drilling residual stress measurement had shown that the residual stress distribution was uneven on the two sides of the specimen and that microhardness measurements had shown that the weld metal had the hardest microstructure.

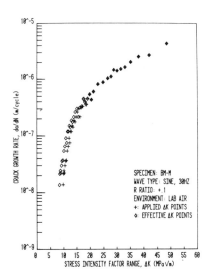 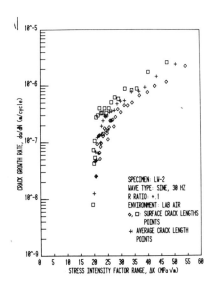

Figure 3. (Left) Typical base material da/dN vs. ΔK_{app} and ΔK_{eff} plots.

Figure 4. (Right) da/dN vs. ΔK_{app} plots for a longitudinal weld specimen.

The da/dN vs. ΔK_{app} data for one of these specimens is shown in Figure 4. The upper and lower bounds of the scatterband were derived from the two surface traces of the crack. The results from the average lengths of the crack's surface traces are also shown. Since the crack-opening displacement gauge was mounted in front of the specimen, any particular crack-closure load can only be regarded as corresponding to that particular average crack length. The crack-closure correction can, therefore, only be applied to the da/dN vs. ΔK data derived from the average crack lengths. The da/dN plotted as a function of both ΔK_{app} and ΔK_{eff} is shown in Figure 5. The effect of the crack-closure correction was quite significant. No difficulties were encountered during testing of the stress-relieved specimen.

Transverse Weld Specimens

As with the as-welded longitudinal weld specimens, erratic crack growth was also observed with the as-welded transverse weld specimens. However, in these specimens corner cracks nucleated at the machined notch and propagated on two different planes at 45 degree to the plane of the notch until one crack became dominant and propagated onto the other plane. The residual stresses in one specimen close to the machined notch were determined by blind hole drilling on both sides of the specimen. The measurement confirmed the earlier finding that the residual stresses were not equal on the two sides of the specimen. Furthermore, at the location of measurement, the values of the transverse residual stresses were greater than the longitudinal residual stresses. The transverse (perpendicular to the weld) residual stresses were tensile: 95 MPa on side 1, 49 MPa on side 2. The longitudinal residual stresses (parallel to the weld and the loading axis) were compressive: -37 MPa on side 1, -88 MPa on side 2. The large lateral stresses and the compressive longitudinal stresses were responsible for causing the corner cracks to nucleate initially and subsequently propagate away from the specimen's plane of symmetry.

The data from the as-welded specimens could not be meaningfully reduced to the da/dN vs. ΔK form. However, the load vs. COD traces showed that as the crack grew into the weld, the crack closure load decreased dramatically.

No difficulties were encountered during testing of the stress-relieved specimen. Its crack growth behavior were similar to those of the stress-relieved longitudinal weld specimen and the base material specimens.

Fractography

Macroscopically, the fractured metal had a very different appearance as compared to the base material. The weld metal fracture surface was more granular and showed definite grain orientation which was a reflection of the solidification pattern and the large grains present.

Microscopically, the base-material fracture surface at low magnifications exhibited transgranular cleavage features (Figure 6). Higher magnification examination of the same area revealed fine striations. With increasing ΔK, the striations become better defined, as shown in Figure 7,

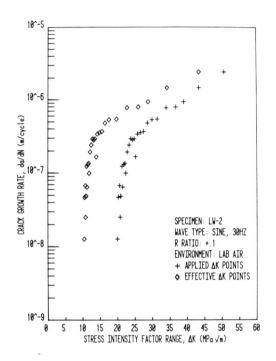

Figure 5. Longitudinal weld specimen da/dN vs. ΔK_{app} and ΔK_{eff}.

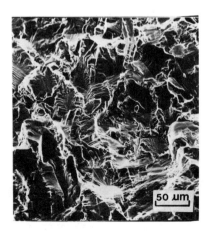

Figure 6. Base material specimen fracture surface at low ΔK.

Fatigue-Crack Growth Behavior / 467

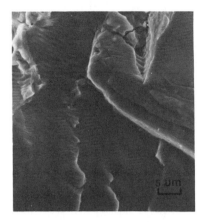 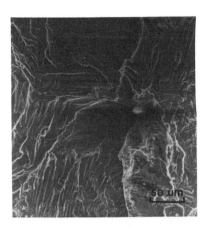

Figure 7. (Left) Well-defined striations on the base material fracture surface at higher ΔK.

Figure 8. (Right) Weld metal fracture surface; note dendritic features.

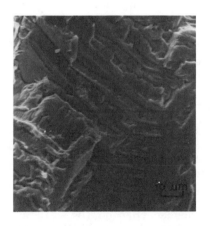 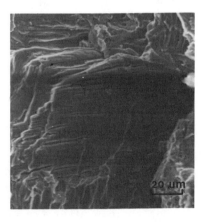

Figure 9. (Left) Weld metal fracture surface; stepped fracture.

Figure 10. (Right) Higher magnification view of Figure 8; striation-like features.

while the cleavage features diminished. The weld metal
fracture surfaces, on the other hand, showed features such
as fractured dendrites and stepped fractures (Figures 8
and 9) associated with the microstructural characteristics
of the weld metal. Fatigue striations were not as prominent
as in the base metal fracture surfaces; where visible, as
shown in Figure 10, these appeared different from the
striations found on the base material fracture surfaces.

DISCUSSION

In the specimens tested where the crack-growth data
could be meaningfully reduced to the da/dN vs. ΔK form, the
da/dN vs. ΔK_{app} results showed marked variations amongst
the different specimens; however, the da/dN vs. ΔK_{eff}
plots of the stress-relieved specimens had a tendency to
coincide, as shown in Figure 11. This indicated that residual
stresses and the hypothesized residual stress-induced crack
closure were major influences in determining the apparent
crack-growth behavior of the specimens tested.

The longitudinal and transverse weld specimens
showed different characteristics in the manner the crack-
closure load varied with crack length. The crack-closure
data from the stress-relieved specimens (in which sufficient
residual stresses were still present to yield distinct
crack-closure patterns) can be used to illustrate the
point. Figure 12 shows a plot of the crack-closure load
(P_{cl}) expressed as a ratio of the P_{cl} at the particular
crack length of interest to the P_{cl} at a/w = 0.25 versus
a/w for the transverse and longitudinal weld specimens.
The variation in P_{cl} with a/w for the case of a hypothetical
constant clamping moment about the crack tip, which can be
expressed as a compressive residual stress intensity factor
K_{IR}, is also shown in this figure. It can be seen that
in the transverse weld specimen the crack-closure load
decreased rapidly after the crack passed through the weld.
The variation of the crack-closure load with crack length
can be rationalized by considering the position of the
crack with respect to the dominant cause of crack closure,
i.e., the source of residual stresses in this investigation.

Although the macroscopic mode I crack-growth rates
correlated on the basis of the effective stress intensity
factor range were not noticeably affected by the different

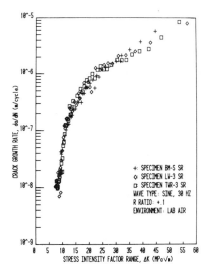 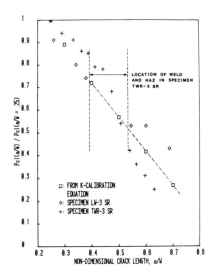

Figure 11. (Left) Composite plot of the da/dN vs. ΔK_{eff} points for the three types of stress-relieved specimens.

Figure 12. (Right) $P_{cl}(a/W)/P_{cl}$ (a/W = 0.25) vs. non-dimensional crack length (a/W) for two stress-relieved specimens and the case of the constant K_{IR}.

microstructures, the differences in the macroscopic and especially the microscopic fracture surface appearances between the base and weld materials were important clues that under certain circumstances, such as in pre-mode I crack growth and/or at near-threshold crack growth rates, microstructural alterations in and around the weld joint could play an important role in determining the fatigue life of a weld joint.

CONCLUSIONS

The results of this investigation showed that over the range of fatigue crack growth rates attained in welded commercially-pure titanium, the following conclusions could be made:

1) The welding-induced residual stresses and their redistribution/relaxation played a major role in influencing the crack growth behavior.

2) Where applicable, the mode I crack-growth rates could be most effectively correlated with the effective mode I stress-intensity factor range.

3) Correlated on the basis of the effective mode I stress intensity factor range, the macroscopic mode I crack-growth rates were not noticeably affected by the different microstructures in the neighborhood of the weld although fractographic examination of the fracture surfaces showed definite microstructural influences in the cracking behavior.

REFERENCES

1. L. A. James, "Crack Propagation Behavior in Type 304 Stainless Steel Weldments at Elevated Temperatures," Welding Journal, Vol. 52, pp. 173s-179s (April 1973).

2. M. Parry, et al., "Fatigue Propagation in A514 Base Plate and Welded Joints," Welding Journal, Vol. 51, pp. 485s-490s (Oct. 1972).

3. S. M. El Soudani, R.M.N. Pelloux, "Anisotropy of Fatigue Crack Propagation in Aluminum Alloy Butt Welded Joints," Welding Journal, Vol. 54, pp. 144s-152s (May 1975).

4. K. R. Dowse and C. E. Richards, "Fatigue Crack Propagation Through Weld Heat Affected Zones," Met. Trans., Vol. 2, pp. 599-603 (Feb. 1977).

5. B. M. Kapadia, E. J. Imhof, Jr., "Fatigue Crack Propagation in Electroslag Weldments," Flaw Growth and Fracture, ASTM STP 631, pp. 159-173 (1977).

6. G. Glinka, "Effect of Residual Stresses on Fatigue Crack Growth in Steel Weldments under Constant and Variable Amplitude Loads," Fracture Mechanics, ASTM STP 677, pp. 198-214 (1979).

7. W. Elber, "Fatigue Crack Closure Under Cyclic Tension," Engineering Fracture Mechanics, Vol. 2, pp. 37-45 (1970).

8. W. Elber, "The Significance of Fatigue Crack Closure," Damage Tolerance in Aircraft Structures, ASTM STP 486, pp. 230-242 (1971).

9. R. J. Bucci, "Effect of Residual Stresses on Fatigue Crack Growth Rate Measurement," Fracture Mechanics: Thirteenth Conference, ASTM STP 743, pp. 28-47 (1981).

10. G. E. Nordmark, et al., "The Effect of Residual Stresses on Fatigue Crack Growth Rates in Weldments of Aluminum Alloy 5456 Plate," Paper presented at the ASTM Committee E9 Symposium on Residual Stress Effects in Fatigue, Phoenix, AZ (1981).

SHEAR BAND FAILURES IN THREADED TITANIUM ALLOY FASTENERS

George Hopple*

ABSTRACT

Failure analysis was performed on threaded Ti-6Al-4V fasteners that had fractured in the threads during installation. Scanning electron microscopy (SEM) and optical metallography revealed that the fractures initiated in circumferential shear bands present at the thread roots. The fractures propagated by microvoid coalescence typical of that observed in notched tensile specimen fractures of the same material. For comparison, Ti-6Al-4V fasteners from various commercial sources were tested to failure in uniaxial tension and examined in the SEM. In all cases, the fracture appearances were similar to that exhibited by the fasteners that failed during installation. In addition, results of optical microscopy indicated that the geometry and extent of the shear bands appeared to depend on the fabrication process employed by the individual manufacturers. Causes of shear band formation are discussed along with potential methods to eliminate these microstructural inhomogeneities.

---

* Lockheed Missiles and Space Company, Inc., Sunnyvale, CA 94086 USA.

(a)   (b)   (c)

Figure 1. Photomacrograph of the fastener combination investigated in this paper (a). During normal installation, the collar separates (b). For a number of fasteners, the pin fractured instead of the collar separating (c).

INTRODUCTION

Threaded Ti-6Al-4V fasteners are used extensively by the aerospace industry for joining structural components. One popular type of fastener is designed so that during installation, the nuts (collars) separate in half (torque off) leaving a portion of the collar tight against the bolt (pin) at a specified torque (Figures 1a and 1b). Pins from three different commercial sources were evaluated in the present investigation and are referred to as those procured from manufacturers A, B and C.

Shear Band Failures / 475

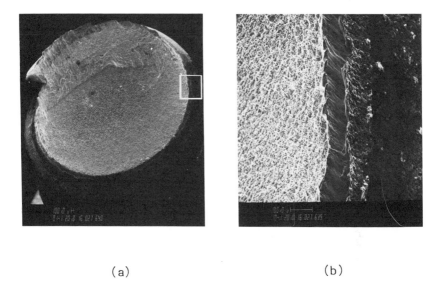

(a) (b)

Figure 2. Electron micrographs of a typical fracture surface of a failed pin (a). A featureless band (arrows) was found on the circumference of the fracture surface (b).

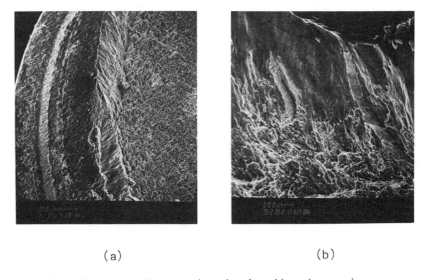

(a) (b)

Figure 3. Electron micrographs showing the change in orientation of the band (a) and faceted appearance of the band surface (b).

During installation of a number of pin collar combinations, the pins fractured before the collars torqued off (Figure 1c). Only the pins from manufacturer A failed in this manner and have done so infrequently over a number of years. In the past, little time was spent on the analysis of the failures because the parts were relatively inexpensive and the failed parts were easily replaced. However, when the frequency of the failures demonstrated a marked increase, greater emphasis was placed on providing a detailed understanding of the problem. This paper reports and discusses the results of a failure analysis directed to provide this information.

PROCEDURE AND RESULTS

Failed Pins

Since the mating collars to the failed pins were intact, they were torque-tested in accordance with the required specification. The torque values were within the required range.

The failed pins were examined in a scanning electron microscope (SEM), and in every case, their fracture appearances were similar. The fractures were flat and had propagated across the pin through a single thread root (Figure 2a). A distorted, band-like region measuring 150 micrometers in width and angled 45 degrees to the pin axis (Figure 2b), was present at the circumference of the fracture surface. At low magnifications, the band appeared featureless for practically the entire circumference on some pins. For others, the band was featureless only in isolated segments which constituted as little as 10% of the total circumference. In addition, the slope of the band alternated between an up or down orientation in approaching the flat fracture surface (Figure 3a) and exhibited a faceted appearance in some locations (Figure 3b). The dimples found on the shear band surface were small, partially formed and generally equiaxed (Figure 4a). The remaining portion of the fracture surface was composed of numerous, partially formed dimples similar to those of a notched, uniaxially tested tensile bar of a similar material (Figure 4b) [1].

A number of the failed pins were sectioned, polished and etched for optical microscopy. Standard metallographic techniques were used in grinding and coarse polishing the failed pins. Final polishing was performed on a semi-automatic unit employing Kroll's etch and a rare earth polishing compound.

Shear Band Failures / 477

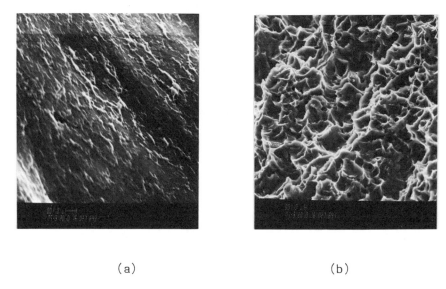

(a) (b)

Figure 4. Electron micrographs of the band surface (a) and the flat fracture surface adjacent to the band (b). The dimples on the band surface were small, poorly formed and widely spaced (a).

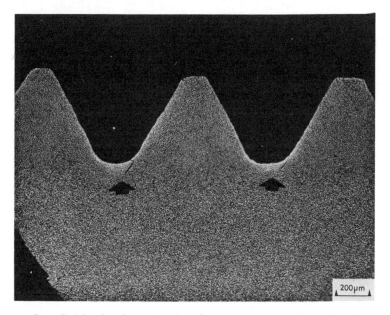

Figure 5. Optical micrograph of a cross-section of a failed pin. The thread root areas are heavily cold worked and contain cracks (arrows).

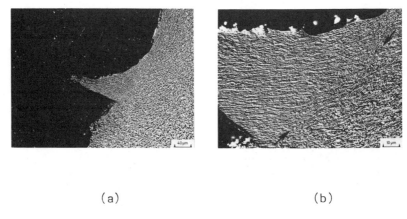

(a)   (b)

Figure 6. Optical micrographs of the metallographically prepared, failed pin detailing the fractures in the thread root (a). An enlargement of this region revealed the presence of shear bands (b, arrows).

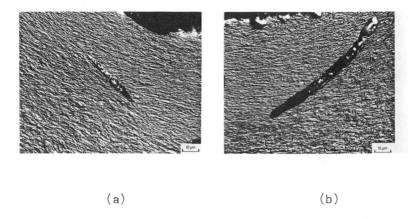

(a)   (b)

Figure 7. Optical micrographs of the failed pin showing the initiation of voids (a) which grow into cracks (b) in the shear bands. Note the voids are round (a).

Application of a duplex etching treatment consisting of Kroll's solution followed by a hydrogen peroxide-potassium hydroxide mixture, combined with interference microscopy, was needed to delineate the shear bands.

Figure 8. Photomacrographs of the fracture surfaces of the pins from the three manufacturers that were uniaxially tested to failure. Note the manufacturer A pins fractured through one thread root, while the manufacturers B and C pins fractured through a number of thread roots.

The threads were heavily cold worked and in one pin, cracks were detected in the roots (Figure 5). The cold worked region adjacent to one of the pin fractures contained numerous shear bands (Figure 6). Voids appeared to have nucleated in several of the shear bands at the thread roots adjacent to the fracture (Figure 7). The voids were round and appeared to have been enlarged by the etching solution. Generally, the width of the shear bands ranged from 60 to 100 micrometers, while their orientation relative to the pin axis varied from 40 to 60 degrees.

480 / Microstructural Science, Volume 12

 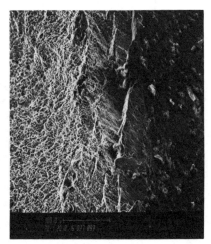

(a)   (b)

Figure 9. Electron micrographs of the shear band regions of the pins tested to failure for comparison. The manufacturer A shear band (a) is at a much lower angle to the pin axis than the other manufacturer's shear band (b).

Comparison of the Pins from Various Manufacturers

Pins from the three commercial sources were uniaxially tested to failure. Although all of the pins tested passed the minimum load requirement, pins from manufacturers B and C were consistently about 10% higher in strength values than those fabricated by manufacturer A (producer of the failed parts).

The macroscopic fracture appearance of the comparison A pins was similar to the failed A pins: a flat fracture through a single thread (Figure 8). On the other hand, the pins from manufacturers B and C fractured through a number of threads creating a more uneven surface appearance (Figure 8). Circumferential shear bands were present on all samples.

A SEM examination of the pins from the three manufacturers that were tested to failure revealed that the orientation of the shear bands varied with each manufacturer. The large featureless regions found on the failed A pins were rarely detected on the comparison test pins. Dimples were generally interspersed within the featureless regions (Figure 9). The orientation of the shear bands on the comparison A pins were similar to those observed on the failed A pins. The B and C pins had shear bands oriented at much lower angles to the pin axis than the A bands (Figure 9). In all of the comparison pins, the center portions of the fractures were similar to the centers of the failed pins.

Untested pins were metallographically prepared in the same manner as the failed pins. The thread root radius of the A pins tended to be more square (Figure 10) than the root radius of the B or C pins (Figure 11). Shear bands present in B and C pins were generally at a much lower angle (20 to 30 degrees) to the axis than the A pins (40 to 50 degrees). Shear band widths for all of the comparison pins were similar to those observed for the failed A pins (50-100 micrometers).

DISCUSSION

The failure of pins from manufacturer A is related to the existence of shear bands in the thread roots. The fractures initiate in the shear band as clearly evidenced by the metallographically prepared pins (Figure 7). The dimples on the shear bands were partially formed and widely spaced, indicating a lack of adequate adhesion (Figure 4a). The fracture appearance of the shear band is similar to a channel fracture [2]. The mechanism for this type of fracture is theorized to be a pile-up of dislocations in a narrow band of crystallographic planes forming a dislocation channel. In the present case, channel formation (shear band) is probably related to tool geometry and deformation texture induced during thread rolling. The result of this channeling of dislocations is a faceted fracture appearance as observed in some locations on the shear bands of the fractured pins (Figure 3b).

The shear bands form during the rolling of the pin threads. Titanium alloys are known to have a high propensity for shear band formation owing to their high strength and low thermal conductivity [3]. It has been reported that shear band formation in Ti-6Al-4V alpha-beta alloys is extremely temperature dependent [4]. Below 704 C (1300 F) plastic deformation is highly localized (forming shear bands); while above this temperature, uniform deformation increases dramatically. The evidence of cold work in the thread roots of the failed pins

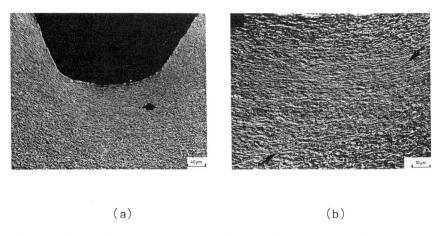

(a)   (b)

Figure 10. Optical micrograph of the thread root of manufacturer's A pin. The broad arrow in photograph (a) denoted region where photograph (b) was taken. Shear bands (b, arrows) were oriented at about 45 degrees to the pin axis.

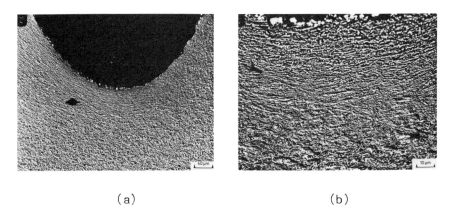

(a)   (b)

Figure 11. Optical micrographs of the thread root of a manufacturer's B or C pins. The broad arrow in photograph (a) indicated the region enlarged in photograph (b). Shear bands (b, arrows) were oriented about 25 degrees to the pin axis (b).

indicates that the threads were probably rolled below the reported annealing range for the alloy. In a similar alpha-beta alloy, shear band formation has been shown to be a function of processing temperature, microstructure, and deformation rate for a given tool geometry [5].

Considering the above evidence, it was not surprising that all of the pins examined in this study contained shear bands in the thread roots. Therefore, the mere existence of shear bands does not entirely explain the pin failures. However, the extent of shear band formation is related to the observed failures. All of the failed pins had rather large, featureless segments on the band (at least 10% of the circumference); whereas the pins uniaxially tested for comparison generally did not. Without extensive microstructural analysis, quantification of this observation would be extremely difficult and was considered beyond the scope of this investigation. Nevertheless, the extent of shear band formation does appear to correlate with the observed pin failures.

Perhaps a more important factor in initiating the pin failures is shear band geometry which may be the reason that the pins from manufacturer A are more susceptible to fracture than those from the other suppliers. The angle of the shear band (to the pin axis) was relatively constant for each manufacturer and probably reflects the various thread processing techniques employed by them. This hypothesis is supported by the observed difference in thread root geometry between the A (Figure 10a), and B and C (Figure 11a) pins, which results from a difference in thread tool geometry. Shear band formation generally occurs under compressive stress; whereas fracture typically is promoted by an applied tensile stress [6]. Consequently, regardless of how extensive shear band formation is, it will only open up and appear as a crack or void as a result of an applied tensile stress. In our case, the tensile stress, shear band fracture initiation mechanism appears to be operating because the voids in the band were round, not elongated (Figure 7a). In addition, the dimples on the shear bands were equiaxed (Figure 4a). During installation, the pins are subjected to uniaxial tension which can be resolved into two stress components acting on the shear band, a shear stress exerted parallel to the shear band coupled with a tensile stress exerted perpendicular to the shear band. Because the tensile component is initiating fracture, it alone need only be considered. Therefore, due simply to geometry, the 45 degree shear bands in manufacturer A's pins would have a much greater tensile stress component of the uniaxial stress (70% of the applied load) and exerted normal to the band than the 25 degree shear bands in manufacturer B's and C's pins (42% of the applied stress)*.

In addition to microstructural inhomogeneities such as shear bands, material properties will have a profound effect on the fracture behavior of this material. The only universally accepted test performed on titanium alloys is the uniaxial tensile test. Unfortunately the remnants of the failed pins were too small for tensile testing. Although not reported earlier in this paper, materials analysis (using tests that could be performed) was implemented on the failed pins as well as the comparison pins from the different manufacturers. Elemental composition, hardness and bulk microstructure were identical for all pins.

Numerous options exist for minimizing shear band formation in the pin thread roots. Altering the tool-workpiece geometry may be the most effective alternative since this parameter appeared to be a major difference between manufacturer A's pins and the pins produced by manufacturers B and C. Related to the thread tool geometry is the processing method for producing threads. Generally, threads are rolled by two common methods, flat die rolling and radial infeed rolling [8]. One of these rolling methods may be more likely to form shear bands than the other. Of course, altering any one or more of the previously mentioned, more fundamental parameters (temperature, deformation rate and microstructure) could also be employed to minimize shear band formation in titanium alloys.

CONCLUSIONS

The pin failures are related to the orientation and extent of shear bands present in the pin thread roots. The pins from manufacturer A are more susceptible to failure because their shear bands are at a higher angle to the pin axis than the other manufacturer's pins.

---

* Stress components can be determined using the simple relationship alpha = P/A cos Θ where alpha is the applied stress, P is the load and A is the area for a specific plane, and Θ is the angle between the plane and the applied stress [7].

ACKNOWLEDGEMENTS

The author is grateful to Dr. E. H. Rennhack for his suggestions in preparing the manuscript and to the many members of the Materials and Process Laboratories who assisted in preparing this publication.

REFERENCES

1. SEM/TEM Fractography Handbook, Air Force Material Laboratory Publication MCIC-HB-Q6.

2. W. J. Mills, Fractography and Materials Science, ASTM, p. 108 (1981).

3. R. F. Recht, J. Appl. Mech., Vol. 31E, p. 189 (1964).

4. C. M. Mataya, et al., J. Applied Met. Working, Vol. 2, No. 1, p. 36 (1982).

5. S. L. Semiatin, et al., Met. Trans., Vol. 13A, p. 287 (1982).

6. H. C. Rogers, Annual Review of Materials Science, Vol. 9, p. 304 (1974).

7. G. E. Dieter, Mechanical Metallurgy, McGraw-Hill, p. 15 (1976).

8. Metals Handbook, 8th ed., Vol. 3, American Society for Metals, p. 130 (1967).

EFFECT OF SULFIDE INCLUSIONS ON THE FRACTURE OF A LOW-CARBON ALUMINUM-KILLED STEEL

R. Chattopadhyay*

ABSTRACT

Elongated sulfide inclusions are known to cause lower transverse ductility, lower Erichsen values, and poor electrical resistance welding (ERW) characteristics of low-carbon aluminum-killed hot-rolled strip when compared to a similar steel without stringer-type sulfide inclusions. The fracture characteristics of this steel were studied using scanning electron microscopy and rapid ductile fracture was found to be related to stringer-type manganese sulfide inclusions of mainly type-II morphology.

INTRODUCTION

The stringent requirements for highly stressed welded structures have renewed the interest in the transverse properties of hot-rolled flat products. It was recognized quite early that the presence of elongated inclusions in hot-rolled plates was a source of gross mechanical anisotropy. Normal manganese sulfides, particularly in hot-rolled strip, form elongated stringers. Such stringers have a deleterious effect on formability, especially with respect to bends made parallel to the rolling direction. In steel strips for thick walled tubes formed by 'U' and 'O' techniques, good transverse properties, emphasizing not only transition temperatures but

---

* EWAC Alloys Ltd., Saki-Vihar Rd., Post Bag 8933, Powai, Bombay 400072 India.

also shelf energy levels, are usually demanded. The relevance of inclusion shape and content with respect to transverse shelf energy levels have been investigated by numerous authors [1,2] and it has been shown that here too, elongated inclusions play a leading part in the mechanism of transverse fibrous rupture. The object of the present work is to study by scanning electron microscopy [3] the effect of stringer-type inclusions on the fracture characteristics of a hot-rolled aluminum-killed C-Mn steel which has been used for ERW tube production.

EXPERIMENTAL RESULTS

The average chemical composition of the steel used for the present investigation is given in Table 1.

Table 1. Chemical Analysis, wt. %

C	Mn	Si	P	S	Al	Cr	Ni	Mo	Cu
0.11	0.52	0.36	0.025	0.030	0.053	0.17	0.30	0.04	0.16

On a transverse bend test, cracks appear in the outer surface and the fractured surface has a dull grey appearance. The mechanical properties of the strips were evaluated and are given in Table 2.

Table 2. Mechanical Properties of Aluminum-Killed C-Mn Steel

	Yield Point (kg/mm^2)	Tensile Strength (kg/mm^2)	Elong. (%)	Erichsen Value on 3.25 mm thick strip (mm)
Longitudinal	36.5	46.0	34.7	12.5
Transverse	43.8	54.0	31.0	

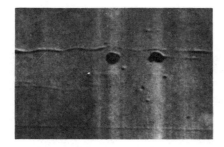

Figure 1. Micrograph of stringer plus globular inclusions. 1000X (Reduced in reproduction).

 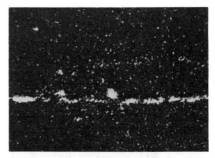

Figure 2. (Left) Scanning electron microscope x-ray image of aluminum in the globular inclusion. 1000X.

Figure 3. (Right) Scanning electron microscope x-ray image of manganese in the stringer inclusion.

Metallographic specimens were prepared from the strips and preliminary observations revealed stringer-type inclusions (Figure 1). The globular part of the inclusion contained aluminum (Figure 2) as aluminum oxide. The stringer portion of the inclusions contain manganese (Figure 3) as manganese sulfide.

490 / Microstructural Science, Volume 12

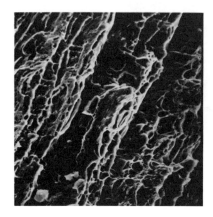 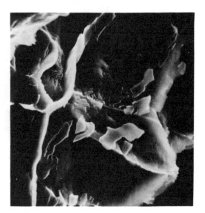

Figure 4. (Left) Fractured surface of the Al-killed steel strip. 497X.

Figure 5. (Right) Fractured surface showing the presence of inclusions. 1000X.

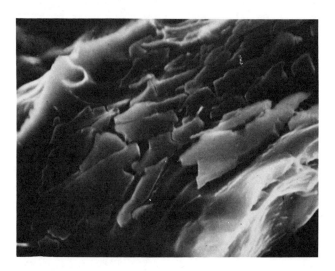

Figure 6. Fractured surface with inclusions. 3700X.

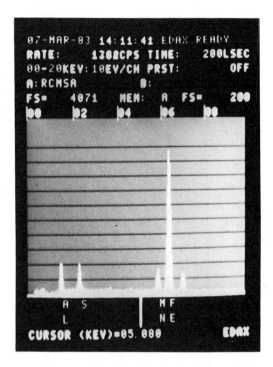

Figure 7. Identification of elements in inclusions by EDAX.

The ERW tubes made from the above steel failed in drift and flattening tests. The strip itself failed in the 180-degree bend test. The fractured surface (Figure 4) exhibits dimples characteristic of ductile fracture excepting one central region where micro-void coalescence seems to have occurred around elongated inclusions. The inclusions present in these regions are shown in Figures 5 and 6 and are aluminum oxide and manganese sulfide. The identification of elements in the inclusion was carried out by energy-dispersive x-ray analysis (EDAX) as shown in Figure 7.

DISCUSSION

The major factor controlling fibrous fracture energy is the dispersion of non-metallic inclusions, of which manganese sulfides are by far the most important inclusion. Although it is in the type-II morphology that MnS is usually precipitated in aluminum-killed steels, the presence of high concentration of C, Al, Si and P results in the formation of the type-III morphology of MnS [4]. Type II and III sulfides are more deformable than the type-I morphology and with inclusion elongation, the toughness of the steel decreases dramatically in the transverse and short-transverse directions. Simpson and Allen [5] reported that there was no relationship between the stringer length and loss in bendability, i.e., the only requirement for a detrimental effect was the presence of stringers regardless of their length.

The loss in bendability in this steel can be thus ascribed to stringers as shown in Figure 1. Dewsnap et al. [6] reported a small drop in deep drawability with increased inclusion content, particularly with higher strength steels. They attributed the loss of formability to the presence of sulfide stringers. The present investigation also showed a loss of formability. Compared to normally obtainable Erichsen values for similar steels of the same thickness (15 to 16 mm), a low value of 12.5 mm was obtained for the steel under investigation.

In the steel under investigation, the inclusions consist of elongated MnS stringers with globular aluminum oxide. The sulfide inclusions are originally of the types described by Sims [4], predominately type-II along with some type-III sulfide. Farrar and Dolby [7] have stated that for carbon levels up to about 0.2% in C-Mn fully-killed steels with aluminum additions and with minimal alloying additions, the dominant inclusion types are type-II manganese sulfide and aluminum oxide. In strips, the deformation can be sufficient to produce MnS inclusion stringers with aspect ratios in the range of 100 to 1000 and lengths of several millimeters parallel to the major rolling direction. The alumina particles, although sometimes found close packed to give dense beds of particles, may give rise to locally poor short transverse ductility properties, but may not pose a serious lamellar tearing problem in this type of steel.

It is difficult to confirm the presence of type-III sulfide inclusions in the rolled product since positive morphology identification relies on observations made in the as-cast state. However, in the stringer type shown in Figure 1, it is quite possible that the more deformable type-III sulfide may be associated with these inclusions [8].

The general mechanism of ductile failure is agreed to be one of void nucleation, void growth, and finally void coalescence. Voids may nucleate by cracking of a second phase particle [9] or by decohesion at the metal/particle interface [10]. The growth of a void formed around an inclusion leads to gross void linking and, with increased deformation, to final tearing. The fracture surfaces of broken short-transverse tensile specimens exhibit a classical terrace fracture pattern evidencing extensive dimple formation around inclusions [11] similar to those in Figure 4. At higher magnifications, Figures 5 and 6, numerous inclusions are found in these regions and the inclusions are of MnS associated with alumina.

Therefore, it is most likely that decohesion at the matrix/inclusion interface has occurred at these regions followed by void formation around inclusions and, at the last stages of deformation, gross void linking and final tearing. Equations have been prepared to describe the growth of voids [10]. The theory assumes a uniform particle spacing, which is not usually observed in practice, and because the criterion for failure is a critical ratio of inclusion width to inclusion spacing, the importance of local segregation of inclusions on the ductility at failure is clearly apparent. The importance of this factor with regards, for example, to lamellar tearing in welding, is obvious.

Pickering [12] calculated the critical void fraction for the onset of rapid ductile failure for the given type of steel; an average ratio of inclusion width to void diameter of 0.21 was obtained. Actual measured ratios obtained by Pickering [12] were in the range of 0.3 to 0.6, and the slight disagreement between the theory and observation has been ascribed as due to the heterogeneity of the sulfide distribution. In the steel under investigation, the measured ratio of inclusion width to void diameter has been found to be within a range of 0.23 to 0.64 which is quite consistent with the results of Pickering [12]. Therefore, it can be reasonably concluded that the fracture characteristics of this steel correspond to that of rapid ductile fracture due to stringer-type manganese sulfide inclusions of mainly type-II morphology.

REFERENCES

1. H. Vogels and F. Bruning, Arch. Eisenhuttenwesen, Vol. 35, p. 115 (1964).

2. L. Luyckx et al., Metal. Trans., Vol. 1, p. 3341 (1970).

3. D. A. Melford and P. Duncumb, Metallurgia, Vol. 57, pp. 159-161 (1958).

4. C. E. Sims, Trans. AIME, Vol. 215, p. 367 (1959).

5. I. D. Simpson and D. V. Allen, "The Influence of Sulfide Inclusions on Transverse Bend Ductility," B.H.P. Bulletin, Vol. 16, pp. 27-30 (May 1972).

6. R. F. Desnap, D. M. Keane and J. R. Brenson, "Factors Affecting the Formability of Hot Rolled Strip Steels," BISRA Open Report MG/34/71, p. 25 (1971).

7. J. C. M. Farrar and R. E. Dolby, Sulfide Inclusions in Steel, American Society for Metals, Metals Park, Ohio, pp. 252-268 (1975).

8. T. J. Baker and J. A. Charles, Effect of Second Phase Particles on the Mechanical Properties of Steel, The Iron and Steel Institute, Special Report 145, pp. 79-87 (1971).

9. C. T. Liu and J. Gurland, Trans. ASM, Vol. 61, pp. 156-167 (1968).

10. J. Gurland and J. Plateau, Trans. ASM, Vol. 56, pp. 442-454 (1963).

11. G. Bernard, M. Grumbach and F. Moliexe, Metals Technology, pp. 512-521 (November 1975).

12. F. B. Pickering, Towards Improved Ductility and Toughness, Kyoto International Conf., October 25-26, 1971, Climax Molybdenum Development Co., pp. 9-31 (1972).

MICROSTRUCTURE-FRACTURE CORRELATIONS FOR PERMANENT MOLD GRAY IRON

Daniel C. Wei*

ABSTRACT

The fracture, solidification structure, graphite morphology, and tensile properties of hypereutectic gray iron cast into permanent molds were studied. The relationship of graphite morphology versus solidification orientation was investigated. The effect of graphite morphology on the tensile properties as well as the tensile-overload fractures are presented. The samples from the initial ten pours of the permanent mold casting were analyzed. The mold temperature increased from room temperature for the first pour to approximately 400 F at the end of the tenth pour. The measured dendrite spacing and the graphite morphology of these samples were correlated with the pour sequence. The scatter of the data in the relationship of dendrite spacing versus pour sequence was explained by examining the graphite morphology. It was concluded that the existence of fine-tipped undercooled graphites was detrimental to the performance of the permanent mold gray iron.

INTRODUCTION

It has been recognized that the as-cast mechanical properties of gray iron can be effectively influenced by variations in both the solidification structure and the size, shape and distribution of gray iron graphite [1-4]. Permanent mold casting of gray iron, used for relatively high volume

---

* Kelsey-Hayes Corporate Research & Development Center, 2500 Green Rd., Ann Arbor, MI U.S.A.

production of small castings (usually, no heavier than 30 pounds) [5], is particularly sensitive to the rate of solidification as well as to graphite morphologies. Gray iron used for this process should be hypereutectic, so as to obtain maximum fluidity and minimum chill. Its typical microstructure shows undercooled or type-D graphite which normally results from the rapid cooling rate obtained in permanent mold casting.

A number of studies [6-9] have discussed the effects of chemical composition, mold temperature and section thickness on the microstructure of permanent mold gray cast iron. In this paper only one composition of gray iron has been cast to a specific size casting. Therefore, the effects of chemical composition and section thickness on the gray iron graphite morphologies are beyond the scope of this paper. As for the mold temperature effect on the permanent mold gray cast iron, it is noted that [9] casting in a relatively cold mold can only have disadvantages; for a negligible savings in solidification time, one creates very high temperature gradients on either side of the casting surfaces. Structural heterogeneities are aggravated, running the risk of defects such as cracking, sweating and excessive internal stresses. Thermal fatigue takes place earlier as the high temperature gradients occur in between the interface and the mold mass. However, it is reported [6] that a mold temperature above about 700 F is not desirable when using gray iron permanent molds because it may cause more rapid thermal cracking; and it is objectionable in graphite permanent molds because it may accelerate the rate of oxidation.

In this paper, the effect of a cold mold on the dendrite spacing and the graphite morphologies of the air-cooled permanent mold gray cast iron was studied. It is noted that the dendrite spacing increases linearly with the increasing solidification time in a logarithmic plot. In addition, this paper discusses the effect of graphite morphologies on the tensile properties of the air-cooled permanent mold gray cast iron samples. The correlations between the impact fracture pattern and the solidification structure were also demonstrated. The objectives of this study were: (1) to reveal the correlations between the gray iron graphite morphologies and the fracture pattern, (2) to reveal the effect of the gray iron graphite morphologies on the tensile properties; and, (3) to reveal the correlations between the dendrite spacing and the gray iron graphite morphologies.

MATERIAL AND EXPERIMENTAL PROCEDURES

Material

The permanent mold gray cast iron used in this study was hypereutectic. The range of carbon equivalent value was 4.5 to 4.8%. The representative chemical composition of this gray iron is shown in Table 1. The typical microstructure was undercooled or type-D graphite in a ferritic matrix. It is noted that [10] transformation to ferrite plus graphite in gray iron is most likely to occur with a high value of carbon equivalent, the presence of fine undercooled flake graphite, and high silicon contents which favor the formation of graphite rather than cementite.

Table 1. Chemical Composition of Permanent Mold Gray Cast Iron

Sample	Element, wt. %						
	C	Mn	Si	P	S	Cr	CE*
A	3.75	0.73	2.59	0.21	0.10	0.11	4.68
B	3.82	0.73	2.58	0.21	0.10	0.11	4.75

* Carbon Equivalent Value

$$CE = \% C + 1/3 (\% Si + \% P)$$

The cast samples were 152.4-mm (6-in.) high, 25.4-mm (1-in.) diameter cylinders. The pouring temperature of these samples was 2350 to 2450 F. The normal mold temperature was about 450 F and the mold was coated with acetylene soot before each pour.

Tensile Testing

Tensile specimens with gage sections 25.4-mm (1.00-in.) long, 6.35-mm (0.250-in.) diameter were prepared from the selected cast samples based on the ASTM E8 Standard [11]. The tensile properties were evaluated on an Instron universal testing machine using a cross head speed 5.08-mm (0.2-in.) per minute to failure. A 25.4-mm (1.00-in.) Instron strain gage extensometer was used for recording load-elongation curves.

Fractography

Fracture surfaces of the cracked gray iron castings were examined using a Cambridge Stereoscan 600 scanning electron microscope. This fractographic work focused on the examination of the interfacial fracture patterns, which related to the undercooled graphite morphologies. In addition, fracture surfaces of the tested tensile specimens were examined using a Nikon low magnification stereomicroscope. This work focused on the overall fracture pattern examination and measurement of the depth of the outer circumferential ridges on the fracture surfaces.

Metallography

Samples for microstructural examination were cut from the area near the as-cast surface of the gray iron castings. Most of this metallographic work focused on the gray iron graphite morphologies. All the gray iron samples were etched in 2% nital. A Bausch & Lomb Research II metallograph was employed for the microstructural examination. The dendrite spacing was measured and averaged on a micrograph at 100X magnification.

RESULTS AND DISCUSSION

Observations on Fracture-Microstructure Correlations

Two selected permanent mold gray iron castings, one with nil-ductility which was cracked during casting or machining operations, and one with normal ductility which had no cracks, were studied. In order to compare these castings in terms of the correlations between the fracture pattern and microstructure, a fracture of the casting with normal ductility was generated artificially through an impact force. Figure 1 shows the fracture surface of the permanent mold gray cast iron sample with nil-ductility. This fracture was initiated and propagated from graphite-ferrite interfaces. This phenomenon reveals that a serious weakness existed at the interface of the undercooled gray iron graphites and the ferrite cells. Figure 2 shows the solidification structure and the graphite morphologies of the

Permanent Mold Gray Iron / 499

(a)   (b)

Figure 1. Fracture surface of the permanent mold gray cast iron sample with nil-ductility. (a) Low magnification. (b) High magnification.

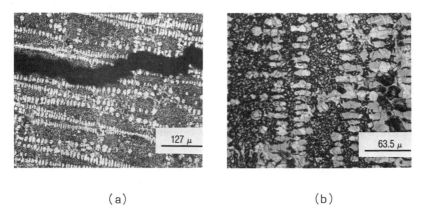

(a)   (b)

Figure 2. Micrographs of the permanent mold gray cast iron sample with nil-ductility, etched 2% nital. (a) Low magnification. (b) High magnification.

permanent mold gray iron sample with nil-ductility. Evidently, crack propagation followed the preferred orientation of solidification, and the dendrite spacing is very narrow, approximately 50 microns. This correlates very well with the findings from the fracture surface as shown in Figure 1. Furthermore, its graphite structure consists of fine-tipped undercooled graphite which is isolated from the ferrite cells. This situation seems to be a major contributor to the graphite-ferrite interface cracking. The fine-tipped undercooled graphite is more-or-less cylindrical in shape, indicating more rapid growth in one crystallographic direction [8].

Figure 3 presents the artificially generated fracture surface of the permanent mold gray iron sample with normal ductility. A significant fracture pattern of resisting the graphite-ferrite interface cracking can be traced. Figure 4 presents the solidification structure as well as the graphite morphologies of the permanent mold gray iron sample with normal ductility. By comparison to the sample with nil-ductility shown in Figure 2, the solidification structure of the ductile sample shown in Figure 4 is more isotropic and the dendrite spacing is larger, 85 versus 50 microns. In addition, its graphite structure contains interconnected medium-sized undercooled graphite, which well-penetrate into ferrite cells. This type of graphite morphology will most likely provide higher resistance to graphite-ferrite interface cracking. In summary, the gray iron with fine-tipped undercooled graphite has lower impact resistance and lower ductility than the gray iron with interconnected medium-sized undercooled graphite. A similar result was revealed during the comparison between the undercooled graphites and the random flake type-A graphites in gray irons [1,12].

Microstructure and Tensile Properties

Undercooled graphite structures usually show much lower ductility than normal graphite structures [12]. This implies that the ductility of undercooled graphite structures are effectively influenced by graphite morphology. In order to correlate microstructure with tensile properties, two selected permanent mold gray iron samples coded A and B were studied. The chemical composition of these samples was shown in Table 1. Their chemical compositions were very similar.

Permanent Mold Gray Iron / 501

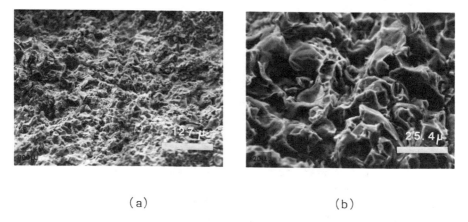

(a) (b)

Figure 3. Artificially generated fracture surface of the permanent mold gray cast iron sample with normal ductility. (a) Low magnification. (b) High magnification.

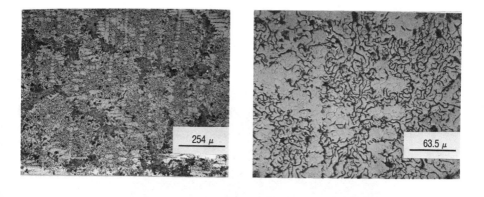

(a) (b)

Figure 4. Micrographs of the permanent mold gray cast iron sample with normal ductility, etched 2% nital. (a) Low magnification. (b) High magnification.

Table 2 presents the tensile properties of gray iron samples A and B. It is noted that the tensile strength and Young's modulus of samples A and B were almost identical, while sample B had higher ductility than sample A.

502 / Microstructural Science, Volume 12

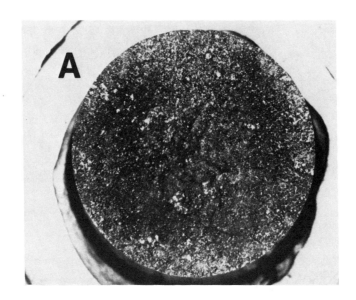

Figure 5. Tensile-overload fracture in the permanent mold gray cast iron Sample A with 12.7-mm (0.005-in.) diameter.

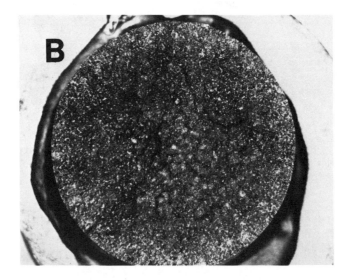

Figure 6. Tensile-overload fracture in the permanent mold gray cast iron Sample B with 12.7-mm (0.500-in.) diameter.

Table 2. Tensile Properties of Permanent Mold Gray Cast Iron

Sample	Young's Modulus GPa (Ksi)	Tensile Strength MPa (Ksi)	Ductility %
A	105 (15.3)	186 (27.0)	0.109
B	105 (15.3)	188 (27.3)	0.336

Figures 5 and 6 show the tensile-overload fracture in the tested samples A and B, respectively. The circumferential fibrous ridges appeared in the area near the outer edge of the fracture surfaces. In general, the area of the circumferential fibrous ridges decreases with increasing degree of brittleness [13]. By comparing the degree of the circumferential fibrous ridges in samples A and B in terms of the percentage of the gage cross-section area, it is found that sample B had higher percentage than sample A, 15.1% versus 11.6%. This implies that sample A was more brittle than sample B.

Figures 7 and 8 present the microstructures of samples A and B, respectively. It is evident that sample A had fine-tipped undercooled graphite, while sample B had interconnected medium-sized undercooled graphite. As such, the fine-tipped undercooled graphite had lower ductility and a higher degree of brittleness than the interconnected medium-sized undercooled graphites, although their tensile strength and Young's modulus were almost identical.

Microstructure and Dendrite Spacing

In order to correlate the dendrite spacing during the solidification process and the graphite morphology of the permanent mold gray iron, the castings from the initial ten pours of the permanent mold casting were studied. The mold temperature increased from room temperature for the first pour to approximately 400 F at the end of the tenth pour. The cycle time was 3.5 minutes. Figure 9 presents mold temperature versus pour sequence for the permanent mold gray iron samples. The mold temperature increased gradually with the increasing pour sequence.

504 / Microstructural Science, Volume 12

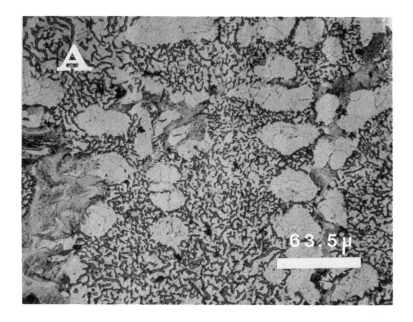

Figure 7. Micrograph of the permanent mold gray cast iron Sample A, etched 2% nital.

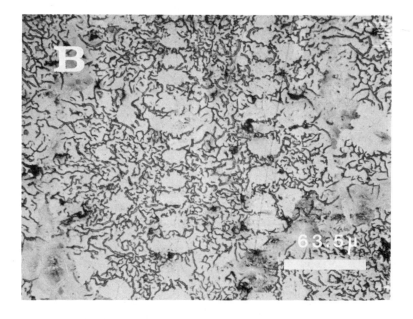

Figure 8. Micrograph of the permanent mold gray cast iron Sample B, etched 2% nital.

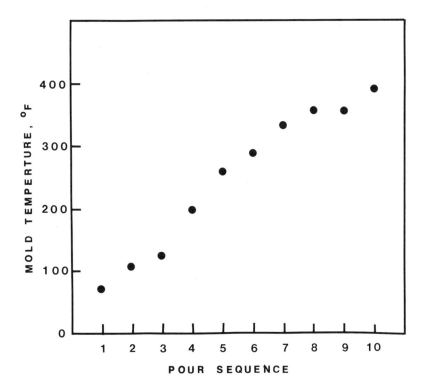

Figure 9. Mold temperature versus pour sequence for permanent mold gray iron samples.

Figure 10 shows dendrite spacing versus pour sequence data for the permanent mold gray iron samples. The regression line indicates that the dendrite spacing increased with increasing pour sequence. Scatter of the data points is apparent. This scatter might result from a number of factors; e.g., mold coating, air cooling, mold temperature, mold cleaning, and cycle time control. Nevertheless, this data scatter might be explained by examining the microstructure, particularly the graphite morphology.

Figure 11 presents the microstructures of the permanent mold gray iron samples for the ten pours. The individual pours with narrower dendrite spacing, i.e., the first, third, fourth and fifth pours, had fine-tipped undercooled graphite. The

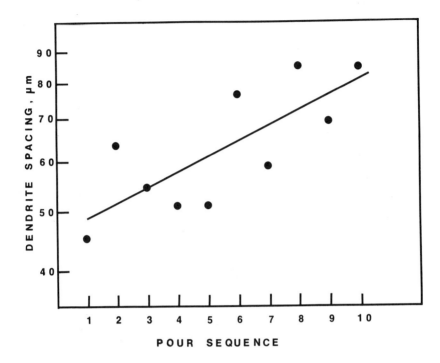

Figure 10. Dendrite spacing versus pour sequence for permanent mold gray iron samples.

pours with larger dendrite spacing, i.e., the second, sixth, eighth, ninth and tenth pours, had interconnected medium-sized undercooled graphite. The graphite of the seventh pour was in between these two distinct types of graphite.

In summary, the dendrite spacing of the permanent mold gray iron casts were highly sensitive to graphite morphology. More specifically, the increasing dendrite spacing was reflected by the transition changing from the fine-tipped undercooled graphites to the interconnected medium-sized undercooled graphite.

Permanent Mold Gray Iron / 507

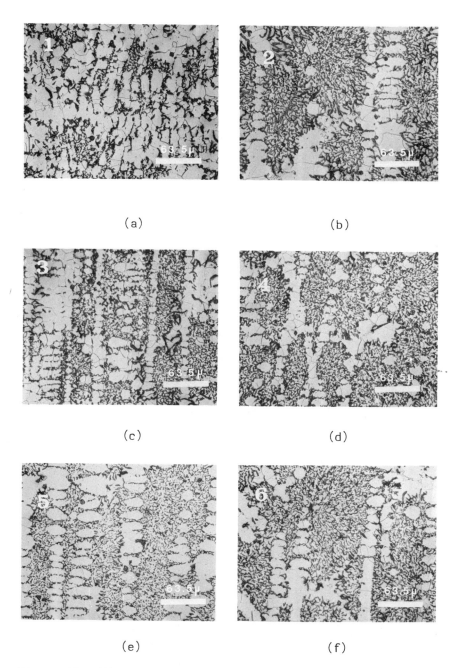

Figure 11. Micrographs of the permanent mold gray cast iron samples of the initial ten pours, etched 2% nital. (a) First pour, (b) Second pour, (c) Third pour, (d) Fourth pour, (e) Fifth pour, (f) Sixth pour, (g) Seventh pour, (h) Eighth pour, (i) Ninth pour, (j) Tenth pour.

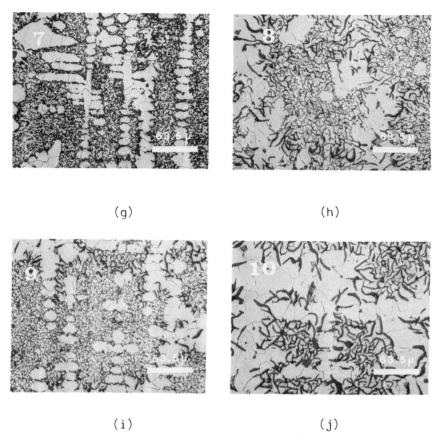

(g)　　　　　　　　　　　　(h)

(i)　　　　　　　　　　　　(j)

Figure 11. (Continued)

CONCLUSIONS

Based on the experimental results and discussion, the following conclusions have been drawn:

1. For permanent mold gray iron, fine-tipped undercooled graphite appear to be a major contributor to graphite-ferrite interface cracking.

2. For permanent mold gray cast iron, those casts with fine-tipped undercooled graphite had lower impact resistance, lower ductility and a higher degree of brittleness than the casts with interconnected medium-sized undercooled graphite.

3. For permanent mold gray iron, tensile strength and Young's modulus were not sensitive to changes in graphite morphology.

4. For permanent mold gray iron, the casting with fine-tipped undercooled graphite had narrower dendrite spacings than the casting with interconnected medium-sized undercooled graphite.

ACKNOWLEDGEMENTS

This study was performed under the support of the Research and Development Center of Kelsey-Hayes Company, a subsidiary of Fruehauf Corporation. The author wishes to thank Mr. L. Roe, Mr. N. Waldecker and Mr. M. Dunn for their expert assistance in the experimental aspects of this work. The author is also grateful to Dr. J. Rhodes for his valuable discussion on solidification processing.

REFERENCES

1. G. F. Ruff and J. F. Wallace, "Effects of Solidification Structures on the Tensile Properties of Gray Iron," AFS Transactions, Vol. 85, pp. 179-202 (1977).

2. G. F. Ruff and J. F. Wallace, "Control of Graphite Structure and Its Effect on Mechanical Properties of Gray Iron," AFS Transactions, Vol. 84, pp. 705-728 (1976).

3. T. Carlberg and H. Fredriksson, "Influence of Silicon and Aluminum on the Solidification of Cast Iron," Solidification and Casting of Metals, The Metals Society, London, England, pp. 115-124 (1979).

4. E. Hornbogen and J. M. Motz, "Fracture Toughness of Gray Cast Irons," AFS International Cast Metals Journal, Vol. 2, No. 4, pp. 31-36 (December 1977).

5. Metals Handbook, 8th ed., Vol. 5, pp. 265-284, American Society for Metals, Metals Park, Ohio (1972).

6. C. A. Jones, J. C. Fisher, and C. E. Bates, "Permanent Mold Casting of Gray, Ductile and Malleable Iron," AFS Transactions, Vol. 79, pp. 547-559 (1971).

7. R. R. Skrocki and J. F. Wallace, "Control of Structure and Properties of Irons Cast in Permanent Molds," AFS Transactions, Vol. 77, pp. 297-302 (1969).

8. R. R. Skrocki and J. F. Wallace, "Control of Structure and Properties of Irons Cast in Permanent Molds, Part II," AFS Transactions, Vol. 78, pp. 239-250 (1970).

9. S. Parent-Simonin and J. Parisien, "Permanent Mold Casting of Irons: Gray Iron and Malleable Castings with Good Machinability in the As-Cast Condition," AFS Transactions, Vol. 81, pp. 260-267 (1973).

10. Metals Handbook, 9th ed., Vol. 1, pp. 11-32, American Society for Metals, Metals Park, Ohio (1978).

11. 1979 Annual Book of ASTM Standards, Part 10, pp. 160-180, American Society for Testing and Materials, Philadelphia, PA (1979).

12. H. T. Angus, Cast Iron: Physical and Engineering Properties, 2nd ed., Butterworth & Co. Ltd., London, England (1976).

13. Metals Handbook, 8th ed., Vol. 9, pp. 36-48, American Society for Metals, Metals Park, Ohio (1974).

FIELD METALLOGRAPHY AIDS NDT OF EVALUATION
OF INDICATIONS IN TURBINE MAIN COLUMN HORIZONTAL
PLATE WELDS AT POWER PLANT

David J. Diaz*
and
Steve E. Benson*

ABSTRACT

An evaluation of indications in the main turbine building column horizontal plate welds was conducted by the joint efforts of field metallography and nondestructive examinations. The turbine building main column horizontal plate welds were selected at random and were inspected to find discontinuities, metallurgical evaluation of the discontinuities, analysis of any failure modes, and determination of the best repair techniques.

The welds were made with prequalified joints in accordance with AWS D1.1-77 and required only visual inspection. More sensitive inspection methods were applied to the welds in order to better define the indications found with the visual inspections.

Cracks were found in 17 field welds and in two test plate welds. The causes of the cracking are related to the weld design and installation procedure. Three field welds were rejected because of the depth of the cracks. The NDT inspections,

---

* Lawrence Livermore National Laboratory, University of California, P.O. Box 808, L-351, Livermore, CA 94550 USA. (D. J. Diaz is now at FMC Corp., 1185 Coleman Avenue, P.O. Box 580, Santa Clara, CA 95052 USA.)

Plates Installed in Main Columns

 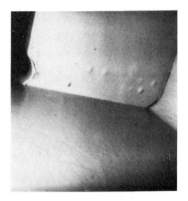

Top side of plate **Bottom side of plate**

Figure 1. Seismic modifications to the turbine building main columns.

evaluations, method of field metallography, analysis and conclusions are discussed with recommendations for corrective actions in the following report.

INTRODUCTION

Seismic modifications to the turbine building main columns included welding horizontal (stiffener) plates between

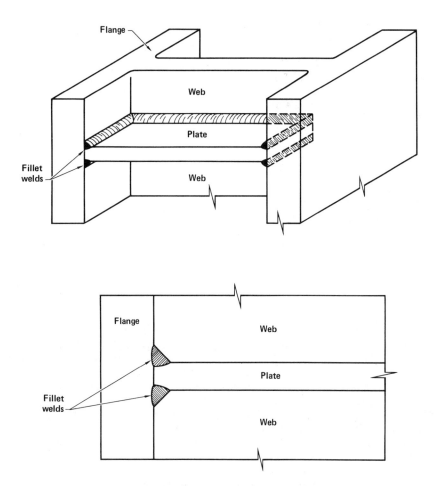

Figure 2. Top and bottom fillet welds specified in original horizontal plate design.

the flanges and along the web of the turbine building main columns (Figure 1). The purpose of these plates is to allow a higher design loading of the main columns between the turbine building buttresses and the overhead crane rail. The plates are not subject to any calculated stresses, but provide extra resistance to buckling of the columns.

The original specification required fillet welds along the top and bottom edges of the 3/4-inch plates (Figure 2). This requirement was later changed to a 5/8-inch partial penetration weld from the top edge of the plate (Figure 3) in accordance with the American Welding Society (AWS Code, Section D1.1, Rev. 2-77). The partial penetration design requires a 1/8-inch land at the bottom edge of the plate, with a 45-degree bevel to the top surface of the plate.

After the plates were welded to the columns, a visual inspection of 400 exposed weld termination points revealed indications in 30 plate welds. An evaluation program was developed which incorporated nondestructive examinations (NDE) to find discontinuities, metallurgical evaluation of these discontinuities, analysis of the failure modes and determination of the best repair techniques. A recommended resolution of the problem was to be prepared based on the results of the evaluation program.

EXAMINATION TECHNIQUES USING NONDESTRUCTIVE SPECIFICATIONS

This procedure is applicable to the magnetic particle examination of ferromagnetic components using prods and dry powder with direct current (DC) according to the dry continuous inspection method. This procedure is limited to the detection of discontinuities open to and immediately below the surface in materials greater than 1/4-inch thickness in the least dimension, and under 600 F in temperature. This complies with the examination requirements of the ASME Boiler and Pressure Vessel Code, Sections V and XI, 1974 Edition with addenda through W76.

Surface preparation by metal removal may be necessary in areas where surface irregularities may mask discontinuous indications. Prior to the examination, areas plus the surface within at least one inch of the examination areas shall be dry and free of dirt, grease, lint, scale, welding flux, spatter, oil, or other extraneous matter that would interfere with the examination.

The normal examination shall consist of orienting the prod in two directions to produce approximately perpendicular magnetic lines of force over each inspection area. (See Figure 4). When applicable, expected flaw location and/or orientation may determine the need for a single examination of an area which shall be noted as such on the data sheet. The examination shall be performed by the continuous application

Field Metallography Aids NDT / 515

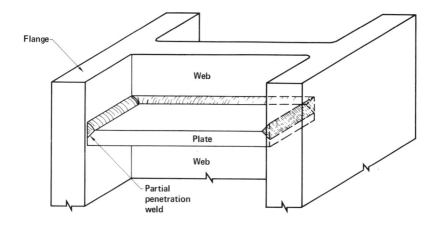

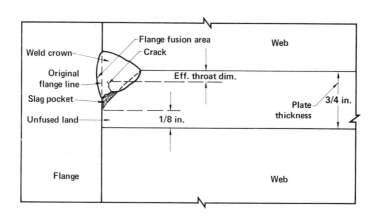

Figure 3. Partial penetration weld specified by revised horizontal plate design.

of direct current. For sections that are 3/4-inch thick or greater, a magnetizing current of at least 100 amps, but not more than 125 amps, per inch of prod spacing shall be used. For sections less than 3/4-inch, the amperage shall be 90 to 110 amps per inch of prod spacing. The inspection powder shall be lightly dusted or floated over the inspection area while the prods are energized. When an indication is observed, the examination may be repeated for further evaluation by optimizing the magnetic flux density and field direction relative to the orientation of the indication. The examination results are summarized in Table 1; all of the field welds

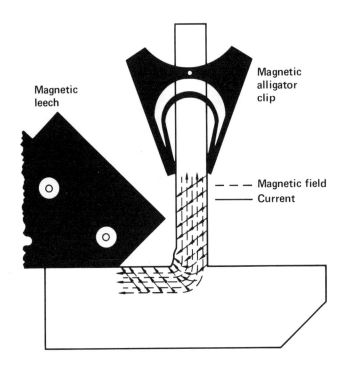

Step 1. A magnetic alligator clip shall be fastened to the stiffener plate two inches from weld.
Step 2. A magnetic leech is then attached to the flange two inches from weld.
Step 3. Current is then passed from alligator clip to leech creating a longitudinal field.
Step 4. Powder shall be applied using the continuous method.
Step 5. Form shall be used to document examination record.
Step 6. A magnetic penetrameter shall be used to determine orientation of field and field strength.

Figure 4. Procedure for magnetic inspection of stiffener plates.

with magnetic test (MT) indications had been previously accepted by the dye penetrant test (PT) examinations. The slightly subsurface detection capabilities of the MT procedure provides a more effective examination of as-ground surface

Table 1. MT Inspection Summary

Line	Column	Girder Face	Plate Side	Field Welds ** Plate No. from El.140 ft.	Unit No.	MT Result	Effective Throat (in.)	Reject-able	Remarks
A	6	S	E	4	1	RI	5/16	yes	Crack*
A	8	S	W	4	1	RI	15/32	–	
A	9	N	W	2	1	RI	1/2	–	
A	9	N	W	4	1	RI	7/16	–	
A	12	S	W	3	1	NRI	–	–	
A	16	S	W	2	1	RI	1/2	–	
A	18	N	W	1	1	RI	1/2	–	
A	28	S	E	3	2	NRI	–	–	
A	28	N	E	6	2	RI	7/16	–	Crack 2 Exams*
A	28	N	W	6	2	RI	5/16	yes	2 Exams
A	29	S	E	3	2	RI	7/16	–	Crack*
G	5	S	W	5	1	RI	9/32	yes	
G	6	N	W	4	1	RI	7/16	–	
G	6	N	W	3	1	RI	1/2	–	
G	6	N	E	2	1	RI	15/32	–	
G	6	S	W	2	1	RI	3/8	–	Crack*
G	6	S	W	1	1	RI	17/32	–	
G	14	S	E	3	1	NRI	–		
G	22	S	E	5	2	RI	7/16	–	
G	27	S	E	2	2	RI	1/2	–	Crack*

* Metallographic analysis   RI = reportable indication   NRI = non-reportable indication
** All field welds previously accepted by fabricator's Liquid Penetrant Examination

Plate	Section	MT Result	Test Plate Welds Effective Throat (in.)	Rejectable	Remarks
W	1	RI	1/2	–	
W	2	NRI	9/16	–	
W	3	RI	7/16	–	
G	1	RI	23/64	Yes	Crack
G	2	RI	7/16	–	Crack
G	3	RI	23/64	Yes	Crack

that PT, due to the possibility of metal being smeared over the discontinuity by the grinding process.

Demagnetization following the examination is required where residual magnetism can interfere with the subsequent use of the component. The examination area may be demagnetized by reducing the magnetic field strength, reversing the field polarity, and repeating until the part is demagnetized. A substantially greater demagnetization force should be used at the start of the process, and the polarity should be reversed approximately once per second. Demagnetization may be verified with a field strength indicator.

FIELD METALLOGRAPHIC EVALUATION OF NDT RESULTS

Field metallography was performed on five representative welds. All welds contained an unfused root, and a large slag pocket at the base of the weld. Most welds contained cracks which initiated at the slag pockets and propagated either along the column side fusion line into the column flange, along the column side fusion line into the weld, or directly into the weld metal. Other discontinuities such as porosity were noted in several of the welds.

The surface to be examined was prepared using the following materials:

- a. 180 grit silicon carbide
- b. 240 grit silicon carbide
- c. 320 grit silicon carbide
- d. 400 grit silicon carbide
- e. 600 grit silicon carbide
- f. 6 micrometre diamond paste with microcloth
- g. 0.05 micrometre aluminum oxide with microcloth
- h. Acetone
- i. Replicating tape
- j. Etching solution (3% nitric acid in alcohol) or portable etcher
- k. Distilled water
- l. Cotton balls
- m. Air grinder with medium rubber pad 2"

The effective throat dimensions of the five metallographically evaluated welds were measured. The effective throat is measured from the top surface of the plate vertically downward to the top end of any crack or mechanical opening (slag pocket) that may be present in the weld (Figure 3). The welds, group numbers, effective throat measurements, and the results of a comparison of each measurement to the 3/8-inch minimum required by design, are shown in Table 2. Four field welds and one test plate weld were acceptable, but one field weld and one test plate weld were rejectable because they lacked the required minimum effective throat.

Two of the plate welds with reportable indications were also evaluated. These welds were fabricated in a less restrained configuration than the field welds and would, therefore, be expected to show less severe cracking than the field welds. However, the test plate welds contained cracks similar to the field welds.

Table 2. Effective Throat Measurements

Weld	Group	Effective Throat (in.)	Result*
A6SE4	1	11/32	Rejectable
A28NE6	2	7/16	Acceptable
A29SE3	3	29/64	Acceptable
G6SW2	4	3/8	Acceptable
G27SE2	5	29/64	Acceptable
W3	4	7/16	Acceptable
G1	4	23/64	Rejectable

Weld Symbol Designation

A6SE4, for example, means: W3, for example, means:

A - plate number
6 - plate side
S - girder face
E - column line number
4 - column line letter

W - section
3 - test plate

Group Designation

1. Linear indication parallel to a weld fusion line.
2. Flat linear with a vertical linear rising towards the weld crown.
3. Flat linear with a linear parallel to a weld fusion line.
4. Vertical linear.
5. Linear parallel to a weld fusion line with a broad linear rising vertically to the weld crown.

* Comparison with 3/8-inch minimum effective throat.

ANALYSIS OF INDICATIONS

All well-defined indications were discontinuities which adversely affected the effective throat dimensions of welds. The linear indications parallel to the weld fusion lines in Group 1, 3, and 5 were confirmed as cracks. The flat linear indications in Group 2 and 3 were obliquely ground slag pockets. The vertical linear indication in Group 2 was an extended slag pocket with some porosity, and the vertical indication in Group 4 was a crack. The broad indications in Group 5 were caused by distortion of the MT magnetic field around the unfused portion of the weld when the magnetic leech was placed on the bottom side of the weld.

The analysis of the Group 5 indications revealed the importance of adhering closely to the specified examination method. Only sharp, well-defined MT indications should be interpreted as discontinuities.

FAILURE ANALYSIS

A weld failure analysis disclosed three traits of this weld design and procedure that were found in all the welds. First, the initial weld bead (root pass) does not fuse the bottom edge of the bevel to the main column, Second, the large slag pockets and associated discontinuities are inherent in the joint design. And third, the normal weld metal shrinkage induces tension stress on the weld, often resulting in cracks.

The cracks originating from the slag pockets tend to relieve stresses. The cracks initiate immediately or shortly after the plate weld is completed. Further propagation should not occur as a result of the remaining residual welding stresses. Stresses induced by heating, cooling, or external loading may cause further propagation of the cracks.

SUMMARY AND CONCLUSIONS

The welds hold the horizontal plates in place and are not subject to calculated design stresses. The welding was performed in accordance with a fabricator's procedure and required a visual inspection. More sensitive inspection techniques were later specified during the resolution of the visual indications.

AWS Code D1.1, Rev. 2-77, allows discontinuities provided the minimum effective throat dimension of the weld is not violated. The minimum required effective throat dimension of the weld has been established as 3/8-inch.

All true magnetic particle indications are caused by discontinuities (slag pockets, unfused lands, and cracks) that decrease the effective throat dimension of the plate welds. Large slag pockets and associated discontinuities exist in most of the partial penetration welds. Since similar discontinuities were found on five of the six test plate cross sections, the discontinuities present at the weld termination points probably exist throughout the lengths of the welds.

The specified welding procedure used with the partial penetration weld joint design did not consistently provide the minimum required effective weld throat thickness. The formation of the slag pockets and other discontinuities result primarily from the partial penetration joint design. The welding procedure and weld joint design resulted in tension stresses which caused many welds to crack.

Three field welds and one test plate weld had less than the required 3/8-inch effective throat thickness and are therefore rejectable. Other rejectable welds may exist in the 810 turbine building plate welds.

RECOMMENDATIONS

To insure that the design requirements and acceptance criteria for the horizontal plate welds are not overly conservative, it is recommended that they be reviewed by the responsible design engineer. This review should include a determination as to whether through-thickness cracking in limited areas can be tolerated. The possibility of crack initiation and propagation due to stresses caused by heating, cooling, or external loading should also be addressed. The minimum acceptance criteria for the welds should be specified in the review.

Following the review, the results of this investigation should be re-evaluated. If the responsible design engineer determines that the reported defects in the welds are not significant, then no further work should be required. If defects in the welds are significant, then further examinations would be necessary. One of the sampling programs shown in Table 3 could be used to determine the number of randomly selected welds to be inspected in order to achieve the specified confidence level.

Table 3. Recommended Sampling Programs*

Basis	Program 1 Weld termination points, only		Program 2 Entire plate welds	
Population	1620		810	
Confidence level	95%	96%	95%	96%
AQL	6.5	4.0	6.5	4.0
Sample size (n)	40	40	35	35
Maximum allowable percent defective (M)	11.85	8.09	11.87	8.10
Number defects that will reject sample	5	3	5	3

* Based on ANSI Standard Z1.9-1972, Inspection by Variables for Percent Defective.

If the weld termination points are selected for inspection, then Sampling Program 1 as shown in Table 3 would be used. With this program, 40 randomly selected weld termination points would be inspected. The subsequent analysis and corrective actions would be identical to the procedure for Sampling Program 2, but the quality of the weld between the two weld termination points would not be known.

If discontinuities are found, then the acceptance criteria would be applied to determine the weld's acceptability. If welding repairs are required, they should be approved and reviewed for acceptance, carefully outlining the method and procedures.

ACKNOWLEDGEMENTS

This work was performed under the auspices of the U.S. Department of Energy by the Lawrence Livermore National Laboratory under contract number W-7405-ENG-48. The authors are grateful to W. C. Ham, F. J. Dodd, E. A. March, R. S. Blackman, P. M. Tibals and R. D. Adamson.

Advances in Metallographic Techniques

Chairpersons: W.J.D. Shaw and J.D. Braun

THE MICROCOMPUTER IN THE METALLOGRAPHIC LABORATORY

A. S. Holik[*] and J. C. Grande[*]

INTRODUCTION

A microcomputer has been employed to facilitate several quantification procedures in a metallographic laboratory. An Apple II Plus personal computer (PC) has been interfaced with a Zeiss MOP-3 digitizing tablet to improve the memory capacity and the data manipulation capabilities of the system. A computer program was also written to drive an ink plotter to inscribe dimensional labels onto micrographs. Further, a system has been developed in which the PC, interfaced to the filar eyepiece of a Kentron microhardness testing machine, directly processes impression length measurements and generates a final report with tabulated microhardness values. All three of these applications - employing "user friendly" programs - save time, improve the ease of operation, and significantly enhance the quantification capabilities of the laboratory at a modest capital equipment investment.

DIGITIZING TABLET

A Zeiss MOP-3 semi-automatic digital image analyzer with a digitizing tablet (Figure 1) has been in use for several years in our laboratory as an aid in the quantification of microstructural parameters such as phase volume fraction, particle size, and particle size distribution. In a typical application, a micrograph/macrograph is secured to the surface of the digitizing tablet. Electromagnetic signals emitted from a wire grid array embedded beneath the surface of the tablet are sensed by an operator-directed stylus or cursor. These

---

[*] General Electric Corporate Research and Development, P. O. Box 8, Schenectady, NY 12301 USA.

Figure 1. Zeiss MOP-3 system consisting of digitizing tablet and microprocessor with keyboard and LED display.

signals are translated into x and y coordinate positions by an integral microprocessor. These data are placed in the system's memory and then processed to obtain information such as area, perimeter, length, count, angle, and shape. The lateral resolution is of the order of 0.1 mm.

The memory storage and data manipulation capacities of the digitizing tablet system, however, are limited. For example, in order to generate a particle size distribution histogram (Figure 2), the system requires the operator to predict and input the histogram interval width (size range) prior to making the measurements. A histogram of different interval widths from these same data cannot be generated after the measurements have been made; the measurements must be retaken. It is obviously desirable to have the capability to modify histogram interval widths after the measurements have been taken since it is usually difficult to predict what will be a suitable or acceptable interval width.

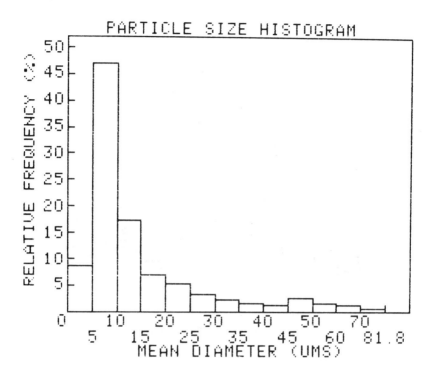

Figure 2. Particle size histogram of aluminum oxide particles in a ceramic matrix.

To achieve this, as well as to increase memory capability, the digitizing tablet system was interfaced with a PC (Figure 3). The hardware required for the interface - a Zeiss 'V24' circuit board - was installed in the MOP-3. This circuit board permits two way data communication between the PC and the digitizing tablet system. Measurements made with the digitizing tablet are sent to the PC via an RS-232 serial interface and stored in an array. Once the measurements are made, the data then can be manipulated and/or permanently stored on a 5-1/4" floppy disk. The format for data transmission from the digitizing tablet system to the PC is in the basic form:

$$0.000000 \quad E\underline{+}00 \quad 00$$

$$\text{Mantissa} \quad \text{Exponent} \quad \text{Code}$$

Figure 3. Microcomputer system consists of an Apple II Plus personal computer linked to (a) Zeiss MOP-3 digitizing tablet instrument, (b) printer, (c) plotter, and (d) monitor.

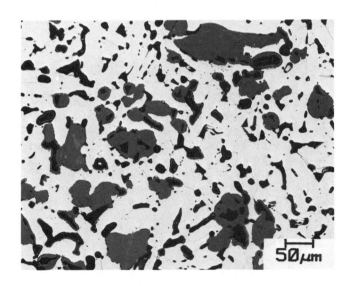

Figure 4. Dimensional label on micrograph of calcium-doped aluminum oxide ceramic.

The code number serves to identify the particular measurements (i.e., area, perimeter, etc). Multiple measured quantities can be sent to the PC simultaneously. This format enables the operator to discern data sent to the PC without a priori knowledge of the measurements made on the digitizing tablet.

DIMENSIONAL LABEL

There is a need to inscribe micrographs with a dimensional label (as shown in Figure 4) in order to facilitate rapid estimations of the size of microstructural features such as inclusions, phases, coating thicknesses, and grains. The use of the dimensional label avoids the mental gymnastics required to convert from a magnification value to a length when a micrograph is under quantitative assessment. Existing methods to apply dimensional labels have shortcomings when evaluated on the basis of speed, quality and ease.

A system has been developed in which a Hewlett-Packard #7220A Plotter coupled to a PC is used to inscribe micrographs with the appropriate dimensional labels. The interface was effected through a standard RS-232 serial interface board. The software was written to provide only round number labels such as 10, 25, 75 and 250 micrometer and to forbid labels like 13, 27.2 or 102 micrometer. A "turnkey" system was developed which asks the operator for the print magnification and print size after the appropriate floppy disk has been inserted into the disk drive of the PC. The PC then determines the dimensional label parameters and the size of the lettering. On command, the plotter inks the legend onto the micrograph. For the first micrograph of a given magnification, this sequence requires approximately 15 seconds of time; labelling each additional micrograph of the same magnification takes 10 seconds. The legend is inked into a small rectangular area of the micrograph that is left unexposed during the darkroom photographic printing by an appropriately positioned opaque stop (Figure 5).

A more sophisticated program is also available which provides opportunity to customize the label by offering the operator the choice of any specific numerical label, such as 100 micrometer, for which the PC computes the appropriate scale length. Further, it offers the option of matching the scale length, or an increment or multiple of the length, with a microstructural feature of the micrograph.

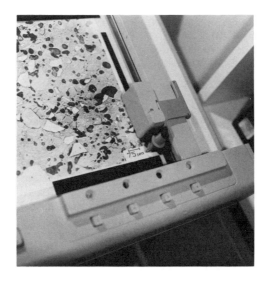

Figure 5. Dimensional label is inked into a small, clear area of the micrograph by graphics plotter that is under command of PC.

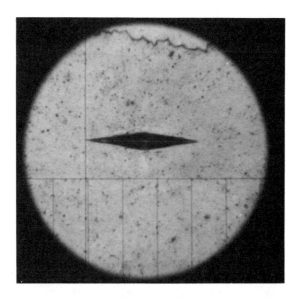

Figure 6. Hairline of filar eyepiece positioned at end of a Knoop microhardness impression.

MICROHARDNESS TESTER

The usual method for measuring a micro-indentation hardness impression is to read the diagonal length(s) of the impression with the filar eyepiece of a microscope. The hairline of the filar eyepiece is positioned at the diagonal end of an impression and a reading in filar units is made (Figure 6). A second reading is taken at the opposite end of the impression to obtain, by subtraction, a diagonal length in filar units. This value is converted to micron units and a hardness number is read from a chart according to the load that was used for the test.

This procedure has been considerably simplified and made less prone to error by the use of a PC that was linked to the filar eyepiece of the testing machine. Under the new arrangement, the operator sets the hairline of the filar eyepiece at the edge of the microhardness impression, as before, but does not take a filar reading. Instead, a key on a module is pressed. Another key is pressed after the hairline is moved to the opposite end of the impression. The PC does the rest, i.e., it calculates and prints the microhardness value. Parameters such as the objective lens used (for length calibration), the load, and the type of indentor (Knoop or Vickers) are input to the computer via a user friendly program prior to the diagonal length measurement step. The program also asks for other information necessary for a final report printout that is generated after all the impression measurements have been made (Figure 7). All of the accumulated data and other pertinent information are permanently stored on a floppy disk file for future reference.

Appendix 1 lists the interactive portion of the program. Appendix 2 provides details of the hardware and procedure for linking the various components. A block diagram of the various components of the entire system is shown in Figure 8.

SUMMARY AND FUTURE WORK

These three applications of microcomputers are examples of how computer technology can be effectively utilized in the metallographic laboratory. Significant benefits of improved quantification capabilities, time savings, and quality of

```
****************************************************************
                                    //
                                    //   MATERIALS CHARACTERIZATION
         GENERAL ELECTRIC           //          OPERATION
                                    //
    CORPORATE RESEARCH AND DEVELOPMENT //
          SCHENECTADY, NEW YORK      //  LIGHT MICROSCOPY
                                    //
****************************************************************
```

```
        TO:  DR. COFFIN                    FILE NO.:  55643

        BLDG:  K-1  RM:  1C22              DATE:  5/12/83
```

* MICROHARDNESS TEST *

THE FOLLOWING RESULTS WERE OBTAINED WITH THE SAMPLES OF ZIRCALOY TUBE
WHICH YOU SUBMITTED ON 5/9/83:

        SAMPLE #56656A1: BREAKDOWNPASS - SECTION, #1
        SAMPLE #56656B1: BREAKDOWNPASS - SECTION #2

HARDNESS IMPRESSION NUMBER	AVERAGE DIAGONAL LENGTH	AVERAGE DIAGONAL LENGTH	KNOOP HARDNESS NUMBER
(#)	(FILAR UNITS)	(MICRONS)	(KG/MM2)
	* SAMPLE #56656A1 *		
1	510	261	105
2	500	256	109
3	520	266	101
4	490	250	113
5	505	258	107
	* SAMPLE #56656B1 *		
1	525	268	98.9
2	520	266	101
3	522	267	100
4	524	268	99.2
5	527	269	98.1

```
LOAD USED: 500GMS          OBJECTIVE LENS USED: 20X
REMARKS: IMPRESSION INTERVAL - 1MM

ANALYST: JIM GRANDE
EXT. 5-8531, K-1/1C22
```

Figure 7. Typical microhardness report produced by the microhardness program described.

presentation have resulted from these usages. Further, an additional degree of technical sophistication and professionalism have been added to the discipline of metallographic laboratory practice.

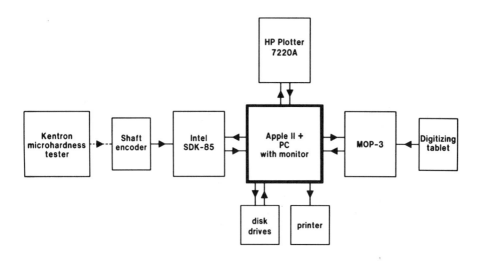

Figure 8. Block diagram of the components of the system.

Other applications of computer technology are under study. We are currently developing software to store an etchant file. The program will have the capability to access etchants by a number of different keywords. For example, recommended etchants for Rene 41 alloy could be accessed by the input of any one of several keywords such as "Rene 41", "nickel-chromium alloy", or "nickel-base heat resisting alloy". Proper names of common etchants, such as Vilella's reagent, and uncommon etchants, such as Valenta's reagent, also will be access keywords. Other possible metallographic applications of computer information storage and retrieval are chemical, mechanical and electrolytic polishing recipes for specific alloys and ceramics.

The applications of computer science to the metallographic laboratory promise to be manifold. It is hoped that the successful examples presented will help to inspire others to exploit this technology to its limits.

ACKNOWLEDGEMENTS

We gratefully acknowledge the technical assistance of D. B. Sorensen and J. J. Nasadoski for the implementation of the Intel SDK-85 microcomputer and shaft encoder.

APPENDIX 1

Interactive Portion of Microhardness Program

```
Type the customer's name ...
Customer's location ...
     Bldg. #:
     Rm. #:
What is the metallography file number of this job?
Type today's date (month/day/year) ...
What is the specimen material?
When was the job submitted?
Type your name and the name or names of others who are
responsible for this report ..
Are there any remarks to be included in this report?  (Y/N) ...
Type remarks ...
What load (in grams) will you use?
Which objective lens will you use?
     Type 10, 20, or 50:
Will you use Knoop (press 1) or Vickers (DPH) (press 2)?
How many samples will you test?
What is the metallography file designation for specimen
number 1?
     (54020A1 for example)
What is the customer designation of sample number 1?
You are now ready to start your microhardness test.
```

APPENDIX 2

Hardware and Procedures for the Implementation
of the Microhardness Program

An Intel SDK-85 microcomputer kit was used as an intermediate data communication link between the Kentron microhardness tester and the PC. A shaft encoder, coupled to the filar eyepiece of the microhardness tester (Figure 9), was interfaced to the PC via the SDK-85 microcomputer. The shaft encoder was chosen to correspond one-to-one with the 100 filar units per revolution of the filar eyepiece. The SDK-85 supplies an input potential of 5 volts to the shaft encoder. A separate pulse counter circuit receives a 5-volt pulse for every 1/100th of a revolution on the shaft encoder. Two channels from the encoder to the pulse counter circuit provide an independent path for the 5-volt pulse depending on the rotation direction of the shaft encoder. Once the pulses are totaled, they are sent to the PC for further processing.

The microhardness program sends the appropriate "handshaking" signals through an RS-232 serial interface to the SDK-85 thereby initializing the programming sequence. The operator aligns the filar eyepiece hairline with the end of the impression diagonal. A press of the "B" key of the SDK-85 keypad module (Figure 10) simultaneously zeros out the LED display and signals the PC that the initial filar hairline alignment has been made. The PC then sends a character "b" to the SDK-85 (which appears on the LED display) to confirm two-way data communication. The LED display on the SDK-85 continuously updates the number of filar units accumulated by the pulse counter as the impression diagonal is traversed. After the filar hairline is set onto the other end of the diagonal, the "E" key on the keypad is pressed to end the measurement. A string containing the filar unit value and the character "E" is then sent to the PC and stored in an array for further data manipulation that results in the generation of a Vickers or Knoop microhardness number.

Figure 9. Shaft encoder mechanically coupled to the filar eyepiece of the microhardness tester.

Figure 10. Keypad and LED display of the Intel SDK-85 microcomputer.

FIELD METALLOGRAPHY
THE APPLIED TECHNIQUES OF IN-PLACE ANALYSIS

J. F. Henry*

ABSTRACT

Due to utility interest in extending the life of older steam generating units, attention has been focussed on creating a non-destructive, empirically based method of estimating the residual creep life of high-temperature pressurized components. Such a method is currently being developed as part of a joint CE/CEGB effort sponsored by EPRI. The primary tool of material investigation for this method is replication. It is shown that replication, as an adjunct of basic field metallographic techniques, is capable of accurately recording microstructural features whose size approaches the limits of the resolving power of the light-optical microscope. In particular, replication has been used successfully to uncover evidence of the initial stages of creep-induced cavitation in damaged boiler components.

INTRODUCTION

Over the last several years, the basic techniques of field metallography have become fairly well established; however, there have remained questions regarding the limits of its capabilities. In particular, there has been no comprehensive study of the extent to which in-place surface

---

* Combustion Engineering, Inc., C. E. Power Systems, 911 W. Main St., Chattanooga, TN 37402 USA.

preparation techniques are able to reveal, and replicating methods are able to record, relatively subtle microstructural features, such as the grain boundary cavitation that is the earliest stage of creep-rupture damage detectable by the light-optical microscope.

Traditionally, the primary function of field metallography, as it has been employed throughout industry, has been to investigate nondestructively relatively gross microstructural phenomena, either as part of a routine quality control procedure, or in the course of failure analysis. A typical field metallographic examination might have as its object the determination of a material's grain size, the characterization of the morphology of a fracture, or the identification of major microstructural phases in a critical area of a heat-treated component. While these uses of field metallography are eminently beneficial, we have found that they do not exhaust the investigative capabilities of the technique.

In 1982 an EPRI-sponsored program, supported jointly by CE and CEGB, was initiated to develop a nondestructive method of empirically determining the remaining life of high-temperature boiler components. As part of the development work, it was necessary to fully evaluate current field metallography practices. Particular attention was given to the replication of the metallographically prepared surface, to insure that the replica could serve as the primary analytical tool in detecting the early stages of intergranular creep cavitation that precede failure by stress rupture. This evaluation showed that, with certain minor refinements, basic field metallographic methods are fully capable of exposing and recording evidence of structural features, the discrimination of which approaches the resolving power of the light-optical microscope.

At this point, a brief explanation of the EPRI Boiler Life Estimation Project (Contract 14282) is in order. As utilities have attempted to extend the service lives of older steam-generating units, for obvious economic reasons, questions have been raised concerning the amount of useful life remaining for critical pressure parts that have operated for prolonged periods within the creep regime. Design calculations based on stress-rupture curves are, in this case, of marginal value because they cannot account for variables in material and unit operation, and frequently lead to pessimistic life predictions. Consequently,

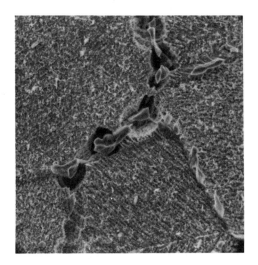

Figure 1. Grain boundary cavitation (SEM micrograph, 725X).

an empirically verifiable, nondestructive method of determining how much of a component's useful life has been expended is essential if a valid estimate of residual creep life is to be made. Such a determination is thought to be possible because of the nature of the creep process. Very simply put, as a pressurized component operates at elevated temperatures, damage due to creep accumulates in the material.

Creep damage tends to concentrate at grain boundaries and is oriented normal to the maximum principal stress. When a substantial amount of the total useful life of the material has been consumed, small cavities, whose visible size is on the order of magnitude of 0.1 micrometer, appear in the boundaries (Figure 1). With further aging, the cavities coalesce and become grain boundary fissures which, with still further service exposure, form a network of cracks through the material. Remaining life studies using interrupted creep-rupture test specimens currently are being performed to obtain not only qualitative, but also quantitative values correlating the various stages of degradation with the creep curve [1,2].

The task of field metallography thus becomes the detection of the earliest visible stage of creep-damage -- the cavities -- and the subsequent monitoring of the component to aid in the run, repair, or replace decision.

EXPERIMENTAL PROCEDURE

As previously mentioned, the basic techniques of field metallography are well established. Briefly, they involve surface preparation, replication, and documentation. Stage I, surface preparation, may be broken down into a rough grinding step, an intermediate polish, and a final polish/etch/polish sequence of surface treatment. Rough grinding involves treatment with successively finer grits of a silicon carbide or similar type of paper, starting with 80-grit, and progressing through 180-grit, 320-grit, and 600-grit papers.

The intermediate stage of polishing is accomplished using a 6-micron and 1-micron diamond paste to remove all but a very thin layer of deformed metal from the surface to be examined.

The final step, involving a sequence of polishing with a suspended solution of 0.05 micron gamma-alumina, and etching with a standard 3% nital etchant for ferritic materials, is the most critical in the surface treatment process for revealing early cavitation damage. We have found that a minimum of three polish-etch cycles is required to effectively remove material disturbed by the previous polishing steps. This polish-etch repetition has the additional effect of slightly enhancing the size of the creep cavities so that they are more readily detected in the replica. The importance of the final preparation step is emphasized in Figure 2, which shows a surface at various stages in the polish-etch sequence.

Having disclosed the cavities, it now remains for the field metallographer to record the structure, so that a meticulous examination of the grain boundary features can be conducted in the laboratory. The mechanics of replication are schematically illustrated in Figure 3. For this purpose a standard cellulose acetate replicating film, approximately two mils thick, is used with a methyl acetate solvent. The methyl acetate is superior to acetone and other solvents, because it is less prone to form bubbles in the film as the replica dries. The softened replicating film is placed on the etched surface and allowed to dry; it is then removed, mounted on a glass slide, and stored. Although a Leitz field microscope is used

Field Metallography / 541

After 1-micron diamond

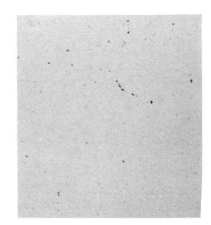
After one polish/etch/polish cycle

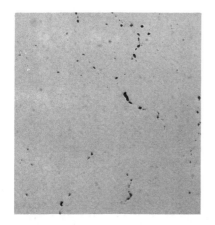
After two cycles

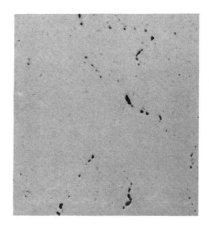
After three cycles

Figure 2. Effect of the final polish/etch sequence on the revelation of cavitation damage.

to make a preliminary evaluation of replica quality, final analysis and documentation of the structure recorded on the replica is deferred until the replica is returned to the laboratory, where it can be shadowed with a vapor-deposited layer of chromium to enhance reflectivity.

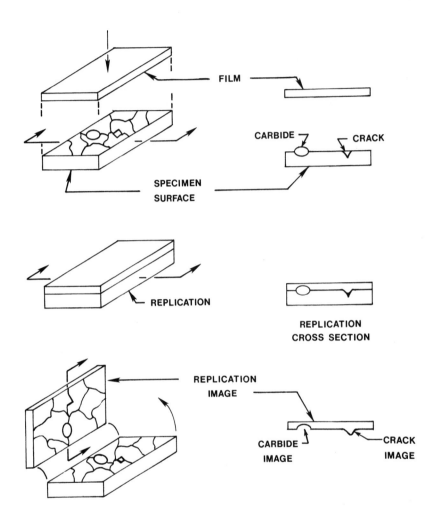

Figure 3. Schematic illustration of replication.

RESULTS

To substantiate the efficacy of the replicating method, the results of field metallographic analyses of several components suspected of having sustained creep-related damage will be presented.

Field Metallography / 543

Figure 4. Inlet desuperheater "T" section with three OD surface cracks.

Figure 5. Transverse section of desuperheater inlet T showing cracking in weld repaired region.

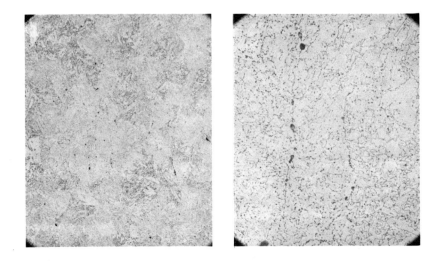

Figure 6. Metallographic mount of weld repair in inlet desuperheater section (left: 60X, right: 300X).

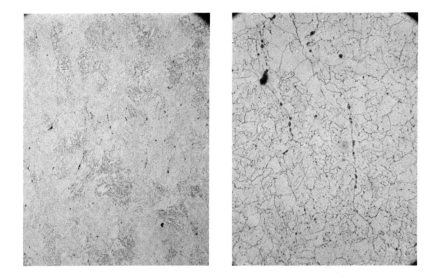

Figure 7. Replica of weld metal repair in T section showing oriented nature of cavitation at low magnification and grain boundary cavities at high magnification (left: 65X, right: 320X).

The first analysis involved the inlet "Tee" on a header that had operated for approximately 28 years and was known to have undergone several prolonged overtemperature excursions. A section of the "Tee" was removed from service because of visible cracks in the O.D. surface (Figure 4). In this case interest focussed not on the failure mode, itself, which was readily identified as stress-rupture cracking (Figure 5), but on the ability of the replication technique to detect the early stages of creep cavitation remote from the major cracks. This sample was particularly useful because it was possible to compare directly the replicated structure with the actual structure prepared under laboratory conditions, and thereby evaluate the fidelity of the replicated image. The results, illustrated in Figures 6 and 7, were satisfactory, confirming the belief that any structural feature that could be resolved in a metallographic mount using a light-optical microscope could be successfully reproduced on a replica.

The next example of successful replication involved the evaluation of an AISI 316 stainless steel main steam line piping. This pipe was 9.05-inch O.D. x 2.525-inch minimum wall thickness (MWT), and had been designed for supercritical service at 1200 F and 5500 psi (Figure 8). One section of the pipe had developed extensive stress-rupture cracks (Figure 9), and it became necessary to investigate the full run of the steam line for any evidence of incipient creep-rupture damage that would indicate a need for replacement. Utilizing the replication technique, it was possible to evaluate the piping in-place, and locate those sections in which cavitation had begun to develop, Figures 10 and 11, and which, therefore, required removal from service. Of equal importance is the ability of the replica to record the absence of damage in a sound component. Figure 12 shows an area of the pipe in which no cavitation had occurred and which was, therefore, deemed serviceable.

The final example to be considered involved the examination of a reheat outlet header that had logged approximately 80,000 h service. This header, relatively young by normal utility standards, showed no visible signs of distress. However, recorded incidents of prolonged overtemperature operation in certain areas of this header led to an interest in the material integrity at the interface between the heat-affected zone (HAZ) and base metal at welded stub tubes. Due to the low creep ductility of coarse-grained bainitic structures, this transition region was selected as a site susceptible to the accumulation of creep damage. Replication of the area at the

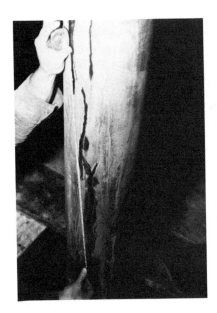 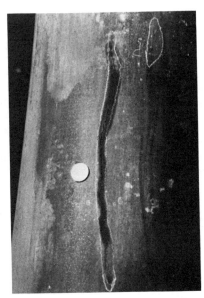

Figure 8. Typical O.D. longitudinal cracking in 4B main steam line.

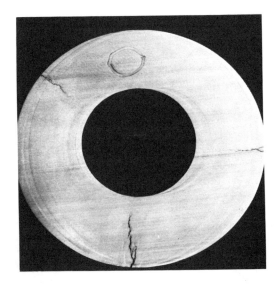

Figure 9. Cracking in section of failed 4B steam line from unit 1.

Field Metallography / 547

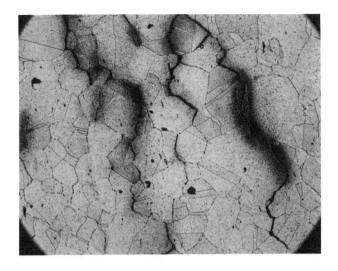

Figure 10. Replica of grain boundary cracks in AISI 316 main steam line (75X).

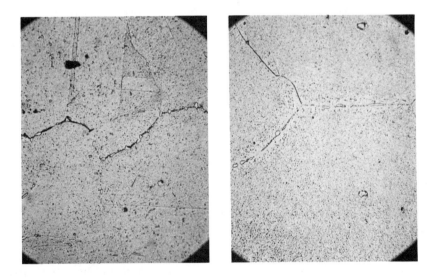

Figure 11. (Left) Replica of oriented cavitation in AISI 316 stainless steel main steam line (300X).

Figure 12. (Right) Turbine inlet pipe with large sigma phase particles in grain boundary and no cavitation damage (300X).

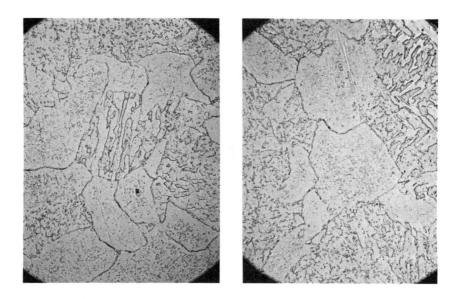

Figure 13. Creep cavitation in header base metal (320X).

edge of the HAZ revealed that cavities had, in fact, begun to form in some grain boundaries, as shown in Figure 13. This form of cavitation, termed isolated cavitation, represents a relatively early stage in the visible creep process; future monitoring of the damaged regions should offer a clearer picture of the advancement of creep-related damage in this type of component.

SUMMARY

The demand for a nondestructive method of detecting the early stages of visible creep damage in high-temperature pressurized components led to a re-evaluation of basic field metallographic techniques and, in particular, replication. We have found that, with a modicum of care and an understanding of the task to be accomplished, the standard field metallographic replica will accurately record all structural features that can be detected in a standard metallurgical specimen with the light-optical microscope.

REFERENCES

1. B. J. Cane, "Creep and Creep Fracture of Low Alloy Ferritic Steels Structure Property Correlations," National Symposium on Creep Resistant Steel for Power Plants, BHEL, Hyderabad, India (January 1983).

2. J. F. Henry and F. V. Ellis, "Plastic Replication Techniques for Damage Assessment," Electric Power Research Institute, RP2253-01 (September 1983).

DETAILED MACROSEGREGATION STUDIES USING A
TANDEM VAN DE GRAAFF ACCELERATOR FACILITY

N. A. Shah*, J. J. Moore*
and J. H. Broadhurst**

ABSTRACT

The Tandem Van de Graaff Accelerator of the University of Minnesota is described in terms of its application and advantages in determining the extent of macrosegregation in cast steel billets. Point analyses 1-mm or larger in diameter for all elements down to helium in the periodic table can be taken to a depth of 1-mm without destroying the surface being analyzed. Therefore, microscopy of the same area can be correlated with the analyses. The alternative is to analyze drilling samples which then have to be produced in a solution form before analysis can be conducted by instrumental chemical analysis techniques.

INTRODUCTION

Segregation of elements is one of the major quality control problems in continuously cast, high-carbon steel products since it persists even after hot-rolling and heat-treatment. Practically, it is impossible to eliminate segregation completely. More generally, segregation influences the form, amount and distribution of the precipitated second phase

---

* Mineral Resource Research Center, University of Minnesota, Minneapolis, MN 55455 USA.
** School of Physics and Astronomy, University of Minnesota, Minneapolis, MN 55455 USA.

with a resultant heterogeneous microstructure. Thermal conditions, casting parameters, chemical composition and the primary phase of the solidifying alloy largely contribute to the extent of the segregation.

Recently, macrosegregation in three continuously cast, high-carbon (0.80-0.85% C), low-alloy steel billets was examined qualitatively/semi-quantitatively using sulfur printing, macroetching, optical microscopy, scanning electron microscopy with respect to the distribution of sulfur, phases, inclusions, primary and secondary dendrite arm spacing and the extent of columnar and equiaxed zones and correlated to the casting conditions [1]. This paper describes the metallurgical application of the Tandem Van de Graaff Accelerator of the University of Minnesota for the determination of macrosegregation of solute elements in these same billets.

LIMITATIONS OF CONVENTIONAL METHODS

Macrosegregation is normally determined quantitatively by wet chemical analysis, which is, unfortunately, time consuming and has limitations regarding the optical resolution of the macrosegregation due to the diameter of the drill and the depth to which it is used [2]. As the volume from which the drillings are obtained increases, there is a greater opportunity for compositional gradients to be reduced in apparent intensity due to averaging effects. A small diameter drill produces a lower weight of drillings so that only one or two elements can be analyzed which makes the simultaneous study of the macrosegregation of a number of solute elements fairly difficult.

Use of the electron probe microanalyzer with a beam scan area of about 10,000 micrometer square for the determination of carbon, nitrogen, oxygen and other elements has been reported in the literature [3]. This method provides compositional mapping ability for elements down to boron and is nondestructive. However, microprobe analyses require a small metallographically polished specimen and the minimum detectable limits and analysis accuracy are not as good as can be obtained by wet chemical analysis.

In order to obtain the best features of each of these methods while minimizing their weaknesses, a method utilizing the Tandem Van de Graaff Accelerator for the quantitative determination of the macrosegregation of elements in continuously cast steel has been developed at the Williams Laboratory jointly with the Mineral Resources Research Center at the University of Minnesota. This technique could also be used for point elemental analysis of other materials.

The novel feature in determining macrosegregation with the Tandem Van de Graaff Accelerator is that, besides providing a nondestructive form of chemical analysis, it is possible to achieve multi-element analyses, including low energy elements, e.g., C, N, O, Al, in a 1-mm diameter region of the sample at ppm levels depending on the operating accelerator voltage, sample current, type of detector, etc. The normal net analysis time for each point is approximately 10 minutes during which a host of elements can be detected in contrast to a selected one or two elements from samples produced via small drillings. Thus, a more extensive study of macrosegregation than was previously achievable can now be obtained with this technique. Moreover, the sample size can be as large as 600 mm in length or width. A comparison between the various techniques for the quantitative study of macrosegregation in steel is presented in Table 1.

THE VAN DE GRAAFF DETECTION TECHNIQUE

Analysis of the steels for solute elements is accomplished by bombarding a spot on the material with protons, resulting in x-ray fluorescence in the steel alloy with energies (and, therefore, wavelengths) characteristic of each solute element. Detection and counting of these x-rays then provides, after corrections for production efficiency and loss before detection, a measurement of the ratios of the number of atoms of each element at the site of the proton bombardment, i.e., the elemental concentrations.

The use of protons at mega electron volt energies allows a focused beam of approximately one millimeter diameter to penetrate a window into the air without appreciable defocusing. The same property of massive energetic particles which allows them to penetrate the material without appreciable deflection also facilitates the acquisition of x-ray data, the competing background process of x-ray production due to deflection by the positive charge of atomic nuclei (bremstrahlung) being suppressed by at least three orders of magnitude relative to energetic electrons.

Table 1. Comparison Between Techniques for Studying Macrosegregation in Steel

	Wet Chemical Analysis	Electron Microprobe Analysis	Tandem Van de Graaff Accelerator
Type of Segregation	Macro	Macro & Micro	Macro & Micro
Accuracy	Limited by diameter of drill. (a) If large: Averaging out problem (b) If small: One or two elements only analyzed	High	High
No. of Elements	A few	Multi elements	Multi elements including low energy elements (e.g., C, O, N)
Sample Preparation	Saw Cut Drillings Drill size: 3-8 mm diameter.	Saw Cut. Sample Size: < 20 mm Mount, grind, polish. Accuracy dependent on sample preparation.	Saw cut or surface ground. Sample size: varied up to 200 mm in height or dia. Spot size 1 mm - 2.5 cm.
Sampling Time	Hours	20 min	10 min. per series of analyses at one point
Estimated Cost	$5-20 per element	$50/hour	$60/hour

The generation of energetic protons is shown in Figure 1. Negative hydrogen ions produced by sputtering a surface of titanium hydride by means of a cesium beam, are injected into a double-ended tandem electrostatic accelerator. Here, a large positive potential, typically several megavolts, attracts the negative ions to the charged terminal. At this terminal, the outer electrons are stripped from the hydrogen ion converting it to a positively charged proton. The proton is repelled by the terminal and emerges from the far end of the accelerator at a kinetic energy proportional to twice the terminal potential. These protons are then transported and focused by a variety of magnetic fields before passing through a 0.10-mm tantalum window into the laboratory. A plan of the facilities at the Williams Laboratory is shown in Figure 2.

The proton beam passes through the exit window at the top and impinges on the billet under examination which has been placed on a screw jack. X-rays generated by the proton interaction are collected by detectors mounted at 45-degree to the sample face. In order to minimize the loss (absorption) of x-rays by the surrounding air, the exit window and detector penetrate a lucite cylinder which is filled with helium. The screw jack enables the impact point of the protons on the samples to be changed as desired.

EXPERIMENTAL WORK

Because high-carbon, low-alloy steels have been found to provide severe segregation problems in continuously cast steel products [2], we decided to study macrosegregation in three production billets, cast under conditions (i.e., superheat > 20 C, casting speed moderately high, section size < 160-mm square) which would produce severe segregation in an attempt to provide a detailed assessment of the physical model of the extent of the problem.

The steel billets used for the present study were continuously cast in a curved mold strand caster under production conditions. The nominal chemical composition and the casting parameters employed are detailed in Table 2. A previous study [4] has shown that Cast A has the highest cooling rate followed by Casts B and C.

Each billet was longitudinally cut through its center and placed in the sample chamber of the accelerator filled with helium. Some of the typical operating parameters are presented in Table 3. The W15 beam line was used (Figure 2).

556 / Microstructural Science, Volume 12

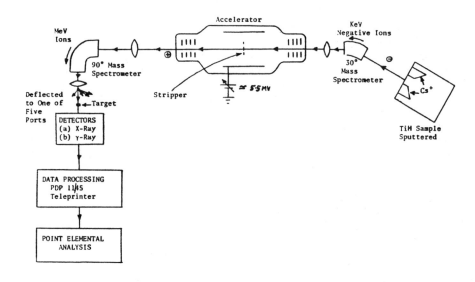

Figure 1. Schematic representation of proton beam generation and data processing.

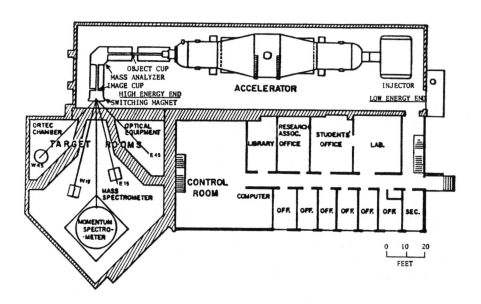

Figure 2. Plan of the first floor of the Williams Laboratory, University of Minnesota.

Table 2. Continuous Casting Conditions and Steel Composition (Wt. %)

Cast* No.	C	Mn	Si	S	P	Cu	Ni	Cr	Mo	V	Mn/S	Liquidus Temp.** (C)	Tundish Superheat (C)	Casting Speed (m/min)
A	.80	.98	.23	.038	.017	.25	.13	.53	.07	.002	25.8	1466.2	57.7	1.40
B	.83	.79	.24	.026	.014	.24	.09	.12	.01	.02	30.4	1466.9	48.6	1.65
C	.85	1.05	.58	.043	.023	.22	.08	.12	.02	.03	24.4	1459.2	34.1	1.52

* Conventional curved mold caster.

** Liquidus temperature was calculated using the following formula [5]:

$$T_L = 1538 - \{f[\% C] + 13.0 [\% Si] + 4.8 [\% Mn] + 1.5 [\% Cr] + 4.3 [\% Ni] + 30 [\% S] + 30 [\% P]\},$$

where for C $0.50 < C < 1.0\%$,

$$f [\% C] = 44 - 21 [\% C] + 52 [\% C]^2.$$

NOTE: Section size: 152.4 × 152.4 mm.

Table 3. Typical Operating Parameters of the Tandem Van De Graaff Accelerator Used for the Experiments

LOW ENERGY END	
Ionizer current	27A
Injector voltage	- 150 kV
Beam current	0.02 μA
Cesium boiler temperature	200 C
ACCELERATOR	
Accelerator voltage	5.86 MV
Corona current	50 μA
Belt current	0.16 mA
Belt speed	50 m/sec
Minimum vacuum in beam line	4 x 10^{-6} Torr
Normal vacuum in beam line	2 x 10^{-7} Torr
HIGH ENERGY END & TARGET LINE	
Proton energy	11.73 MeV
Minimum vacuum in beam line	4 x 10^{-6} Torr
Normal vacuum in beam line	2 x 10^{-7} Torr
Target current	3.0 NA
Colliminator size	0.7 mm
Distance of tantalum window to target	9.5 cm
Distance of target to γ-ray detector	14.0 cm
Distance of target to x-ray detector	4.6 cm
Gas in target chamber	Helium
DETECTOR	
(a) γ-ray photomultiplier detector (for low energy elements) diode bias voltage	1100 v
(b) X-ray Kevex detector (for high energy elements) diode bias voltage	900 v

Point analyses (1-mm diameter) were taken along the centerline and mid-radius of the longitudinal section of each billet. High-energy elements present in the steel (e.g., Si, P, S, V, Cr, Fe, Ni, Cu) were detected by an x-ray energy-dispersive detector. Data on lower energy elements (e.g., C, Al, O, N), which were detected by a γ-ray detector, are not reported here since work is still in progress on developing the software to process the data from these elements. Data from the detectors were recorded on paper tape and later fed into the PDP 11/45 computer through a teleprinter. For the data analysis of high-energy elements, some of the software that was incorporated include: DSP, PEKFIT, FIX3 and CALIB, which were originally developed at the Williams Laboratory.

Optical microscopic examination of the billets was conducted to provide a correlation between the macrosegregation analysis and the microstructural features. This examination was carried out after the segregation analyses were conducted allowing a direct correlation between structure and segregation. This was only achievable due to the nondestructive nature of the accelerator technique.

RESULTS

The number of x-rays counted for each element needed correction before being transformed into a representation of the atomic density. The three components of these corrections are as follows. First, the number of x-rays produced for any given element has to be corrected for x-ray production efficiency as a function of proton energy by taking into account the stopping of protons in the billet. Next, the number of x-rays so produced then requires correction for absorption, which occurs both in the billet itself, and in the intervening helium between the billet and the detector. Theoretically, it was necessary to include the efficiency of the x-ray energy-dispersive detector; however, the latter correction was not made here because the detector efficiency is known to approach 100% at the x-ray energies used in these measurements.

These corrections are simplified by two considerations. The protons travel in an approximately straight line for a well-defined distance into the billet, losing energy at a known rate. Also, the composition of the billet is known to be mainly iron, of which the x-ray absorption rate is well known. Consequently, the iterative search techniques typical of the Bence-Albee correction of electron-induced x-rays can be replaced by calculated correction factors. Calculation of the correction factors is described in Reference 4.

Macrosegregation analyses of Si, P, S, V, Cr, Mn, Fe, Ni and Cu present in the three cast billets along the longitudinal centerline and mid-radius positions are presented in Figures 3 and 4. Macrosegregation is expressed in terms of the segregation ratio S_i, where:

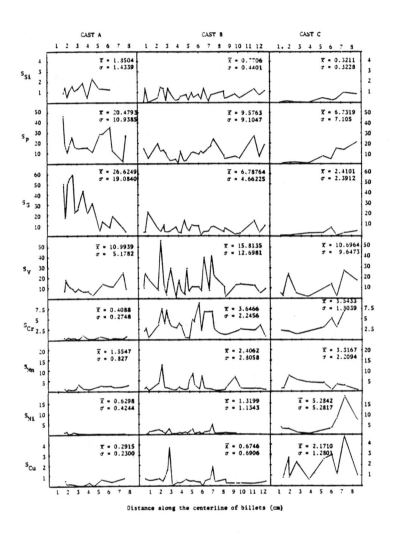

Figure 3. Segregation ratio of solute elements along centerline of various casts.

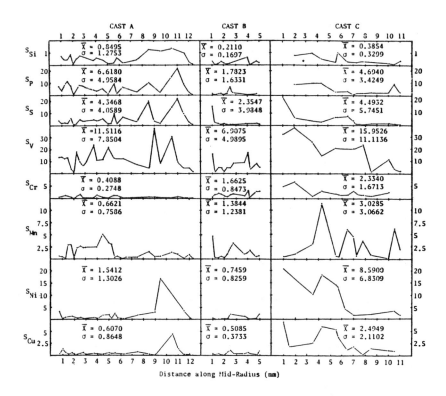

Figure 4. Segregation ratio of solute elements along mid-radius of various casts.

e.g.,

$$S_i = \frac{\text{Composition of solute i at that position}}{\text{Nominal composition of solute i}}$$

$$S_{Si} = \frac{Si}{Si_0}$$

Optical microscopic examination of Casts A, B and C revealed a microstructure of pearlite and cementite, whose amounts vary, depending on the extent of C segregation.

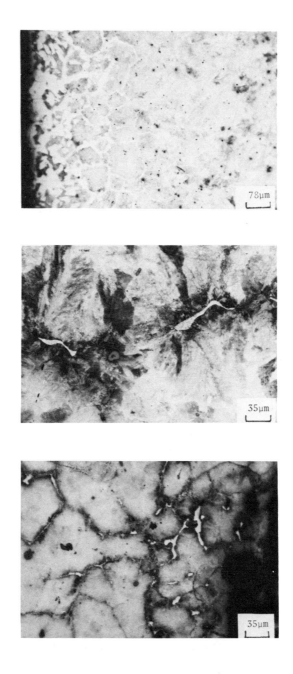

Figure 5. Typical microstructures of billets at: (a) edge, (b) mid-radius, (c) center.

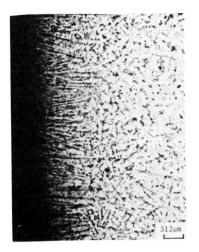

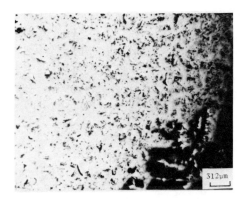

Figure 6. Typical phosphorus segregation in billets at:
(a) edge, (b) mid-radius, (c) center   (Phosphorus etch in which phosphorus-rich areas appear white).

Typical microstructures at edge, mid-radius and center are shown in Figure 5. A high proportion of ferrite is evident in the chill zone while increasing levels of grain boundary cementite are observed from the mid-radius position towards the center of the billet.

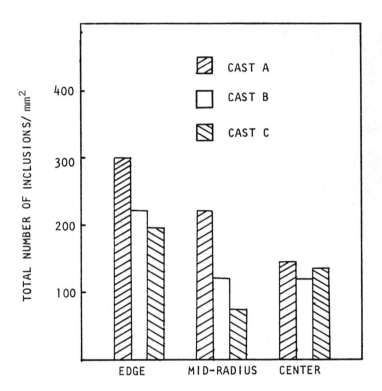

Figure 7. Distribution of total MnS inclusions on transverse section of various casts.

Figure 6 provides a typical microstructure of edge, mid-radius and center, treated with an etchant (composition: 2.5g $CuCl_2 \cdot 2\ H_2O$, 20g $MgCl_2 \cdot 6\ H_2O$, 10 ml HCl, 500 ml ethanol. Etching time: Etch until coppery sheen appears [6]) to show P segregation. The etchant selectively precipitates Cu on the P-depleted regions (dark areas) while the P-enriched regions (white areas) remain comparatively unaffected.

Figure 7 shows that the total number of sulfide inclusions per unit area decreased from edge to mid-radius to center of each of the billets.

DISCUSSION

From the macrosegregation analysis curves presented in Figure 3 it appears that certain elements, e.g., S, P and V, exhibit severe positive centerline segregation while others, e.g., Ni, Cu, Si, exhibit negative centerline segregation under the continuous casting conditions used in these production billets. The fact that some of the positive segregation ratio values are much higher than those previously reported is probably due to the use of a 1-mm diameter beam which is much smaller than drill diameters used in earlier work [7] and, therefore, is more sensitive to changes in elemental composition such as denser areas of sulfide inclusions.

In general, increasing the cooling rate increases the columnar zone resulting in an increase in the extent of positive centerline segregation of S and P in agreement with previously reported research [7]. The same trend is also seen in this present work for Si segregation. Furthermore, the overall level of Si, P and S (within the range used for these billets) does not have a significant effect on the extent of centerline segregation since Cast C contained the highest amounts of these elements, yet provided the least positive centerline segregation. This demonstrates the overriding effect of cooling rate and columnar zone on macrosegregation. The reverse trend is observed for Ni and Cu and, to a lesser extent, Mn and Cr, in that increased positive centerline segregation is produced with a decreased cooling rate and columnar zone. The exception to these trends is V which exhibits high positive centerline segregation for all the three casts, but the highest level with the intermediate cooling rate of Cast B.

The segregation analyses for the mid-radius positions for the three casts, as presented in Figure 4, also show some interesting general trends. The elemental segregation ratios of Cast B at the mid-radius position are less than those at the center. Comparing the three casts, Cast B exhibits the lowest mid-radius segregation ratios while Cast C exhibits the highest.

This work is the first reported on the application of a linear accelerator to determine the extent of centerline segregation and thus, as such, is necessarily speculative until more controlled solidification conditions are used with this technique.

Table 4. Equilibrium Distribution Coefficients, k_o, and Densities for Some Common Elements in Liquid Steel

Element	$1-k_o$ for liquid Fe and γ-Fe	Density (g/cm^3)
Sulfur	0.95	2.07
Oxygen	0.97	1.15
Carbon	0.70	2.25
Phosphorus	0.94	1.82
Silicon	0.50	2.33
Manganese	0.25	7.20
Chromium	0.15	2.19
Nickel	0.05	8.90
Copper	0.30	8.92
Vanadium	-	5.96

Although these billets were cast under production conditions rather than controlled solidification conditions, it may be possible to draw some general postulations on centerline segregation with respect to trends using previously reported research conclusions [8,9] and data from Table 4.

The elements that produce an increasing positive centerline segregation with increasing cooling rate and columnar zone (Si, P and S) are those which possess the lowest combination of values for distribution coefficient (i.e., in liquid iron and γ-iron, $k_o < 0.5$) and density (i.e., < 2.33 g/cm^3). This combination will result in a high rejection of these elements at the solid-liquid interface and increased density-driven fluid flow in the mushy zone towards the central axis of the billet. Increasing the cooling rate will increase the

temperature gradient in the liquid, consequently, producing
a reduction in the width of the mushy zone. This combination
of conditions, i.e., a high density-driven fluid flow in a
shortened mushy zone, may well result in these elements being
transferred to a greater extent towards the billet axis.

On the other hand, those elements which tend to provide
reduced positive (or increased negative) centerline segregation,
i.e., Mn, Cr, Ni and Cu, are those which possess a high
distribution coefficient (i.e., between liquid iron and γ-iron,
$k_o > 0.75$) and a high density (i.e., > 7.19 g/cm^3). This
combination will reduce the level of these solutes rejected at
the solid-liquid interface and considerably restrict the level
of density-driven fluid flow along the interdendritic channels
of the mushy zone. The fact that increasing the cooling rate
increases the extent of negative segregation for these elements
is presently unexplainable and requires more controlled
solidification experiments to be conducted. It may be simply
due to severe restriction in the fluid flow of these elements
along the interdendritic channels associated with a more rapid
solidification rate. Although the distribution coefficient of
V between liquid iron and solid γ-iron is high, its density
(0.5 g/cm^3) falls between the two extremes discussed above.
Therefore, the large positive centerline segregation is
probably due to a high level of density-driven fluid flow.
From thermodynamic calculations it is found that VC will
precipitate in a liquid iron of 0.82% C if 0.2% V is present.
This would necessitate a segregation ratio for vanadium of 10.
Since a segregation ratio of over 10 has been achieved for
vanadium, the precipitation of VC is quite feasible in this
steel. The possibility of forming fine, low-density vanadium
carbide particles is likely to increase the extent of the
density-driven fluid flow providing an increased positive
centerline segregation, the high melting point carbides being
pushed towards the billet center by advancing dendrite tips.

The mid-radius analyses also indicate the influence of the
extent of the columnar zone on the segregation of elements at
that position in that the more extensive the columnar zone,
the lower the elemental macrosegregation. It may be postulated
that this is due to the increased fluid flow occurring with an
increased columnar zone at this position resulting in these
elements being progressively transferred towards the centerline
and lower segregation ratios at the mid-radius position.

As mentioned earlier, the discussion of these analyses is highly speculative and more controlled solidification conditions are necessary in order to confirm these postulations. Nevertheless, the application of the Tandem Van de Graaff Accelerator offers great potential in determining the full extent of elemental segregation in the casting process, especially when the system is extended to the examination of the lower energy elements, e.g., C, N, O, Al, etc., using γ-ray detection techniques.

The microstructure at the edge, mid-radius and center of the billet (Figure 5) revealed an increase in carbon content from edge to center, the initial purer chill zone being negatively segregated in carbon showing the presence of intragranular Widmanstätten ferrite while an increase in the level of nonequilibrium cementite is precipitated towards the center of the billet. The positive segregation of phosphorus also progressively increased towards the center (Figure 6). The extent and position of this positive carbon (increased cementite) and phosphorus segregation will vary to some degree with respect to the position from which the samples have been taken along the billet length. These structural trends support the general conclusion reached from the accelerator analyses.

CONCLUSIONS

This work is the first reported on the application of the Tandem Van de Graaff Accelerator to determine the extent of macrosegregation and, as such, the conclusions drawn from the analyses are necessarily speculative until more controlled solidification conditions are used to produce samples analyzed by this technique. However, it may be possible to draw some general postulations on macrosegregation with respect to trends using the results of past research.

Conclusions regarding the new analytical technique and the data interpretations are:

1. The detection and measurement of point elemental analysis is possible with the accelerator technique.

2. For the study of macrosegregation of cast steel, the use of the Tandem Van de Graaff Accelerator is advantageous over other conventional techniques, especially regarding:

a. Large number of elements to be analyzed,
b. Large-sized samples which can be used in a nondestructive mode,
c. High accuracy (Proportional to $\pm \frac{100}{\sqrt{N}}$ %, where N is the net counts.
d. A relatively short analysis time of about 10 min per point.

3. The disadvantage of using the Tandem Van de Graaff Accelerator as a metallurgical technique for the study of macrosegregation is that the technique requires expensive equipment and installation. This equipment tends to be confined to a few locations in the world, with access limited by program committees and schedules.

4. The Tandem Van de Graaff Accelerator at the University of Minnesota is now available for metallurgical work in addition to nuclear applications. Rapid access to the Accelerator is possible for users paying commercial rates, the typical commercial rate now being $60 per hour.

5. S, P, Si, i.e., elements exhibiting a low distribution coefficient ($k_0 < 0.5$) and density (< 0.23 g/cm^3), produced an increasingly more positive centerline segregation on increasing the cooling rate and, therefore, the extent of the columnar zone (Cast C to Cast B to Cast A), while Ni, Cu, and to a lesser extent Mn and Cr, i.e., elements exhibiting a higher distribution coefficient ($k_0 > 0.75$) and density (> 7.19 g/cm^3) produced an increasingly negative centerline segregation on increasing the cooling rate. The centerline segregation of V seemed somewhat insensitive to the cooling rate in the three casts studied, but exhibited high positive segregation ratios in all the three casts.

6. The segregation ratios for the mid-radius positions for the three casts indicated that a lower segregation ratio could be expected at this position in the billet where there was an increase in the extent of the columnar zone.

7. Optical microscopy confirmed that there was an increase in carbon and phosphorus contents from the edge to the center of the billets.

ACKNOWLEDGEMENTS

The authors gratefully acknowledge the financial support of an AISI research grant under Contract #33-449 for this work.

REFERENCES

1. N. A. Shah and J. J. Moore, "A Metallography Study of Centerline Segregation in High Carbon Steel Billets," Microstructural Science, Vol. 11, Elsevier Science Publishing Co., Inc., NY, pp. 163-192 (1983).

2. J. J. Moore, "Review of Axial Segregation in Continuously Cast Steel," Iron and Steelmaker, Vol. 7, pp. 8-16 (1980).

3. G. Runnsjö, O. Grinder and T. Gustafsson, "The Determination of Macrosegregation of Carbon in Continuously Cast Steels by Electron Probe Microanalysis," Institutet for Metallforskning, Swedish Institute for Metal Research, IM-1466 (June 1980).

4. N. A. Shah, "Segregation in Continuously Cast Steel," M. S. Thesis, University of Minnesota (April 1983).

5. H. Hirai, K. Kanamaru, and H. Mori, "Estimation of Liquidus Temperature of Steel," Tetsu-to-Hagane, Vol. 52, p. S-85 (1969).

6. Metallography, Structures and Phase Diagrams, Metals Handbook, Vol. 8, 8th ed., American Society of Metals, Metals Park, OH, p. 70 (1973).

7. J. J. Moore and J. C. Hamilton, "Axial Segregation in Continuously Cast Steel Billets Used for the Production of High Tensile Wire," Continuous Castings of Small Sections, Pittsburgh, Oct. 1980, TMS-AIME, p. 75 (1981).

8. H. Yamada, et al., "The Critical Conditions for the Formation of A-Segregation in Forging Ingots," TMS-AIME Annual Meeting, A 82-30, pp. 1-6 (1982).

9. T. Fuji, D. R. Poirier, and M. C. Flemmings, "Macrosegregation in Multi-component Low Alloy Steel," Met. Trans., Vol. 10B, pp. 331-339 (1979).

METALLOGRAPHY OF A NOVEL STIRLING ENGINE HEAT RECEPTOR

R. L. Bronnes[*] and R. C. Sweet[*]

ABSTRACT

In a special embodiment of a Stirling type heat engine, thermal energy is transmitted to its working gas through a multi-component receptor consisting of vacuum-cast silver in Inconel 625 containers. The need to effect that transmission with high efficiency required that special attention be given to the metallurgical design of the receptor. The areas of particular interest in this program included the integrity of the Inconel weld joints, the interfacing of fine silver and Inconel, and the interaction of the Inconel and silver with braze metals. This paper describes the optical metallographic examinations associated with the novel receptor design that was developed.

INTRODUCTION

A special application of a Stirling heat engine, of the type shown in Figure 1, required that radiant thermal energy be transmitted to the helium working gas at temperatures between 700 and 800 C. In order to efficiently transfer the thermal energy, special attention was given to the metallurgical design of the twelve Inconel 625 heat receptors, each of which was filled with vacuum-cast fine silver.

---

[*] Philips Laboratories, 345 Scarborough Rd, Briarcliff Manor, NY 10510 USA.

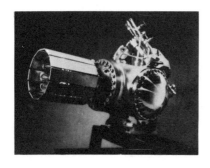 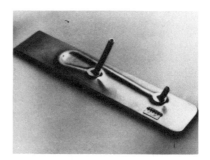

Figure 1. (Left) Stirling engine.

Figure 2. (Right) Completed heat receptor, showing embossed rear section and Inconel tubing.

The receptors, which measured 25-cm long, 5-cm wide, and 8-mm thick, were hydroformed in two parts: the planar front section facing the heat source, and the rear section which was embossed and pierced in two places for the reentrant Inconel 625 tubing that carried the hot working gas to the piston at 100-atmospheres of pressure. The fabrication schedule consisted of vacuum-brazing the tubing to the rear section with Wall-Colomonay Nicrobraz 171 and Nicrogap 108, TIG welding the front and rear sections to form a hermetic container, filling each container with 750-g of vacuum-cast fine silver, and hermetically TIG welding the cap in place. A complete heat receptor is shown in Figure 2.

The successful completion of each step in the fabrication process was critical to the performance of the engine. Defective welds could cause the containers to rupture or permit oxidation of the silver. Incomplete wetting of the Inconel 625 by silver would leave interfacial gaps that would impede the efficient conductive transfer of thermal energy. Unanticipated erosion of the braze metal by silver, or of the Inconel 625 container and tubing by the braze metal, could also cause catastrophic failure.

Optical metallography was selected as the principal method of examining engineering samples of the welded joints, silver-Inconel interfaces, and composite of Inconel,

silver, and braze metal. This assured us that optimal
conditions were achieved at each step in the fabrication
process. This paper describes the procedures for preparing
the metallographic specimens used in the developmental phase
of the program.

SPECIMEN PREPARATION

Specimens were removed from the engineering samples by
means of a conventional water-cooled cut-off wheel. The
specimens were oriented for examination in cross section and
were mounted in thermosetting material to reduce edge rounding
during surface preparation. A final trim cut was made on a
Buehler ISOMET saw to remove surface layers damaged by the
cut-off wheel.

Because of the widely different chemical and mechanical
properties of the Inconel 625 and fine silver, particular
care was required in preparing damage-free surfaces. This
enabled us to show the microstructure with minimal chemical
etching. The surface preparation of these specimens was
accomplished by means of Buehler Mini-Met grinding equipment
and TEXMET polishing pads. Typically, surfaces were lightly
ground on 600-grit silicon carbide papers and polished with
15- through 1-micrometer diamond abrasives. Moderate pressure
was applied to the specimens during each 10-15 minute polishing
cycle. Ultrasonic cleaning in distilled water between polishing
steps prevented the carry over of abrasives from one step to
another.

None of the water-based etchants could be used on this
composite material to show microstructures because of the
severe staining that always occurred. It was found that the
substitution of fresh anhydrous methanol for water gave
excellent results. The two reagents used are:

Ag	50 ml anhydrous methanol 50 ml NH_4OH 20-40 drops of 3% H_2O_2	Immersion etch for 10-20 sec.
Inconel 625	85 ml anhydrous methanol 10 ml HNO_3 5 ml acetic acid	Electrolytic: 2v open circuit; Pt electrodes; 10-30 sec.

Although the silver will etch as the Inconel 625 etches, it is frequently advisable to use the silver etchant to enhance its brightness.

RESULTS AND DISCUSSION

Initially, the TIG welded joints were made by hand. Figure 3 shows the microstructure of one of these early welds in a silver-filled receptor after 100 cycles from 20-750 C in air. The crack shown in the weld will eventually propagate through the Inconel and permit the penetration of oxygen into the silver-Inconel interface, resulting in catastrophic failure as illustrated in Figure 4. Subsequent machine-controlled welds of the type shown in Figure 5 did not fail even after 1000 cycles from ambient to 750 C in air.

The faying surfaces for the Inconel tubing to the rear section of the receptor could not be held to close tolerances. Consequently, large quantities of filler metals were required which increased the possibility of unacceptable erosion of the Inconel by the braze materials, or of the braze materials by molten silver during casting. Fortunately, as seen in Figure 6, the intersection of all three components in this specimen, which was aged at 750 C for 500 h, did not show any evidence of erosion.

Although chemical cleaning of the Inconel improved its wettability by molten silver, some gaps and voids remained in the interface which impeded efficient heat transfer from the source to the working gas. This problem was solved and clean surfaces were consistently obtained by thermally etching the receptors in vacuum at 1160 C for ten min. just prior to casting the silver. Figure 7 shows a typical clean interface obtained in this manner. Examination of creep-rupture specimens indicated no damaging effect of silver on Inconel after aging at 900 C for 1350 h.

SUMMARY

The information obtained from the optical metallographic examination of the novel heat receptors, as illustrated in the photomicrographs, was invaluable to the success of this developmental program and the performance of the Stirling engine. Although there are many sophisticated techniques

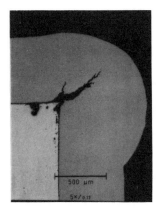 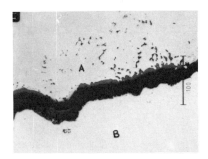

Figure 3. (Left) Cracked weld in Inconel container wall, filled with silver. Unetched.

Figure 4. (Right) Interfacial oxidation and separation of Inconel (A) and silver (B). Unetched.

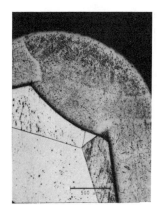 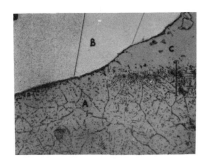

Figure 5. (Left) Sound weld after 1000 cycles from 20 C to 750 C. Etched.

Figure 6. (Right) Intersection of Inconel (A), silver (B), and braze metal (C). Etched.

Figure 7. Clean silver-Inconel interface, showing good wetting of the Inconel by silver. Silver etched.

available today for metallographic examination, optical metallography remains one of the most informative and easily interpreted methods.

ACKNOWLEDGEMENTS

The authors acknowledge with pleasure the technical contributions of D. Christensen, R. Eggleston, J. McKinlay, H. Meehan and J. O'Grady, and the editorial assistance of J. Lebid.

STRAIN INHOMOGENEITY IN ALUMINUM ALLOYS

D. J. Lloyd*
and
M. Ryvola*

ABSTRACT

The deformation structure developed during the cold rolling of an Al-Cu-Mg alloy is revealed by a decoration technique. It is shown that the plastic flow is rapidly localized into shear bands which increase in density as the extent of cold rolling increases. This localization of flow occurs regardless of whether the alloy is in the solution treated or precipitation hardened conditions.

INTRODUCTION

The microstructures developed at large strains have received increasing attention in recent years, with the appreciation that large strain microstructures are not merely an extension of the features formed at low strains [1-4]. As the deformation strain increases there is an increasing tendency for the plastic flow to localize into bands, often referred to as shear bands. This strain localization is poorly understood but has been observed in many different metals and alloys, and can be considered as a general manifestation of large strain deformation. The present paper considers the microstructures developed in AA-2036 alloy, a heat treatable Al-Cu-Mg alloy.

---

* Alcan International Limited, Kingston Laboratories,
  Box 8400, Kingston, Ontario, Canada K7L 4Z4.

Two heat treatments have been considered: (a) T4 -- the solution treated condition, and (b) T6 -- the peak aged, precipitation hardened case. The large strains have been developed by cold rolling the alloy plate. The microstructural features developed have been revealed by first decorating the structure by a low temperature anneal for 1 h at 220 C. This results in precipitation of Θ phase, particularly in those regions of the microstructure having a high defect concentration, which helps in the nucleation and precipitation of the phase.

EXPERIMENTAL PROCEDURE

AA-2036 plate with a chemical composition of 2.4 wt. % Cu, 0.5 wt. % Mg, 0.17 wt. % Mn, 0.37 wt. % Fe, 0.22 wt. % Si, balance Al, was:

(a) solution treated, T4 condition, and
(b) solution treated, water quenched, and aged 8 h at 190 C to the T6 condition.

The plate was then cold rolled to various reductions and given a decoration anneal at 220 C. Microstructures in the rolling plane (RP), the long-transverse plane (LTP), and the short-transverse plane (STP), were examined after etching in an aqueous solution of 0.5% HF with a few drops of H_2SO_4 for about 1 min. The planes examined are shown diagrammatically in Figure 1.

RESULTS

AA-2036-T4

Figure 2 shows that the typical microstructure of the alloy prior to rolling has a fine grain size together with constituent particles produced during casting. Figure 3a shows the LTP section after 20% reduction and some grains exhibit deformation lines or shear bands. These bands can also be seen in the RP sections, Figure 3b, but are more wavy, while in the STP section the bands are not as obvious, but tend to be parallel to the intersection of the rolling plane with the short transverse plane, Figure 3c. Up to 20% reduction there tends to be only one set of lines per grain but as the deformation increases, an intersecting second set of lines develop,

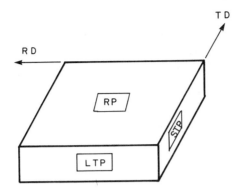

Figure 1. Diagram of examined planes.

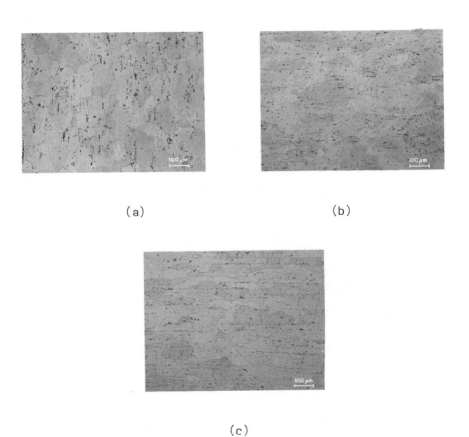

(a) (b)

(c)

Figure 2. Initial microstructures of AA-2036, (a) RP, (b) STP, (c) LTP sections, respectively.

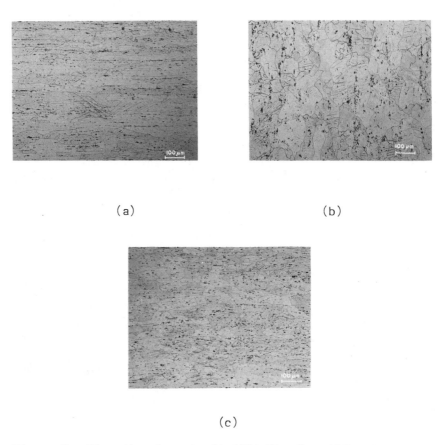

(a)   (b)

(c)

Figure 3. Microstructure in AA-2036-T4 after 20% reduction.

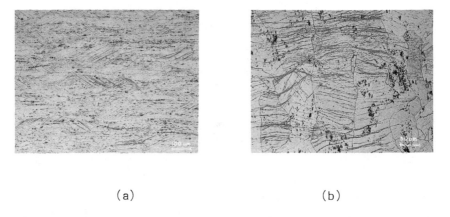

(a)   (b)

Figure 4. Microstructure in AA-2036-T4 after 40% reduction.

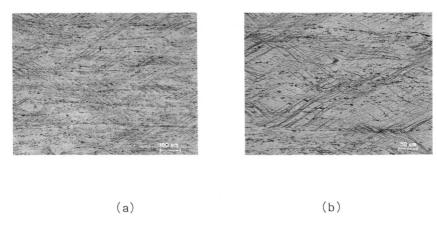

(a) (b)

Figure 5. Microstructure in AA-2036-T4 after 50% reduction.

Figure 6. Microstructure in AA-2036-T4 after 70% reduction.

the total number of lines increase, and the individual lines broaden and become more wavy. This can be seen from Figure 4, which shows examples of LTP and RP sections after 40% reduction. It will also be noted from the figures that many of the lines are continuous with lines in adjacent grains, but reflecting the different orientations across the grain boundary -- this is particularly clear in the RP sections. In the STP section the lines are still predominantly horizontal.

With still further increases in reduction, the deformation lines continue to develop in number, width, and also continuity, extending across many grains. Figure 5 shows the microstructure after 50% reduction and Figure 6 after 70% reduction. At the higher reductions another form of strain localization occurs, a clustering of deformation lines, which extend across a major portion of the cross section of the plate. This feature is equivalent to the macroscopic shear bands observed in Al-Mg alloys. It is interesting to note that macroscopic shear bands are not readily observable in the rolling and short transverse planes, as can be seen from Figure 7.

AA-2036-T6

The deformation features observed in the aged condition are essentially the same as for the T4 condition. However, at low reductions the number of deformation lines appears to be much larger in the aged condition and they also tend to be narrower and straighter, Figure 8. As the deformation increases, shear bands extending across many grains are developed, Figure 9.

DISCUSSION

The present experiments, together with previous work on Al-Mg, have shown that strain localization into bands is a general feature of deformation during rolling in solute- and precipitate-containing alloys. The localization occurs much more readily in the aged alloy. This would be expected since the presence of precipitates inhibits the mobility of dislocations and tends to promote planar slip and band formation. As slip proceeds the precipitates are sheared by the dislocations, reducing the effectiveness of precipitates as barriers to dislocations which results in local softening along the slip planes.

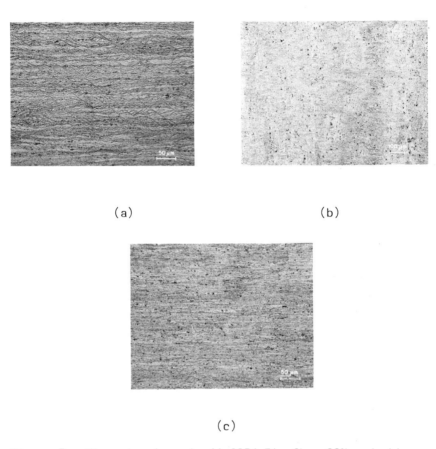

(a) (b)

(c)

Figure 7. Microstructure in AA-2036-T4 after 90% reduction.

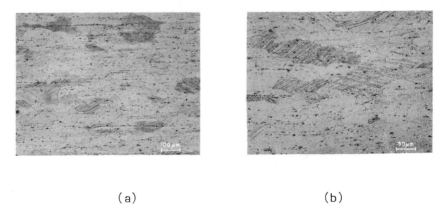

(a) (b)

Figure 8. Microstructure in AA-2036-T6 after 20% reduction.

(a) (b)

Figure 9. Microstructure in AA-2036-T6 after 90% reduction.

Precipitates per se, however, are not a requirement for shear band formation and flow localization, since the T4 condition does not contain precipitates, only solute clusters. In addition, previous work on the precipitate-free Al-Mg alloy [4] also detected shear bands. In fact, shear banding is even more severe in Al-Mg alloys than in the present Al-Cu-Mg alloys. One feature that all Al alloys exhibiting severe shear band formation have is a high work hardening rate, due to the presence of solute or precipitates in the matrix. This suggests that shear band formation is a mechanism of dynamic recovery, because the usual processes of dynamic recovery, such as dislocation cell formation, cannot occur due to the presence of solute and precipitates. Transmission electron microscopy results are consistent with these concepts, in that the deformation structure showed extensive microband formation and an absence of dislocation cells.

CONCLUSIONS

During cold rolling of AA-2036 alloy, shear bands are formed in the first 20% cold reduction and the density of shear bands increases with increasing reduction. At very high reduction, macroscopic shear bands, which extend across many grains, are developed. This localization of flow occurs in both precipitate-free and precipitation-hardened material, but the bands tend to be narrower with precipitates in the matrix.

From this and previous work, it is apparent that strain localization and shear band formation is a general feature of large strain deformation in high-solute and precipitate-containing aluminum alloys.

ACKNOWLEDGEMENTS

The authors are grateful to Alcan International Limited for permission to publish this work.

REFERENCES

1. B. J. Duggan et al., "Deformation Structures and Textures in Cold Rolled 70-30 Brass," Met. Sc., Vol. 12, pp. 343-351 (1978).

2. A. S. Malin and M. Hatherly, "Microstructure of Cold Rolled Copper," Met. Sc., Vol. 13, pp. 463-472 (1979).

3. M. Hatherly, "Deformation at High Strains," ICSMA-6, Vol. 3, pp. 1181-1195 (1982).

4. D. J. Lloyd, E. F. Butryn, and M. Ryvola, "Deformation Morphology in Cold Rolled Al-Mg Alloys," Microstructural Science, Vol. 10, pp. 373-384 (1982).

INTERNATIONAL METALLOGRAPHIC EXHIBIT

1983 Winners

Jacquet-Lucas Award
(Grand Prize)

Veronika Carle and Eberhard Schmid

Max-Planck-Institut für Metallforschung,
Institut für Werkstoffwissenschaften,
Pulvermetallurgisches Laboratorium.

Stuttgart 80/F.R.G.

Application of Interference Layer Microscopy
(from Class 9)

Optical Microscopy--Iron, Steel, Stainless Steel, Nickel and Nickel Alloys (Class 1)

FIRST IN CLASS: V. K. Sikka, C. W. Houck, J. M. Vitek, and K. S. Modrall, Oak Ridge National Laboratory, Oak Ridge, TN, Recrystallization in a Long Term Creep Specimen.

SECOND IN CLASS: M. Calabrese, R. J. Moores, D. S. Gnanamuthu, and J. G. Lumsden, Rockwell International Corporation Science Center, Thousand Oaks, CA, Microstructures of Rapidly Solidified Surfaces.

THIRD IN CLASS: Steven J. Dekanich and Barbara S. Lankford, Oak Ridge Gaseous Diffusion Plant, Oak Ridge, TN, Improved Bending Procedure.

HONORABLE MENTION: G. M. Goodwin, C. P. Haltom, P. A. Sanger, Oak Ridge National Laboratory, Oak Ridge, TN, Unintentional Carburizing Treatment Improves a Superconductor.

HONORABLE MENTION: S. D. Roberts, Stelco, Inc., Hamilton, Ontario, Canada, Failure Analysis of a Piston Rod.

HONORABLE MENTION: R. C. Klug and D. J. Moore, Michigan Technological University, Houghton, MI, Microstructure Refinement in Ausformed/Austempered Ductile Iron.

Optical Microscopy--Metals and Alloys not listed in Class 1 (Class 2)

FIRST IN CLASS: M. Kallfass, C. Weis, Max-Planck Institut fur Metallforschung, Stuttgart 1/F.R.G., Precipitation of Tantalum Suboxides.
SECOND IN CLASS: I. Weiss, F. H. Froes, D. Eylon, and E. Harper, Wright State University, Dayton, OH, Ghost Boundaries.
THIRD IN CLASS: W. G. Hutchings, Atomic Energy of Canada Ltd, Whiteshell Nuclear Research Est. Pinawa, MB, Canada, Overheated Nuclear Fuel.

Optical Microscopy--Petrographic, Ceramographic and Cermet Materials (Class 3)

FIRST IN CLASS: R. H. Beauchamp, N. T. Saenz, J. E. Coleman, D. E. Brownlee, Battelle-Pacific Northwest Laboratories, Richland, WA, Cosmic Deep Sea Spheres.
SECOND IN CLASS: U. Schafer and M. Krehl, Max-Planck-Institut fur Metallforschung, Stuttgart 80/F.R.G., Phase Formation During Diffusion of Oxygen from Nb_2O_5 to Niobium.
THIRD IN CLASS: E. F. Paterson, Stelco, Inc., Hamilton, Ontario, Canada, Metallurgical Coke.

Electron Microscopy--Transmission (Class 4)

FIRST IN CLASS: J. M. Howe, R. Gronsky, H. I. Aaronson, Lawrence Berkley Lab., A HREM Study of Interfacial Structure.
SECOND IN CLASS: H. M. Tawancy, B. E. Lewis, Cabot Corporation, Kokomo, IN, Effect of Long Term Exposure on Tensile Strength.
THIRD IN CLASS: M. H. Crimp, Michigan Technological University, Houghton, MI, Effect of Shock Pulse Duration on Twin Substructure.
HONORABLE MENTION: V. L. A. DaSilveira, R. A. F. O. Fortes, W. A. Mannheimer CEPEL, Electrical Energy Research Center, Rio DeJaneiro, Brazil, The Electroplastic Effect on the Interaction between Dislocations and Electric Current.
HONORABLE MENTION: Dr. P. L. Gai, University of Oxford, Oxford, U.K., Dissociation of Dislocations in III-V Semiconductors.
HONORABLE MENTION: I. G. Solorzano-Narango, McMaster University, Hamilton, Ontario, Canada, Discontinuous Precipitation in Al-22% Zn.

Electron Microscopy--Analytical (Class 5)

NO ENTRIES

Electron Microscopy--Scanning (Class 6)

FIRST IN CLASS: M. McAllaster, Sandia National Laboratories, Albuquerque, NM, Improved Ductility of U-Cr Castings.
SECOND IN CLASS: F. Kurosawa, I. Taguchi, Nippon Steel Corp., Precipitation Behavior of Phosphide.
THIRD IN CLASS: R. J. DeNuccio Gibson Electric Div., G.T.E., Delmont, PA, Oxidation of Silver-Tungsten.
HONORABLE MENTION: R. C. Klug, Michigan Technological University, Houghton, MI, Embrittlement in Austempered Ductile Iron.
HONORABLE MENTION: S. D. Kirchoff, AFWAL/MLLS, Wright-Patterson AFB, OH, Copper 78 Ductility.

Pretty Microstructures (Class 7)

FIRST IN CLASS: D. Downs, Babcock and Wilcox, Lynchburg, VA, Hafnium.
SECOND IN CLASS: B. S. Shabel, T. J. Steinback, R. W. Weleski, ALCOA Research, Continuous Cast Strip Al + 1.0% Mn Unusual Grain Structure.
THIRD IN CLASS: P. Ambalal, Lawrence Livermore National Laboratory, Livermore, CA, Dynamic Recrystallization of Aluminum 5083.
HONORABLE MENTION: M. Moniz, Ontario Research Foundation, Toronto, Ontario, Canada, CrO_3 Crystals in H_2O.
HONORABLE MENTION: D. J. Diaz, Lawrence Livermore National Laboratory, Livermore, CA, Metal Jungle.
HONORABLE MENTION: A. Horata, Y. Sumida, F. Ogata, Daido Steel Co., Nagoya, Japan, Ni Alloy Superalloy Powder Particle.
HONORABLE MENTION: A. Sondergaard, National Institute for Testing and Verification, Copenhagen, Denmark, Cementite in Grey Cast Iron.

Unique, Unusual or Other Techniques (Class 8)

FIRST IN CLASS: S. Kang, P. A. McFarland, F. J. Veltry, Inco Alloy Products Co. Research Center, Suffern, NY, The Microstructure of Inconel Alloy MA6000 Powder.

SECOND IN CLASS: K. Yoshida, H. Tanabe, N. Osakabe, T. Matsuda, J. Endo, T. Okuwaki, A. Tonomura, H. Fujiwara, Central Research Laboratory, Hitachi, Ltd, Tokyo, Japan, Holographic Observation of a Recorded Magnetization Pattern.

THIRD IN CLASS: K. Yoshida, H. Tanabe, N. Osakabe, T. Matsuda, J. Endo, T. Okuwaki, A. Tonomura, H. Fujiwara, Central Research Laboratory, Hitachi, Ltd, Tokyo, Japan, Electron Holographic Observation of a Cross-Tie Wall.

Color Micrographs--from any class (Class 9)

FIRST IN CLASS: I. Taguchi, H. Hamada, N. Fukuoka, Nippon Steel Co., Kawasaki, Japan, Microscopic Internal Structures of an Oxide Inclusion.

SECOND IN CLASS: H. Hamada, I. Taguchi, M. Sasaki, T. Murata, Nippon Steel Co., Kawasaki, Japan, Analysis of Rust on Iron Sword.

THIRD IN CLASS: W. Samells, Ontario Research Foundation, Toronto, Ontario, Canada, Nickel Chromium Abrasion Resistant Cast Iron.

Student Entries--Undergraduate only (Class 10)

FIRST IN CLASS: Randy J. Bowers, Rensselaer Polytechnic Institute, Troy, NY, Interaction of Hydrogen and Sulfide Inclusions in Fracture of 4140 Steel.